FLAX AMERICANA

MCGILL-QUEEN'S RURAL, WILDLAND, AND RESOURCE STUDIES SERIES
Series editors: Colin A.M. Duncan, James Murton, and R.W. Sandwell

The Rural, Wildland, and Resource Studies Series includes monographs, thematically unified edited collections, and rare out-of-print classics. It is inspired by Canadian Papers in Rural History, Donald H. Akenson's influential occasional papers series, and seeks to catalyze reconsideration of communities and places lying beyond city limits, outside centres of urban political and cultural power, and located at past and present sites of resource procurement and environmental change. Scholarly and popular interest in the environment, climate change, food, and a seemingly deepening divide between city and country, is drawing non-urban places back into the mainstream. The series seeks to present the best environmentally contextualized research on topics such as agriculture, cottage living, fishing, the gathering of wild foods, mining, power generation, and rural commerce, within and beyond Canada's borders.

1 How Agriculture Made Canada
   Farming in the Nineteenth Century
   Peter A. Russell

2 The Once and Future Great
   Lakes Country
   An Ecological History
   John L. Riley

3 Consumers in the Bush
   Shopping in Rural Upper Canada
   Douglas McCalla

4 Subsistence under Capitalism
   Nature and Economy in Historical
   and Contemporary Perspectives
   Edited by James Murton, Dean
   Bavington, and Carly Dokis

5 Time and a Place
   An Environmental History of
   Prince Edward Island
   Edited by Edward MacDonald, Joshua
   MacFadyen, and Irené Novaczek

6 Powering Up Canada
   A History of Power, Fuel, and Energy
   from 1600
   Edited by R.W. Sandwell

7 Permanent Weekend
   Nature, Leisure, and Rural
   Gentrification
   John Michels

8 Nature, Place, and Story
   Rethinking Historic Sites in Canada
   Claire Elizabeth Campbell

9 The Subjugation of Canadian Wildlife
   Failures of Principle and Policy
   Max Foran

10 Flax Americana
   A History of the Fibre and Oil
   That Covered a Continent
   Joshua MacFadyen

FLAX AMERICANA

*A History of the Fibre and Oil That
Covered a Continent*

JOSHUA MACFADYEN

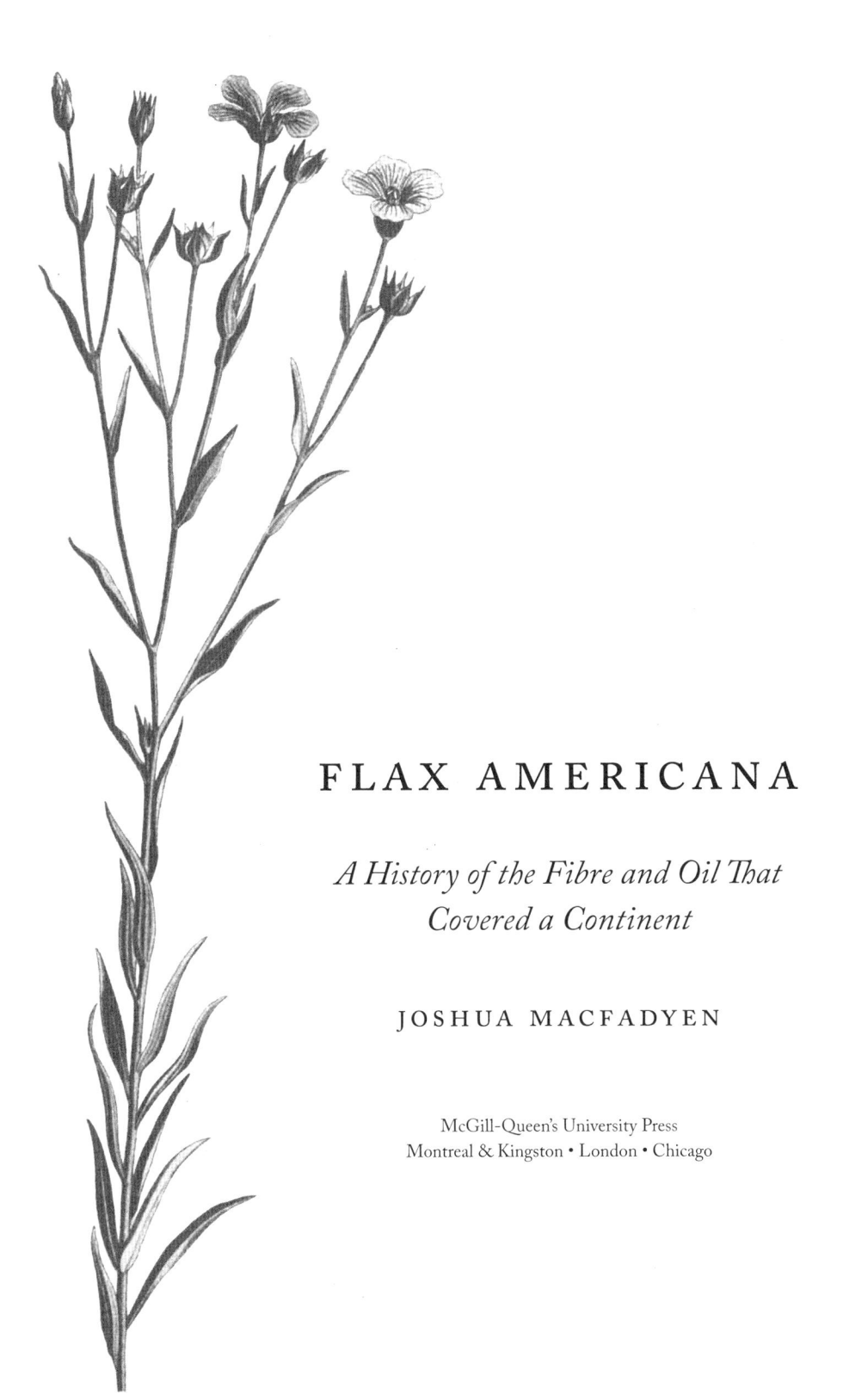

McGill-Queen's University Press
Montreal & Kingston • London • Chicago

© McGill-Queen's University Press 2018

ISBN 978-0-7735-5346-0 (cloth)
ISBN 978-0-7735-5347-7 (paper)
ISBN 978-0-7735-5395-8 (ePDF)
ISBN 978-0-7735-5396-5 (ePUB)

Legal deposit third quarter 2018
Bibliothèque nationale du Québec

Printed in Canada on acid-free paper that is 100% ancient forest free
(100% post-consumer recycled), processed chlorine free

This book has been published with the help of a grant from the Canadian Federation for the Humanities and Social Sciences, through the Awards to Scholarly Publications Program, using funds provided by the Social Sciences and Humanities Research Council of Canada. Funding has also been received from Arizona State University.

Funded by the Government of Canada | Financé par le gouvernement du Canada   Canadä   Canada Council for the Arts / Conseil des arts du Canada

We acknowledge the support of the Canada Council for the Arts, which last year invested $153 million to bring the arts to Canadians throughout the country.

Nous remercions le Conseil des arts du Canada de son soutien. L'an dernier, le Conseil a investi 153 millions de dollars pour mettre de l'art dans la vie des Canadiennes et des Canadiens de tout le pays.

LIBRARY AND ARCHIVES CANADA CATALOGUING IN PUBLICATION

MacFadyen, Joshua, 1979–, author
Flax Americana : a history of the fibre and oil that covered a continent / Joshua MacFadyen.

(McGill-Queen's rural, wildland, and resource studies series ; 10)
Includes bibliographical references and index.
Issued in print and electronic formats.
ISBN 978-0-7735-5346-0 (cloth). – ISBN 978-0-7735-5347-7 (paper). –
ISBN 978-0-7735-5395-8 (ePDF). – ISBN 978-0-7735-5396-5 (ePUB)

1. Flax industry – Canada – History.   2. Flax industry – United States – History.   3. Linen industry – Canada – History.   4. Linen industry – United States – History.   5. Linseed oil industry – Canada – History.   6. Linseed oil industry – United States – History.
I. Title.   II. Series: McGill-Queen's rural, wildland, and resource studies series ; 10

HD9930.C32M33 2018   338.4'7677110971   C2018-902236-1
                                             C2018-902237-X

Set in 11/14 Adobe Caslon Pro
Book design & typesetting by Garet Markvoort, zijn digital

CONTENTS

Tables and Figures
vii

Acknowledgments
xiii

Abbreviations
xvii

Introduction
3

CHAPTER ONE
The Edge of Industrialization:
Finding a Northern Fibre
32

CHAPTER TWO
Everyday Exchanges:
Growing and Harvesting
Flax in Ontario
65

CHAPTER THREE
Flax Fabrications:
Selling the Promoter's Plant
107

CHAPTER FOUR
Covering the Earth:
North American Flax
and Paint to 1878
139

CHAPTER FIVE
Saving the Surface:
Flax in the Urban Industrial Complex
162

CHAPTER SIX
Cover Crop:
Growing Flax for Linseed
Oil and Paint
206

CHAPTER SEVEN
Saving Flax:
Industry, Science, and
the Tariff
244

Conclusion
271

Notes
279

Index
339

TABLES AND FIGURES

Tables

1.1 Value ($) of Perine flax outputs by country, 1857–1865 (part)   50–1
1.2 Value ($) of Perine flax cordage/textile outputs, 1857–1865 (part)   55
2.1 Flax farmers in Waterloo County, 1861   70
2.2 Flax farmers in select townships, 1871   76
2.3 Flax milling in Ontario by workforce, 1871   101
2.4 Cordage milling in Ontario by workforce, 1871   102
4.1 US oilseed product values and GDP (current dollars)   149
5.1 Number and percentage of national output of linseed oil mills by state, 1880–1920   177
6.1 North American flax seed production and price, 1901–1931   223–4

Figures

0.1 The North Dakota Agricultural College Extension Service, "FLAX: *You* can help America win by producing more," *NDAC Special Circular* A-13 (March 1942): 1   4
0.2 Global flax seed production, 1880–1960   7
0.3 The flax commodity web. Diagram by the author and Toggle Media   10

0.4 Flax and two complementary commodity webs. Diagram by the author and Toggle Media   11

0.5 *Linum usitatissimum*, sketch from Franz Eugen Köhler, *Köhler's Medizinal-Pflanzen in naturgetreuen Abbildungen mit kurz erläuterndem* (Berlin, 1887), 88   12

1.1 Perine and Young flax mill, St Thomas, ON, c. 1865. Courtesy of Elgin County Archives, Ian Cameron Collection, R11 S5 SH2 B4 F50   34

1.2 Bowman's flax mill (middle, background) in a mill complex, Floradale, ON. KPL, Woolwich Historical Foundation Collection (WHF), PH590   38

1.3 Flax output and industry in the Great Lakes region, 1850, 1859/60   42

1.4 Map of Doon's built environment in the 1870s, showing the Perine house, mill complex, and mill ponds. Background composition of 1945 historical aerial photographs from the University of Waterloo Library Air Photos Digitization Project, http://www.lib.uwaterloo.ca/locations/umd/project/ (accessed September 2017). Map by the author after Elizabeth Bloomfield, *Waterloo Township through Two Centuries* (Kitchener: Waterloo Historical Society 1995), 202, Figure 33, and *Waterloo Historical Society* 54 (1966): 46–7   44

1.5 Flax output and industry in Ontario and Quebec, 1870, 1890   46

1.6 One of the three central Perine mills, Conestogo, ON. KPL, PH192 WHF Old Flax Mill   47

1.7 Doon Twines, Doon, ON – rear of visible building with drying racks on the right and a tennis court in the foreground. KPL, P001401 WHF, Doon Twines   49

1.8 Scutched flax bales in a rail car. F.W. Stock, BA thesis, OAC   57

1.9 Doon Twines Ltd products. UWA, Special Collections & Archives. Personal Studio fonds. Robert T.G. Nicol, Negative No. 48-1431   62

2.1 Flax in Waterloo County, 1861   71

2.2 Flax fibre, seed, and scutch mills in southwestern Ontario, 1870–1871   75

2.3 Pulling flax, 1910. UGA, Ontario Ministry of Agriculture and Food, Sallows Collection, no. 118, XA1 MS A182. Courtesy of the University of Guelph Archives and Special Collections   86

2.4 Women and children standing in a field, gathering flax near Arthur, ON, c. 1915. Paul O'Donnell and Frank D. Coffey, *Portrait: A History of the Arthur Area*, 1971; photo used by permission of the Wellington County Museum and Archives (WCMA), A1976.87, ph 9535   87

2.5 Tavistock Flax Mill Gang. TDHS, Lemp Studio Collection, Tavistock & District Historical Society, photo #0231, "Flax Gang"   88

2.6 First Nations flax labourers on Drayton train station platform. KPL PH597 WHF, Group on railway platform, Floradale, ON, n.d.   91

2.7 Jesse B. Snyder and two men from the Floradale flax mill at the Drayton train station. KPL PH595 WHF, Three men on train platform, Floradale, ON, n.d.   91

2.8 First Nations camp on field near Arthur, ON, c. 1915. Paul O'Donnell and Frank D. Coffey, *Portrait: A History of the Arthur Area*, 1971; photo used by permission of the WCMA, A1976.87, ph 9536   93

2.9 Peter Hill, St Marys, ON. St Marys Museum, "Peter Hill," 0770ph. Photograph courtesy of St Marys Museum and R. Lorne Eedy Archives, St Marys, ON, Canada   95

2.10 Pulling flax, 1910. UGA, Ontario Ministry of Agriculture and Food, Sallows Collection, no. 93, XA1 MS A182. Courtesy of the University of Guelph Archives and Special Collections   97

2.11 Tavistock flax mill. TDHS, Lemp Studio Collection, Tavistock & District Historical Society, photo #0206, "Flax Mill"   99

2.12 Aboyne flax scutching mill in Nichol Township, Wellington County, ON, c. 1890. Photo used by permission of the WCMA, photo A1993.5, ph 11264   100

2.13 Workers at Keith's St Thomas flax mill, c. 1890, Nathan Ryan on left. Courtesy of Elgin County Archives, Ian Cameron Collection, RII S5 SH2 B4 F53   103

2.14 Chris Yantzi's flax mill work, 1941–1948   105

3.1 Riga flax ordered by agricultural societies, 1866   120

3.2 Oxford Linen Mills: a process of hours instead of months. Oxford Linen Mills, "From Flax to the Fabric" (1908)   128

3.3 Central Experimental Farm, flax fibre pedigree seed plot. Canada, *Fifty Years of Progress on Dominion Experimental Farms, 1886–1936* (Ottawa: J.O. Patenaude 1939), 70; Canada Agriculture and Food Museum, CSTM Agriculture Canada fonds, AGflax1   131

3.4 Pulling flax in Willowdale, ON, 1918. City of Toronto Archives, William James family fonds, Fonds 1244, Item 4511   133

4.1 A house in the St Marys area, c. 1903. St Marys Museum, Carter & Isaac Collection, photo no. 12. Photograph courtesy of St Marys Museum and R. Lorne Eedy Archives, St Marys, ON, Canada   144

4.2 Dominion Linseed Oil and Livingston foundry, Baden, ON, 1920s. TWA, Castle Kilbride Collection, unsorted, scan0030   160

5.1 James Livingston, parliamentary photo (William James Topley Studio, 1883). LAC PA-028383 and MIKAN #3419919   165

5.2 Flaxseed receipts at Livingston Linseed Oil, Baden, ON, 1897   167

5.3 Carl E. Johnson, *Linseed Mill*, 1934. Courtesy of Minnesota Historical Society. MHS, Location No. AV1981.20.56, Negative No. 36584   168

5.4 Alberta Linseed Oil Mills fire (October 1914), Medicine Hat, AB. EMA, Fred Forster family fonds, Accession No. 0206.0015   169

5.5 Alberta Linseed Oil Company, Medicine Hat, AB. EMA, Alberta Linseed Oil Company Fonds, 0494.0009 (n.d.)   173

5.6 "Canada needs more flax." EMA, Alberta Linseed Oil Company Fonds, poster (n.d.)   174

5.7 Minnesota Linseed Oil and Paint Company, Minneapolis, 1915. Photographer Charles J. Hibbard, "Minnesota Linseed Oil Company, 1101–29 South Third Street, Minneapolis." Courtesy of Minnesota Historical Society. MHS, Location no. MH5.9 MP3.1M p. 239   175

5.8 Proof Paints. Pittsburgh Plate Glass Company, *Glass, Paints, Varnishes and Brushes* (Chicago: Lakeside Press 1923), 57, 122   188

5.9 Spencer Kellogg, laboratory, c. 1921. Alexander Schwarcman, *Laboratory Letters: a series of letters written by our chief chemist, Dr. Alexander Schwarcman, and published in Oil, paint, and drug reporter during 1920–21* (Buffalo: Spencer Kellogg & Sons 1921), 8   201

5.10 Battery of twenty French heavy-duty screw presses. Whitney Eastman, *The History of the Linseed Oil Industry in the United States* (Minneapolis: T.S. Denison 1968), 117   202

5.11 Rotocel extractor. Whitney Eastman, *The History of the Linseed Oil Industry in the United States* (Minneapolis: T.S. Denison 1968), 120   203

6.1 Flax as a percentage of all field crops sown on the Northern Great Plains and Prairies   220–1

6.2 Flax at Joe Bellas's homestead shack, Carlstadt (later Alderson), AB, 1911. GMA, NA-4711-3, Calgary, AB   226

6.3 Cutting flax on Canadian Wheatlands fields, Suffield, AB, c. 1912. GMA, NA-587-18, Calgary, AB   227

6.4 Wilcox-area steam tractor ploughing and burning flax, 1907. SA, Photograph No. R-B1004. Courtesy of the Provincial Archives of Saskatchewan   228

6.5 Flax acres seeded and change in improved acreage, 1889–99, 1899–1909   234–5

6.6 Flax acres seeded, 1921, and change in improved acreage in south Saskatchewan, 1920–1921   235

7.1 "Survival of the Fittest Garden," Plot 30, North Dakota Agricultural Experiment Station, Fargo, ND, 1921. HLBP, Photographs,

"Flax – Experiments – Fargo," Negative No. 2705. Courtesy of the Institute for Regional Studies, North Dakota State University, Fargo, ND   246

7.2 "Flax Twine Plant," St Paul, MN, International Harvester Co., c. 1907. Geo. R. Lawrence Co., Copyright Claimant. *Flax Twine Plant, St. Paul, Minn., International Harvester Co.* Saint Paul, MN, c. 1907. Photograph. Retrieved from the Library of Congress, https://www.loc.gov/item/2007663911/. (Accessed September 2017)   250

7.3 "Help save the flax crop, ask us how," L. Simmons Hardware, Foxholm, ND. HLBP, Photographs. Courtesy of the Institute for Regional Studies, North Dakota State University, Fargo, ND   253

7.4 Bolley's office, Propaganda Headquarters, c. 1915. MHS, "Henry Luke Bolley's office in the Science Hall, North Dakota Agricultural College, Fargo, c. 1915," Photograph Collection, Location No. HF3 p122, Negative No. 68351   259

ACKNOWLEDGMENTS

This book is long, and it took a long time to finish. It seems appropriate, therefore, to parrot Charles Darwin, who in 1879 wrote to his friend Asa Gray: "I have written a rather big book – more is the pity – on the movements of plants, and I am now just beginning to go over the MS for the second time, which is a horrid bore." Don't get me wrong, mapping the movement of the flax plant, from Ireland to Ontario and from Alberta to Argentina, has been a fun and fascinating way to explore the people and places at the edges of industrial capitalism. The real trouble of taking so long to finish a project is that I have lived in six cities, nine addresses, and four treaties or traditional territories since starting it, and I have accumulated too many intellectual debts and personal friendships at each one to properly acknowledge here. So, I will proceed, improperly, and say too little about too few of my friends and colleagues.

The idea for *Flax Americana* germinated while I was taking courses in social history at the University of Guelph and working with Dr Douglas McCalla, the Canada Research Chair in Rural History. Other important advisors there included Drs Catharine Wilson, Kris Inwood, Stuart McCook, Richard Reid, Kevin James, and Terry Crowley at Guelph; Suzanne Zeller and Terry Copp at Wilfrid Laurier University; and James Walker, Wendy Mitchinson, Geoffrey Hayes, and Stanley Johannesen at the University of Waterloo. Dr Sterling Evans provided valuable feedback on the first complete draft as external examiner, and he has supported the project ever since. Earlier encouragement and advice came from professors such as Elwood Jones, at Trent University, and Edward MacDonald and Sasha Mullally at the University of Prince Edward Island.

I have gained critical research support from many sources, including the Canada Research Chair in Rural History and the Snell-Murray Ontario Graduate Scholarship at the University of Guelph; a SSHRC Postdoctoral

Fellowship and support from the NiCHE: Network in Canadian History & Environment Strategic Knowledge Cluster at the University of Western Ontario; and the Sustainable Farm Systems SSHRC Partnership Grant at the University of Saskatchewan. Finally, I am grateful for subventions from the Awards to Scholarly Publications Program at the Federation for the Humanities and Social Sciences, and the Institute for Humanities Research at Arizona State University, as well as support from both of my units there. The faculty and staff at the School of Historical, Philosophical, and Religious Studies and the School of Sustainability have been inspirational colleagues, sharpening both my pedagogy and research.

Flax seemed like a manageable research topic for a single-authored monograph, but I quickly realized that "following things around" (as Dr Stuart McCook cleverly calls commodity research methods) requires help from a large network of academics, archivists, and data practitioners. Access to the 1871 Industrial Census database was provided courtesy of Dr Kris Inwood, and I am also grateful for the geospatial data and assistance from the National Historical GIS, the Great Plains Population and Environment Database, and Canadian map and geospatial libraries at the University of Guelph (especially Jenny Marvin) and the University of Toronto (especially Marcel Fortin and Byron Muldofsky). Individuals who have given or directed me to various research materials include Drs Bruce Curtis, Hans Werner, and Jarrett Rudy, and Katelyn MacEachern conducted the data entry for the Tavistock account books. Assistance in Kitchener-Waterloo came in many forms, including a grant from the Waterloo Regional Heritage Foundation, assistance from Elinor Rau and rych mills, and walking tours of "Tow Town" (near Doon) with Jean Haalboom and "Bone Town" (near St Thomas) with Steve Peters. I am especially grateful for Marion Roes, who first unlocked the door to the tiny Woolwich Historical Foundation room containing the Perine Brothers flax mill account books, and for her many contributions since. Many others have offered their time, materials, and even camera batteries including the aptonym-named Gary Flaxbard, William Gladding, and Mrs Yantzi in Tavistock. Staff at all of the archives listed below have assisted me in some way, but special gratitude is due to Tracy Loch of the Sir Adam Beck Archives at Castle Kilbride in Baden, Karen Ball-Pyatt at Kitchener Public Library, Gina Coady at the Elgin County Archives, and the staff at the Institute for Regional Studies in Fargo.

Many friends and colleagues enhanced the intellectual contributions of this book. Dr Douglas McCalla was an excellent supervisor who seemed to have endless hours to dedicate to my intellectual and career development. Now that I'm a professor I realize that he didn't, but because he is a truly generous and principled person he gave (and continues to volunteer) them anyway. Dr Alan MacEachern was a gracious postdoctoral advisor and a generous collaborator and editor. He has shared ideas and indeed entire networks (e.g., NiCHE) with most of the Canadian environmental historians I know. He shared many with me, but the best is obviously the idea for the title of this book, *Flax Americana*. Dr Geoff Cunfer was my second postdoctoral advisor and a fine example of project leadership, grantsmanship, and Vulcan logic. He offered some of my earliest external advice on flax when I visited the US for my first environmental history conference; now I happily visit him in Canada and benefit from similar research advice. Drs Claire Campbell, Claire Strom, and James Murton provided encouragement and comments on various iterations, and I am grateful to Kyla Madden, two anonymous reviewers, and the editors of the Rural, Wildland, and Resource Studies Series at McGill-Queen's University Press, for their patient support and feedback. In particular, Dr Ruth Sandwell dedicated extensive time to an early version of the manuscript, and her comments improved the text a great deal.

Family came first, but book acknowledgments, like the project itself, so often place them last. My parents, Donald and Carolyn MacFadyen, and parents-in-law, Annette and the late Thomas Lamport, supported this project in their own ways for decades. My sisters, Dawn Baglole and Kathleen MacPhee, and their families are always an inspiration. Cameron, Kenya, Siena, and Tori MacFadyen have grown up faster than this book, but not fast enough to recruit as research assistants. It's a delight to be their dad, to have them along for this journey, and to finally increase the number of years they get to spend in one city. The journey began with Colleen MacFadyen – my wife, my travelling companion, and my closest friend. She knows more about, has given more to, this project than anyone. I'm overjoyed to finally dedicate it to her.

ABBREVIATIONS

FDC   Flax Development Committee
NDAC   North Dakota Agricultural College
NDAES   North Dakota Agricultural Experiment Station
NDSU   North Dakota State University
USDA   United States, Department of Agriculture

FLAX AMERICANA

INTRODUCTION

In the century after 1850 the eastern Great Lakes and Northern Great Plains of North America experienced an influx of flax that gave them the largest concentrations of the genus *Linum* in the world, a title that the Canadian prairies still retain. Flax is believed to be the oldest fibre planted and processed by humans, and its Latin name (*Linum usitatisimum*) means "most useful thread."[1] Most people associate flax production with the Irish linen industry or the Belgian flax districts, and indeed the pioneer agro-ecosystems of North America were never a replacement for those famous linen centres. Instead, North American flax was produced in concentrated and contiguous areas for a variety of intermediate products that most consumers would never have associated with the plant. Modern consumers encountered flax in manufactured goods practically every day. Flax grown in the Americas appeared in rough textiles, upholstery, and cordage and in most of the paint, linoleum, and other oil coverings of the Second Industrial Revolution. However, most people knew very little about the plant, where it came from, and what it was capable of.

Even farmers were generally unfamiliar with flax, and those who knew about it were reluctant to grow it. If it seems unusual that most farmers in the Great Lakes and Great Plains regions ignored flax – even as these regions came to host the largest flax crops in the world – they had good reason for the omission. Growers who raised flax for fibre had to pull the crop by hand, and as a result the largest flax producers in the Great Lakes were not farmers at all, but rather millers who grew the plant on rented land. Conversely, farmers who grew flax for linseed oil mainly grew it on the remote edge of frontier grasslands and harvested it as they did wheat. Promoters and farm scientists tended to endorse the fibre side of flax production, and despite centuries of repeated attempts they almost always misunderstood the crop. They failed to see that the fibre was a mill-directed

Figure 0.1   "Flax: *You* can help America win by producing more," 1942

system, and due to periodic wartime cotton shortages, promoters routinely overestimated flax's viability as a substitute fibre for cotton in light textile manufacturing. The average North American farmer simply had no experience with flax, and when he or she did read an expert's advice, it was often propaganda based on alchemy and outright disinformation. Despite the extensive resources dedicated to its promotion, including programs by US and Canadian departments of agriculture, wartime rhetoric, and a plethora of media coverage, farmers remained unconvinced.

US and Canadian governments periodically promoted flax industries, especially in wartime, but first they had to explain what flax was for. Both federal departments of agriculture promoted flax as a replacement for cotton during the American Civil War, but they found that most farmers were unfamiliar with the crop. The Canadian Bureau of Agriculture had to import machinery and experts from the United Kingdom for demonstrations, and Henry Youle Hind, geologist and editor of *British American Magazine*, wagered that Canadian farmers would not know – and would not believe if they were told – that flax was the second most important fibre in the world.[2] During the First World War, the Federal Department of Agriculture distributed free seed from the British Government and explained to farmers that one acre of flax grown would equal one set of linen wings in an Irish airplane factory. And then – since most farmers had seen neither flax nor airplanes – the department hosted a flax festival north of Toronto to introduce them to both.[3] In 1942 the Canadian Agricultural Supplies Board guaranteed farmers $2.25 per bushel for any quantity of flax seed and argued that it was "the most important oil-producing crop now grown in this country." That same year a prairie version of "Uncle Sam" commanded North Dakotans to grow more flax, but also explained that linseed oil came from flax and was "an essential material needed to make America victorious" (Figure 0.1).[4]

Looking for Flax

This is not, however, a book about a plant no one grew. Rather, it is the story of a specialty crop that a concentrated group produced and practically everyone consumed. It is about how plants and humans were co-constituted during the Industrial Revolution – how one plant was shaped by the social-ecological systems it inhabited, and how human society and even human desires were shaped by the plant.[5] It is also a book about a uniquely *northern* commodity and the two Canadian-US regions it occupied. The first is the eastern Great Lakes region of southern Ontario (Upper Canada), Quebec (Lower Canada), Ohio, and Upstate New York. The second is the Northern Great Plains and Prairies where flax was ideally suited to the region's temperate climate and heavy prairie soil.[6] In some ways, the border was one example of how human institutions shaped the plant, but for the most part plants ignore political boundaries. Flax was a transnational commodity in the sense that producers, processors, and pathogens traded, migrated, visited, and corresponded across the border and because economies and agro-ecosystems developed similarly on each side. In the eastern Great Lakes region the flax fibre industry employed thousands of labourers, including First Nations and other mill-directed harvest gangs, and it remained rooted there for almost a century. After military and ecological aggression in Manitoba and the old Northwest, the way was clear for settlers to march west beyond the forest and past the isohyetal line that finally spoke for the last of the eastern precipitation. And they brought flax with them, not for its fibre but for the now valuable oils in its seed.

If North Americans were unfamiliar with flax and its products, it was not for lack of production. In the mid-nineteenth century the crop was modest at no more than 25,000 acres, or less than a fifth of the Irish crop, but later the Northern Great Plains and Prairies became major producers. At its peak in 1912, the US and Canadian flax crop occupied over 5 million acres of land, making it the largest flax-producing region in world history to that point.[7] The next most closely related oilseed crop, canola, did not reach a similar amount until the 1980s. Mapping the plant's location over time shows that it was highly concentrated, both in its production and processing. Wherever the crop was farmed it almost always appeared in contiguous areas. Flax seed was particularly attractive to early grassland farmers because it yielded well on *new breaking* or native grassland that had been recently ploughed. Thus, flax gained a reputation as a pioneer crop.[8] Flax cultivation only managed to make up a fraction of GDP, but *inside* these concentrated districts, it often ranked second or third after wheat (usually after oats) in terms of value. Everyone understood the vital role of prairie wheat for feeding cities. They also knew the importance of oats for powering the horses to feed the cities. But very few people understood that the next most valuable commodity to come off the plains was not for feeding but rather for painting cities. Flax seed processing came to

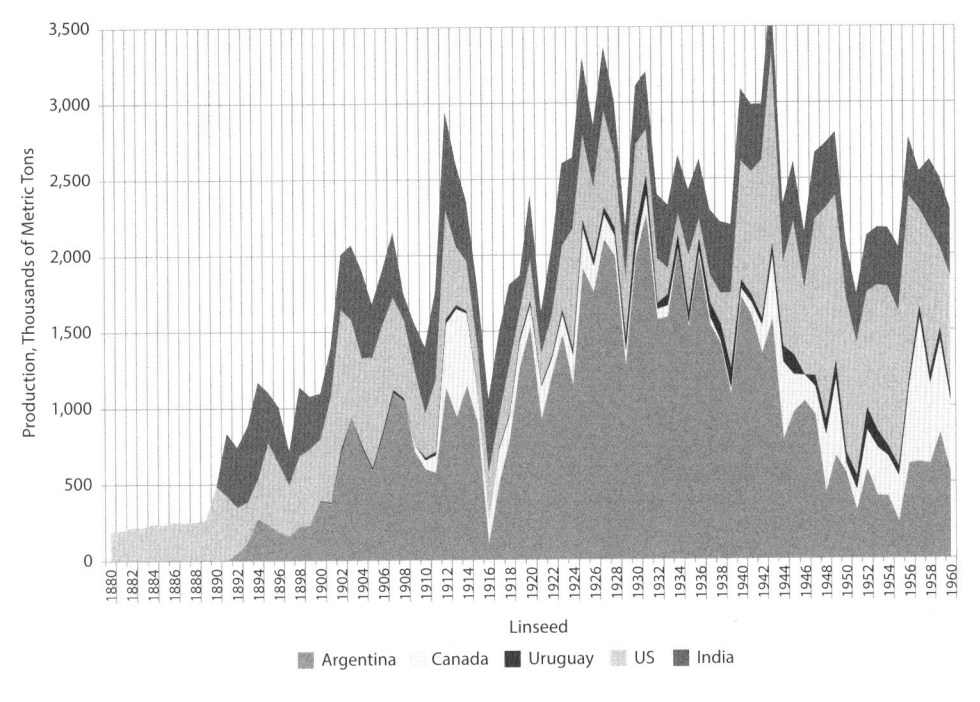

Figure 0.2 Global flax seed production, 1880–1960

Sources: The data for India are suspiciously consistent, and European and Russian flax seed export data are not available. B.R. Mitchell, *International Historical Statistics: The Americas and Australasia* (London: Macmillan 1983), Series D3 and D4, 269–75; B.R. Mitchell, *International Historical Statistics: Europe, 1750–2000* (New York: Palgrave Macmillan 2003), Series D22, 538–40; B.R. Mitchell, *International Historical Statistics: Africa and Asia* (New York: New York University Press 1982), Series C4, 224–5.

be dominated by a few prominent figures in US capitalism, and in a few cities, notably Buffalo, linseed oil manufacturers were among the city's largest industries by value of output.[9] Again linseed was overlooked, in large part because it was an intermediate good. Painting with linseed oil and ground colour pigments was an expensive and highly skilled job and thus out of reach for most consumers. It was not until the development of a new commodity that linseed oil became mass produced. And that commodity was ready-mixed paint.

If travellers journeyed to British North America and the northern United States in 1860 and wanted to see the region's flax and flax products, they would not find much growing in New England, the Maritimes, or along the Lower St Lawrence. Further west, most of Canada's flax seed would have been visible in one factory on the Lachine Canal in Montreal, and the fibre would be found in a handful of mills between Toronto and Berlin (now called Kitchener). Similar fibre production occurred in small clusters of mill complexes in New York State on the upper Hudson River near Troy, and in Ohio on the Cuyahoga, Mahoning, and Miami River valleys. The seed in these states was processed in other mills, similar to the one in Montreal. The semi-processed products would have been used in carriage paint shops in Brantford, upholstery shops in Hamilton, paper mills in Buffalo, and sail cloth factories back in Boston. Consumers bought flax in these forms as well as in cordage, bagging, and a new product called "mixed paint." If the journey continued at the turn of the century, the travellers would see Ottawa's farm scientists promoting flax seed as a crop for new breaking and barn paint as a mark of modern farming. One would find that Manitoba's Mennonites and North Dakota's sodbusters had been sowing the crop on new breaking and that, partly due to a pathogen, the centre of US flax production was moving with the frontier.

A similar "commodity tour" in 1930 would reveal a very different landscape. Flax fibre production continued intermittently in parts of the Great Lakes region, but it was eclipsed by the seed that had become the raw material for one of the most important oils in the burgeoning chemical sector. Virtually all of the continent's flax seed, and a good deal of the world's, was processed by corporations in Minneapolis, Buffalo, New York, and Montreal. Production in the Northern Great Plains and Prairies had again reached 5 million acres, up from an average of 3.4 million across the previous decade. Thanks to strong markets for linseed oil and rising tariffs on foreign flax seed, it looked as though flax seed had found a permanent niche in the Great Plains agro-ecosystem. Houses and vehicles were more colourful now than they had been in the largely unpainted world of 1860, and companies used glossy catalogues and creative reform movements like the "National Clean-up, Paint-up, Fix-up" campaign to increase sales. However, in September 1929, on Black Friday, a major correction in paint consumption had already begun, as with all luxuries.[10] Dust clouds loomed on the grasslands, and flax production began to drop. Rather than a permanent "flax belt," the region attracted a network of processors, chemists, and other specialists to develop the region's oilseed sector. Flax, soybeans, sunflowers, rape, and later canola were becoming the source of Canada's *other* oil: triglyceride. By 1930 the business of converting flax to fibre and oil in small-town mill complexes had been replaced by international networks of managers and chemists. The North American flax crop experienced another boom during the Second World War, and thanks to price controls this continued for several years. However, the global flax seed crop began to decline in the late 1950s when paint and linoleum manufacturers found synthetic alternatives for linseed oil. The Prairies, and to a lesser extent the northern plains, retained a large share of a small crop, and Canada remains the world's leading producer of flax seed. In the entire period of focus in this book, flax was almost never consumed directly as a human food. However, the use of flax seed as an important source of omega-3 fatty acids – discussed in the conclusion – may well add an important new section to this commodity's web.

The Flax Web

The many material stages of the flax commodity web – from cultivation to consumer goods – appear in a simplified diagram in Figures 0.3 and 0.4 ("flax commodity web"). The process began with a seed. Flax had several varieties, all containing what botanists call a constellation of features that characterized the central species, *Linum usitatisimum*. Some varieties were noted for length and quality of fibre, and others had stronger oil characteristics. In the mid-nineteenth century it was difficult to know for certain the variety one purchased, and experts warned that "it is so difficult to get any pure species it is perhaps better to disregard varieties and select seed for its apparent goodness."[11]

Planted in prepared earth and combined with some solar energy, the seed produced a plant (Figure 0.5). Planting in northern climates usually occurred in the last week of April or the first half of May, but the flax calendar was particularly flexible, as spring frosts were not a serious threat to the seed and the crop matured in about ninety-five days.[12] Seeding and weeding were followed by pulling or cutting the crop at harvest. Pulling involved expensive and difficult labour, but cut flax was as much as six inches shorter and less useful for making fibre.[13] Harvesting was followed by separating the fibrous straw from the seed capsules, or *bolls*, at the upper ends of the stalk and branches, a process known as *rippling*. This was done mechanically and by hand. Here the web bifurcated, and the fibre and seed were used for very different purposes.

Retting was a critical process where the straw was exposed to water in order to dissolve and break down the exterior *bast* (waste) fibres, also called woody fibres, which gave the straw its rigidity in the field. This allowed the

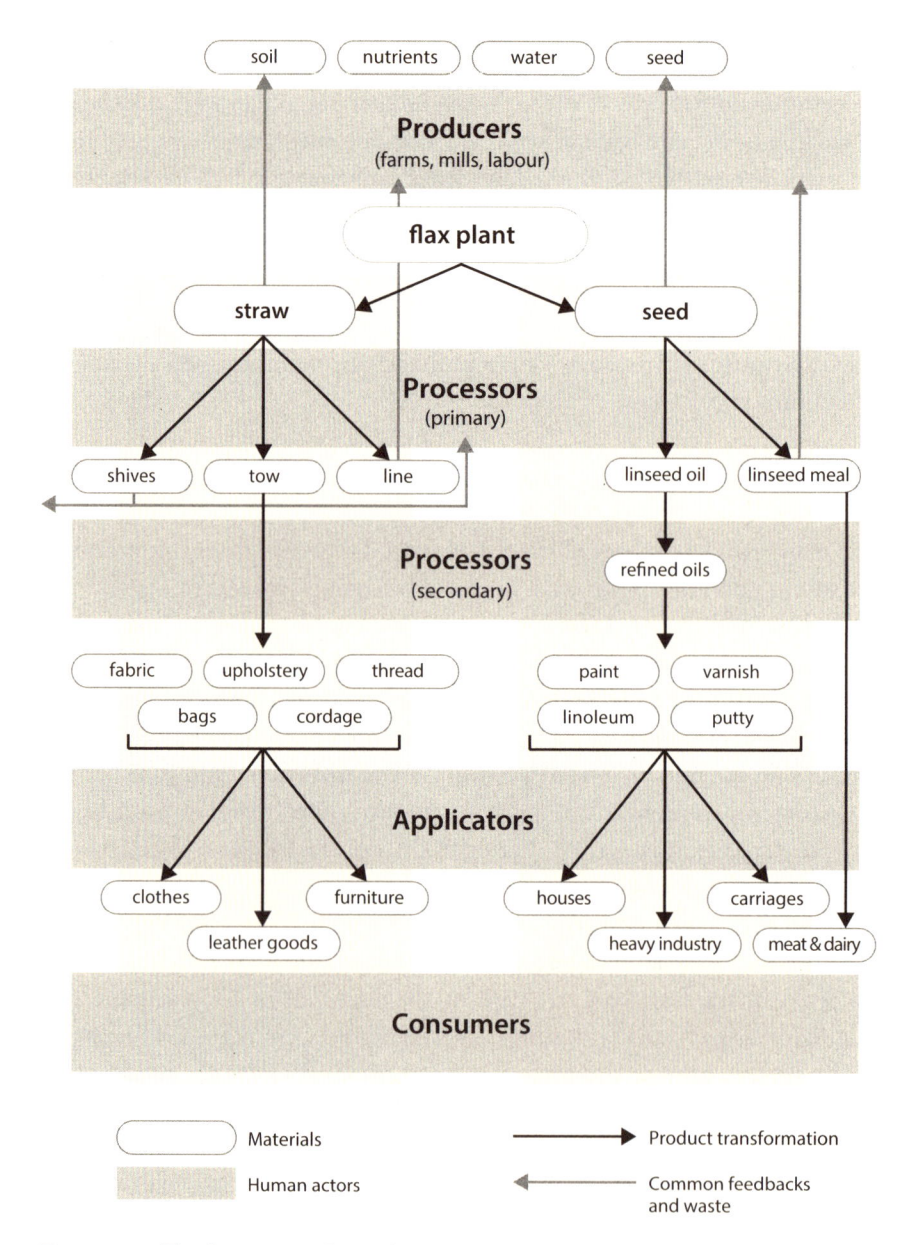

Figure 0.3 The flax commodity web

bast fibres to be removed from the valuable interior fibre (linen) through two processes, breaking and scutching. In most of Europe, flax straw was retted in water and dried over peat fires, but in warmer and drier North American climates these processes occurred in the fields with the dew

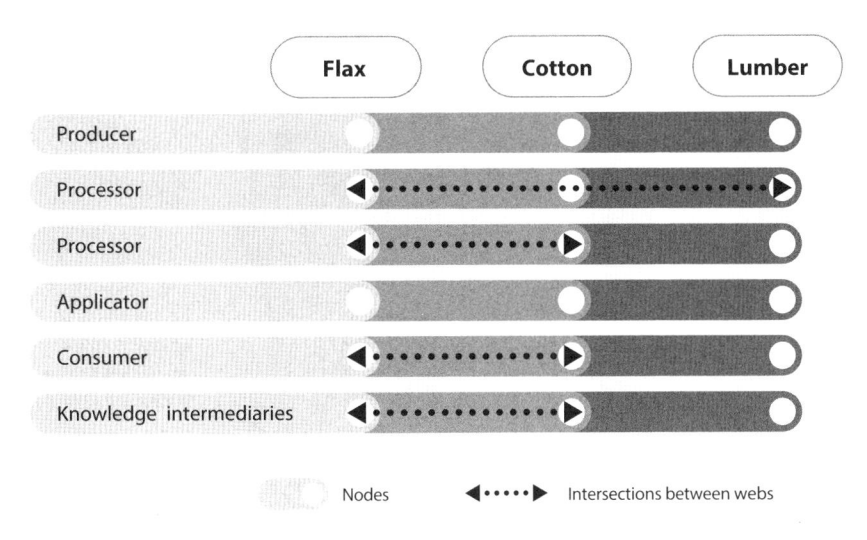

Figure 0.4 Flax and two complementary commodity webs

and sun.[14] Moreover, retting in streams and ponds – sometimes called the Courtrai method – was prohibited in Canada because the rotting straw contaminated the water and threatened both fish and livestock.[15] Retting was a time-sensitive operation; too much exposure to the water would cause the valuable line to deteriorate and not enough exposure would make the ensuing steps difficult.

The second and third stages of processing flax fibre were known as *breaking* and *scutching*. Breaking involved feeding the retted straw through a mechanical or hand-operated brake. A brake was simply a dull horizontal blade that chopped the straw without completely severing it. This broke and loosened the bast fibres but left the linen fibres intact. Scutching was the process by which the retted, broken bast fibres were finally removed from the internal fibres through a process of striking a dull blade down the length of a handful of straw as it hung vertically against a board. This removed the broken bast fibres, which then dropped to the ground as waste (also called shivs). A worker accelerated the mechanical scutching process by inserting the straw into a drum with a set of revolving blades powered by a mill wheel or a small engine. Mechanical scutching often caused more linen fibre to be lost with the waste, but it was about twenty or thirty times faster than scutching by hand. As mills began to switch from water power to steam, millers found that waste fibres were in fact a valuable biofuel for the mill's own engines.

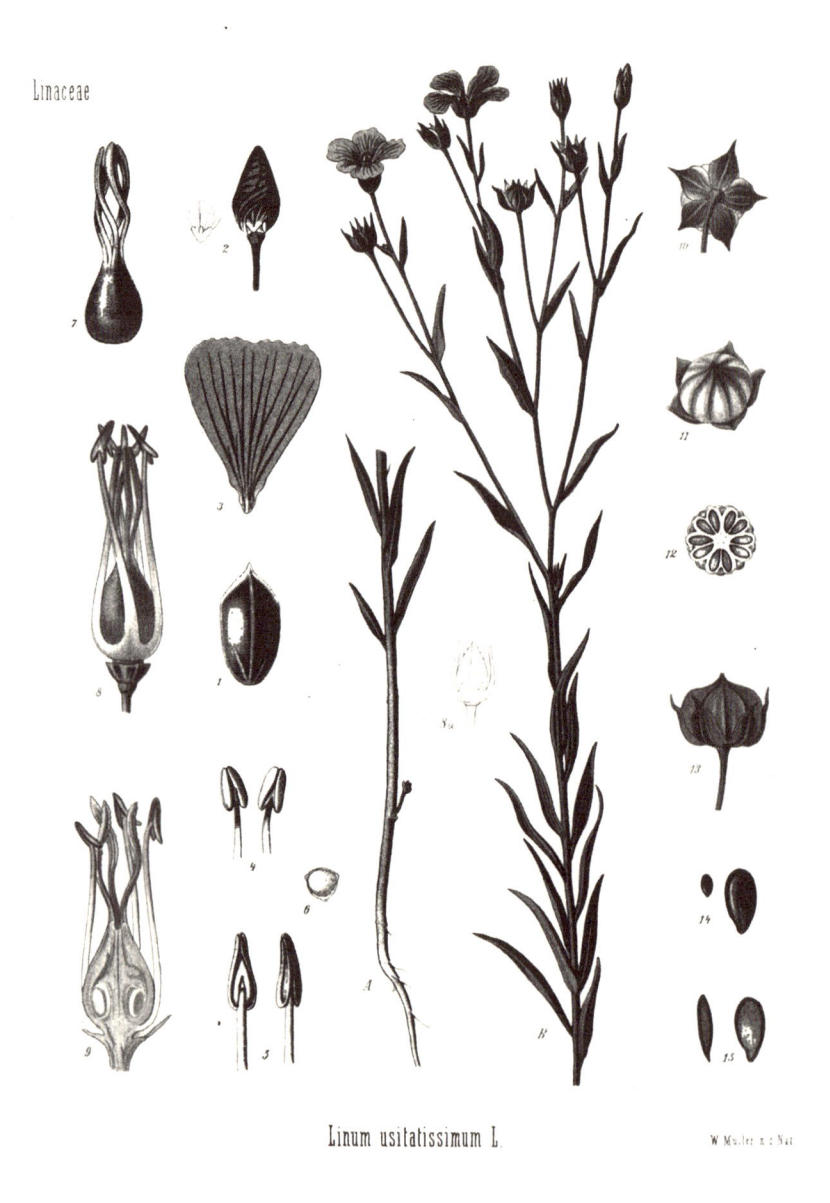

Figure 0.5   *Linum usitatissimum*, sketch from *Köhler's Medizinal-Pflanzen*, 1887

Hackling was the fourth and final step in the primary processing stage. Here, the scutched fibre was pulled through a series of steel combs, or *hackles*, starting with the coarsest and progressing to finer-toothed hackles. This separated the linen fibre into its long fibres (*line*) and short fibres (*tow*). Scutching mills usually performed all four steps in the primary processing operation, and then the line and tow were sold to secondary processors. Secondary processors spun fine linen thread from the line and coarse thread, cordage, and batting from the tow. The threads were used mainly in fine and rough textiles, and the batting was used in everything from upholstery to bedding and medical gauze. These flax fibres were all intermediate products in a variety of highly processed consumer goods. The "applicator" category is meant to represent the dozens of crafts whose members assembled everything from linen shirts and saddles to bed cords and mattresses.

The seed side of the commodity web also required multi-stage processing. After rippling removed the seed bolls from the flax straw, the bolls were threshed, using ordinary threshing equipment, to extract the flax seeds. These seeds were sold to grain merchants, to seed companies, and sometimes directly to linseed oil processors. The primary processors – usually called *crushers* – heated, crushed, and pressed the seed to produce linseed oil and linseed *cake*, or meal. Heating was traditionally performed on open fires, and crushing the seeds was done with rolling wheels or stamping rods powered by wind, water, or steam engines. Later technologies used more efficient heaters and grinders. The purpose of heating and crushing the seed was to create a mash that would release the maximum amount of oil from the seeds. Pressing was done in a variety of ways, from the traditional mechanical presses, resembling wine or olive oil presses, to wedge and screw presses. Wedge presses were literally large wedges that, when stamped down from above, would squeeze a bag of crushed seed against a heavy weight. The oil drained from the bag into a container, leaving the seed husks or "cake" behind in the bag. The cake was ground up as a valuable feed for livestock, and the bag was reused for crushing the next batch of seed. Screw presses squeezed the bag by spinning a large screw, and hydraulic presses used hydraulic cylinders and motors to press the seed. The percentage of oil extracted from the seed cake increased with each improvement in technology.

The raw linseed oil was sent to secondary processors who boiled, refined, and combined it with other chemicals to create the drying vehicle for a wide variety of hardware products, from window putty to paint. There was a range of quality in raw linseed oil, depending on the steps followed in the crushing stage. Raw oil would dry – eventually – but boiling and refining the oil was an important and highly specialized step necessary to produce a drying oil that would perform consistently in the various oil-based consumer goods. After it was refined, linseed oil was combined with driers, or siccatives, such as lead, zinc, or manganese, which accelerated the drying process and produced oil-based varnish and paint. Combined with cork shavings, the hardened oil also produced linoleum. Applying these goods to houses and vehicles was a skilled process that usually required additional inputs from painters, glaziers, and other trades people. As the linseed oil industry expanded in the late nineteenth century, it developed ready-mixed paints and other goods that could be used directly by the consumer.

The actors in the flax industry may be generalized in these five main groups: producers, primary processors, secondary processors (refiners), applicators, and consumers. This book touches on each stage and focuses closely on producers, but identifying flax farmers poses difficulties. They were mostly ordinary, anonymous people who moved frequently, left few records, and grew flax for only short periods. Some made large profits, but few farmers found the crop remunerative over the long term. The processors included millers as well as their workers, including women and First Nations families; many of these actors remained in the business long enough for their dealings with producers and consumers to be examined. Others existed only briefly, and since fraudulent operations abounded, many left a wake of financial debt and ruin. The most successful primary processors in Canada, such as the Perine brothers' flax scutching and cordage mills and James and John Livingston's scutching and linseed oil empires, are more visible because of archival manuscript records and vivid local accounts. Other linseed oil processors, refiners, and paint manufacturers included such well-known names as the Lyman Brothers in Montreal, the Rockefeller family's American Linseed Company, the young Archer-Daniels Linseed Oil Company (now ADM) in Minneapolis, and Spencer Kellogg and Sons in Buffalo, though perhaps they were not commonly associated with flax. Applicators combined processed flax and other materials in a range of goods for the consumer, including upholstery, saddles, carriages, implements, furniture, and houses. The final consumers bought flax in these forms and sometimes more directly, in cordage or linseed oil.

Nature leaned on this commodity web in many ways, including the plant itself, the humans at every stage, and the places where humans and the plant interacted with – and often polluted – the rest of the non-human environment. Environmental factors have been critical to agricultural history, as agro-ecosystems (including climate variation, drainage, pests, and diseases) came to dominate the habitat for plants and animals. Solar energy and the conditions provided by the farmer set in motion a reaction common to all plants – that is, nutrients from the soil were converted to plant mass through the nitrogen and other biogeochemical cycles. This member of the genus *Linum* had several characteristics that directed its growth, and this book stresses environmental adaptation as a form of biological, or non-mechanical, innovation in flax cultivation. First, humans domesticated flax over hundreds of millennia through an unconscious process of transporting varieties with favourable characteristics. In more recent years, flax producers consciously selected varieties that appeared to produce longer stalks, better oil, or better results in harsh climates. Second, farmers developed agricultural practices such as adapting the amount of seed planted per acre. A denser crop would produce more plants, fewer branches, and straighter and taller stalks for more fibre and less seed. Seeding fewer seeds per acre would have the opposite effect, and more branching would allow more seed bolls to develop on the stalk. Third, because flax had a short growing calendar, farmers had the option of planting flax last and still being able to harvest before fall frosts. This made flax more appealing in northern rotations and on new land. Most other crops required such early planting that they were only grown on land ploughed in previous years. Finally, although the fibre and seed came from the same plant, one was usually a by-product or even a waste product of the other. The fibre in the stalk began to degenerate as the seed matured, and there came a point in late summer when a producer could decide between harvesting the flax sooner and producing a high-quality fibre or letting it go to seed and sacrificing most of the quality of the fibre for the seed. Through a combination of the plant's own limits and human adaptation to new agro-ecosystems, environmental conditions and human agency add an important layer of complexity to many strands on the commodity web.

Another set of actors in the flax commodity web has not been included in the diagram or list of five actors. These were the intellectual, or non-physical, members of the web, and their role gained importance over time. They included state officials, farm scientists, and industrial researchers, as well as marketers and field reporters. In the early period, fibre farmers and millers had little use for the prolific writings of flax experts, but research and development eventually became sufficiently important in the oilseed business that some specialists lived purely off the value that their knowledge added, or was supposed to add, to flax seed and its products.

Commodity Webs, Chains, and Frontiers

Sustainability scientists have advanced the concept of telecoupling to model the interactions between human and natural systems in the modern world, and histories of commodities that connected the tropical and temperate worlds, including Sterling Evans's *Bound in Twine* and Sven Beckert's *Empire of Cotton*, make a strong case for the extractive, unequal, and unstable nature of these connections.[16] Twenty-first-century historians and scientists now argue that the impacts of these material extractions on the Earth's systems have been so profound that they threaten to unsteady the ship. One indicator of potentially irreversible changes is the expansion and industrialization of agriculture in the tropical world. In a recent issue of *Conservation Letters*, Nestor Gasparri et al. open with a poignant reminder: "Land-use change is the main driver of ... habitat loss and fragmentation, and the many detrimental off-site effects of industrialized agriculture." And, at current rates of consumption, they argue, "agricultural production will have to continue to increase ... either via agricultural expansion or intensification. Both will further amplify land-use-related pressure on biodiversity ... particularly in the Global South."[17] These are what some sustainability scholars refer to as the "agricultural commodity frontiers," and scientists like Gasparri study some of the mechanisms by which agri-businesses in the burgeoning oilseed sector contribute to the development of ecologically sensitive lands and the conversion of existing small holdings into large soybean monocultures with foreign ownership. They call this mechanism *telecoupling*, which Jianguo Liu et al. defined as "socioeconomic and environmental interactions among coupled human and natural systems over distances." By understanding these interactions through a more systematic framework, these authors hope to integrate a range of fields and encourage more scholars to compare social-ecological systems (SES) over long distances, between diverse systems, and through complex interactions like feedbacks and trade-offs. Ultimately, they believe such an approach could help improve sustainability planning and, indeed, our "ability to predict distant interactions and their consequences in all places." Liu acknowledges that distant socioeconomic and environmental interactions predate our present age, but he argues that where many human needs were met by local resources, they are "now being met by increased global trade."[18]

Environmental historians will be quick to point out that socioeconomic and environmental interactions have been occurring over long distances for centuries, and indeed, they may be one of the defining characteristics of humanity in what many geologists and at least a few historians call the Anthropocene.[19] William Cronon's *Nature's Metropolis: Chicago and the Great West* proposed that when the modern city combined long-distance communication, high-speed transportation, and industrial capitalism, "[g]eography no longer mattered very much except as a problem in management: time had conspired with capital to annihilate space."[20] Other historians will reiterate that these spatial interactions were only possible because of temporal interactions, like the Indigenous displacements, land reforms, energy transitions, and the constructions of Manifest Destiny that went along with US urban growth. And they would be right. A framework like telecoupling faces the crucible of time. Like many frameworks, it awaits a generation of historians who could pass complex case studies through its models as a test of its longevity and rigour. By examining how agricultural commodities expanded from local supply chains to interregional and global exchanges between parts of the temperate world, historians can test how historically coupled land-use systems compared with the commodity frontiers scientists study today.

Scholars have already offered several approaches to commodity studies. The process of moving through the various value-adding stages between producers and consumers has been described by some as following the links in a chain. World systems scholars call this *commodity chain theory*, or the examination of discrete stages of commodity production connecting producers and consumers over time.[21] *Flax Americana* follows a methodology that Matthew Evenden has called "a parallel approach to the commodity chain concept."[22] In business it would be similar to a "life-cycle assessment" history where the entire life of a commodity is considered, from raw material through to waste and disposal. It takes the reader through what Philip Scranton calls the entire "matrix" of environments, materials, knowledge, and consumer tastes that shaped commodity production and consumption.[23] "Commodity chain" can be a useful term, but after examining any commodity in greater detail, it becomes apparent that the linkages between stages are much more complex. In fact, the more we learn about commodities the less they appear to form linear chains. For example, in his study of shipbuilding Y. Eyüp Özveren found that these complex commodities were actually comprised of a whole set of sub-chains. In his initial essay (1992) on the sail sub-chain, Özveren attributes only three major stages to the manufacture of sails from flax, whereas his later essay (2000) recognizes at least eight. Linen historians show that flax producers and processors in the United Kingdom created at least eighteen sub-associations for the various stages and specialties in the trade.[24] Moreover, the fibres used for sail making were not only, and often not primarily, from flax. And, as we have seen, fibre was only one of several markets available to flax producers. There are so many moving parts that the metaphor of a single chain connecting producers and consumers is too limited, static, and unidirectional. Commodities connected people in many ways, power and influence travelled in both directions, and commodities changed over time according to price signals and environments. New commodities were developed out of the waste products of other commodities (such as linoleum from linseed oil and leftover cork), and what was once a by-product could suddenly become the new main output.

The flax industry was not a closed system or a series of boxes, but a web of intersections, a social network of transformative stages connecting people within and between multiple industries.[25] These *commodity webs* were part natural and part human, hybrids of the Holocene. Their organization was not unlike a network of neurons or mycelium if we think of flax not as a standalone chain but as one section of a much larger web of all commodities. In this model, the connecting nodes are the intersections between flax and other webs – places where actors could choose which commodity to invest in. If we focus only on the strands between actors, we see that the interconnections between flax's web and the commodity webs it overlapped are just as interesting as the ways flax connected producer and consumer.

Many environmental historians have recently argued, like Tina Loo, that "following commodities, and the people whose hands they pass through around the world, has become a powerful way to tell stories."[26] There are pitfalls to avoid along the way. Many recent commodity historians write about how salt, potatoes, corn, coffee, and so on changed the world.[27] They embrace a technological and environmental determinism popularized by writers like Jared Diamond and Michael Pollan.[28] Conversely, social constructionists argue that all societies, even industrialized commodity consumers, set out "paths and diversions" to establish the terms by which objects are circulated and exchanged. These terms may have nothing to do with the use value of an object, because all value is ultimately subjective and representative.[29] This story of flax is, by contrast, about how humans shape and are shaped by commodities over time and in relation to the places they cohabit with others. It follows a middle path set out by scholars like Graeme Wynn who interpret historical actors through a "constrained constructivism" and examine humans and nature "through the agency of work within the totality of 'habitat, economy and society.'"[30]

Substitutions are an important example of the constrained constructivist approach that illustrates the usefulness of the commodity web. The flax processes described above produced many intermediate goods, and some of them could be substituted with other materials. For example, if flax tow was retted, scutched, and hackled in a very precise way, it could be spun directly on cotton spindles or mixed in with cotton fibres for a hybrid textile made from tow, one of flax's less valuable fibres. This formed an intersection or overlap between the flax web and the cotton web. Since cottonseed cake could also be crushed and fed to livestock after being pressed for oil, linseed oil meal formed a second intersection with the cotton web. However, fibre and oil processors were usually different people, and only the latter were ever likely to switch between raw materials like flax seed and cottonseed. Oil crushers sometimes did this because they were "set up" for crushing, they had customers for livestock feed, and the only "retooling" they had to do was to find a new customer for the oil (linseed oil was purchased by paint makers and cottonseed oil by food companies). The cotton fibre millers sometimes experimented with flax tow, but since the flax-cotton substitution was very difficult to perform under normal mill conditions, it was not usually a viable exercise. Unless the price for cotton rose and the cost of making flax tow fell, the millers would likely continue spinning the fibre they were set up to spin – cotton. This is precisely what happened to prices during the American Civil War, and Canadian flax millers briefly enjoyed added markets for their tow.

Price usually influenced whether a miller would try to develop a substitute, but cultural attitudes about substitutions (often called "adulterants") were another factor to consider. All the actors in the web could substitute materials and processes. Paint manufacturers could use different driers, or they could dilute linseed oil with another drying oil, such as tung. Painters could decorate with whitewash or wallpaper. Home buyers might begin favouring brick walls over painted wood or carpet over linoleum. Just as consumer tastes changed over time, prevailing attitudes about localism, traditional process, and product purity changed as well. In some periods, finding substitutes for manufactured goods was celebrated, while in others substituting was a crime. Sometimes customers approved and preferred imported materials, while at other times it was deemed unpatriotic to buy raw materials from overseas. These attitudes also depended on the size of the commodity web and consumers' familiarity with raw materials. Canadian textile consumers knew the difference between Irish linen, Belgian Courtrai linen, "North River" linen from New York, and local coarse cloth from Ontario. Oil was much more difficult to detect by place of origin, partly because the whole aim of scientific linseed-oil refining was to create a standardized commodity that looked and performed the same way every time. Substitutions were thus one mechanism of the flax web, and they became the intersections where major changes could begin.

Farmers could choose among any number of crops that would grow under their environmental conditions. This meant that flax overlapped with many commodities at the producer level, but like the cotton miller, the farmer was only set up to grow a few of them. Retooling was possible there too, and like all the other actors, farmers had to consider the cost–benefit ratio. The cost of entering the flax web was the land, labour, and other inputs required to grow the crop, and the benefit was the amount the farmers expected to get in return. Other benefits, however, came in the form of various material and social feedback loops. The rippling could provide seed for next year's crop, or linseed meal from the mill could be fed to livestock on the farm. Dealing with a flax fibre miller might open labour opportunities for the entire farm family, and farming a specialty crop might increase a farmer's experience, knowledge, and social network. Commodity webs did not even have to intersect to influence other webs. Most famously, ship ballast was often an artificial source of coal, salt, or even immigrants in the transatlantic trade. In revolutionary New York and Philadelphia, the September salt market coincided with flax threshing time and meant, according to Thomas Truxes, that "it was largely the demand for salt that brought flaxseed to market."[31]

There was one plant but two primary branches of its commodity web, the flax fibre web and the flax-paint web, and these webs were connected in many ways with the various other fibres, oils, and even other agricultural commodities. In 1850 the Canadian webs were very small both in terms of value and materials and compared to the US mills. But then Canada's flax production began to grow, first with an operation in Montreal and second through a small group of US millers who had emigrated to Waterloo County, Upper Canada. Already growing rapidly, these industries were encouraged by liberal trade policy and the fibre shortages caused by the American Civil War, and production exploded in the 1860s. Suddenly southern Ontario had a vibrant network of flax millers with a range of outputs and many connections to businesses south of the border. But then, as with the 1930 "flax tour" mentioned above, the flax-paint web changed completely, growing almost overnight into a world-class industry and relocating from the small mill complexes to cities like Minneapolis and Montreal. What explains the transition?

Turning Points: Rifts and Shifts

Biographers often speak about turning points in an individual's life and then reflect on key before-and-after moments, patterns, and behaviours in support of their conclusions. Historians know that even the most dramatic changes occur through cumulative, syncretic processes. However, historical changes are stochastic and cyclonic, as well. Pivotal moments do occur. If a turning point was responsible for the massive differences between the flax webs of 1860 and those of 1930, it likely occurred around 1877. Like a person facing a turning point, the plant was the same organism before and afterwards. It had the same DNA and very similar varieties, but it lived a very different life. The new life of flax was defined by its enormous scale, environmental impact, and, most of all, its new main product: ready-mixed paint.

Fibre processing would continue for another fifty years, but the scale remained the same and oil became the area of explosive growth. Many other developments occurred in 1877 and 1878, but one critical event was the establishment of a trading relationship between James Livingston, one of the original Waterloo County flax millers, and the new Mennonite settlement at the West Reserve. The partnership began, and Mennonites grew the prairie's first few acres of flax, introducing it to a region that would soon grow many millions more.

In this study I differentiate between two kinds of temporal changes – rifts and shifts – in the commodity web. Rifts in a commodity web are structural changes that occur in the way capital or materials flow between key nodes, whereas shifts are simply a significant change in the quantity of those flows. Major rifts are turning points in the commodity's history.

A rift can decouple or recouple two nodes, often changing fundamentally the way actors perceived the industry and sometimes developing spinoff effects such as increased intersections with other commodities. The growth of textile factories was widely touted as a rift that shattered subsistence relationships and created many new stages between producers and consumers. But most Canadian textile consumers would not have perceived such a transition. In flax's commodity web, there was never a time when a significant proportion of Canadians produced a significant amount of their own textiles or linseed oil from flax. Most pioneers in the lower Great Lakes were well integrated with urban and regional markets, and flax production was too specialized for home manufacturing.

Nevertheless, major restructuring did occur. The rifts in the *material* flows were threefold in the period examined here: between the products and by-products that the plant yielded, between the rural generalist millers and the urban specialist millers, and, later, between primary and secondary processors like the oil and paint trades. Many changes occurred at the consumer level that reshaped the commodity web to varying degrees, but these were not rifts. Sometimes these changes could cause rifts, as middle-class house and car consumers did when their demand for paint caused producers to focus on the seed, and sometimes consumer tastes were driven by rifts in other webs: for instance, the decoupling of livestock systems, from local feed supplies to imported concentrates, including linseed meal. A fourth, *non-material*, rift occurred between so-called flax specialists. As enthusiasts attempted to create a fibre industry (or save an existing one), they published copious amounts of essentially useless or even potentially ruinous knowledge. Others who recognized the value and viability of the seed specialized in environmental and trade knowledge that could be used in the mill, the market, and the Experimental Farm. Their publication record was much lighter, but the value they added to the web was significant. These actors were semi-professional engineers, chemists, and biologists as well as trade and marketing professionals. Some specialists went to work for oil-press manufacturers, while others worked in the mill laboratories. Those with experience in biology could find work in the experimental stations and seed companies that developed flax varieties. The traders and sellers found work helping linseed oil and paint companies navigate the increasingly complex world of buying and selling flax. And these oilseed specialists could move between fields and between different commodity webs as well.[32]

The *shifts* in the commodity web were more in scale than kind. Shifts occurred when new actors entered the commodity web in significant ways – such as Argentine farmers during the First World War – or when the location of processing moved geographically between regions. These geographic shifts explain why the commodity web approach allows, and requires, historians to follow the plant from one environment to another, often across provincial and international boundaries. Telecoupling frameworks are designed to help scholars follow sectoral and geographic shifts, but they fail to explain the rifts or turning points. These occur over the long run, and their causes are often obscured by the language of industry and other actors, many of whom fail to understand the rift as it develops. For instance, when flax fibre's by-products, linseed oil and meal, became the main products, the mixed focus of the rural mill complex was replaced by the specialized focus of the urban industrial complex. There was no significant rift between producer and consumer, and there was no significant shift in the geography of production. That occurred later as the flax crops of the Great Lakes region were supplemented and then supplanted by the western flax frontier. But the flax web's new shape – or its new oleaginous consistency – helped expand several complementary webs, including other triglyceride and processed-feed industries. Rifts are temporal events with complex causes and unpredictable outcomes, and telecoupled frameworks that take long-run perspectives and perhaps even commodity webs into account will be better prepared "to predict," as Liu posits, "distant interactions and their consequences in all places."[33]

One of the fundamental differences between East and West – between the rural mill complexes and the urban industrial complex – is how millers or processors moved between different commodity webs. In the rural mill complexes of eastern Canada, processors changed their outputs so that they could more easily remain in the same town or region. As flax became a more western commodity and millers moved to the urban industrial complex, processors like the Archer-Daniels-Midland Linseed Company substituted their inputs so they could compete in the same general sector.

Flax East: The Rural Mill Complex
The first large concentrations of North American flax fibre and seed were processed in rural mill complexes as part of intermittent activities ranging from making bricks and carding wool to milling grist and distilling spirits. The intersections between the flax web and other webs were very active

in these places; a single mill or group of mills could invest in fibre and oil machinery if they sensed a market for these products, and they could also use the materials in the production of other commodities.

In the mid-nineteenth century, immigrants from the flax milling region near Troy, New York, converted a few sawmills to flax-processing factories in the rapidly expanding farm regions of western Upper Canada. They combined the motive power, expansive yards, and dry storage of the lumber mills with a few hundred acres of rented farmland and began to produce large amounts of the fibre for US markets and local manufacturing. A pre-existing commodity web, sawmilling shared enough similarities with the fibre-scutching business to repurpose the mills for flax. Farmers were too cautious to grow the quantity required by the mills, and they lacked the time and expertise to orchestrate large flax crops, so the millers became farmers, ploughed the soil, and assumed most of the costs of production. The networks established through these exchanges show the importance of place, primary manufacturing, capital and credit, and the changing relationship between urban consumers and rural agro-ecosystems.

By the 1870s these operations had expanded to dozens of small towns in what was by then Ontario, and some of the first movers were beginning to change in both scale and scope. Some invested in the equipment necessary for spinning flax fibre, and others began to crush oil from the seeds. But still these commodities were intermediate goods – the simplest versions of the consumer goods people created from the flax plant. There were no fine cambric shirts or colourful paints, only rough cords and raw oils. The finer fibres (scutched flax) were shipped to the urban textile mills on some of the region's first trains, and the rough-fibre goods (rags, bags, and broom twine) were almost immediately integrated into the region's vibrant watershed of makers. Flax millers were responding to the growing markets for fibre and oil, and they saw in the eastern Great Lakes region an increasingly interconnected world of suppliers and consumers.

Flax West: The Urban Industrial Complex
As urban paint and other oil-based markets expanded, they outpaced the mill complex and the miller–farmer relationships established there. The city became the new mill complex, the world its new hinterland. As demand for oil increased in eastern cities, millers abandoned farming, moved to town, and hired scientists. The new urban factory replaced the rural mill

complex, scientist-managers replaced millers, and corporate organizational structures replaced local networks built on interpersonal trust. Managers still had to gain the trust of a large network of actors, but farmers were not among them. Now millers dealt mainly with grain elevators, crop scientists, and other commodity specialists. Cities were the preferred site for these factories because they contained the largest markets for goods and the lowest costs for transportation and energy. Paint and varnish played an important role in the development, and the pollution, of London's industrial suburbs.[34] In North America, the oil business gravitated toward eastern cities with excellent water access, and New York, Buffalo, and Montreal became important metropoles.

Flax seed was purchased from a densely clustered area of producers located, until the 1880s, in the Midwest. The centre of flax seed production moved distinctly with the frontier of new cropland, and production on older farms was problematic. Most North American producers were relatively diversified; flax rarely accounted for more than 10 per cent of cropland and then only in counties in the centre of the flax "belt." Canadian millers benefited from this robust supply chain as well, but by 1909 even the United States began producing more linseed oil than its flax farmers would supply. As managers acquired seed from world markets, farmers in North America began to abandon flax for less problematic crops with greater and more consistent returns on investment. Production was outsourced to other grassland countries with fewer alternatives, and producers in Argentina and Uruguay became the new world leaders for the entire interwar period.[35]

The centre of flax production returned to the Northern Great Plains and Prairies after the Second World War, albeit in much smaller amounts. However, the oilseed crop and the urbanites who consumed it impacted the region in multiple ways. Through their demand for varnished carriages, white kitchens, and linoleum floors, a mainly urban base of paint consumers contributed to an injection of capital into the frontier and the ploughing of arid land that may not have been cropped otherwise. And, after decades of growing, trading, and studying the flax plant, those who abandoned flax formed the support base for a series of new but similar commodity webs built on oilseeds. Transborder regions were still important under the urban industrial system, and the network of seed suppliers, scientists, and other experts often operated across the border. The flax-paint web was driven by a new form of consumer taste, permitted by industrial restructuring, and accommodated by the dislocation of indigenous plants and people from the Northern Great Plains.

Placing the Industrial Revolution

This book is about the ascendancy of the urban industrial complex and how commodities reveal a new relationship between urban consumers and rural people and their environments. It is the city–country story told through the radials of a commodity's web. It also examines elements of industrial capitalism as it developed in these two regions. Flax might appear to have followed the "transition to capitalism" in the countryside where self-provisioning gets displaced by mechanized factory goods, farmers come under the control of urban manufacturers, and consumers are separated from producers and ultimately from nature.[36] This was capitalism at work. Cut. Print. Check the gate. However, this book argues, like Arjun Appadurai, that commodities are as complex as the people who produce and consume them. Like the items of great value given away or even destroyed at a potlatch, commodities cannot simply be categorized as "precapitalist" or "capitalist." Similarly, it is difficult to determine where phenomena such as self-provisioning and separation from nature worked in a web of such complexity and flux.[37]

Although flax cultivation is often presented as the pioneer's source of homespun linen, it was rarely grown for this purpose in the nineteenth century. Pioneer villages are particularly eager to share the story of self-sufficient fabric production using flax fibre's visually captivating processes, and recent research has shown that these museums are among Canadians' most trusted sources for historical information.[38] But the fibres, oils, and relationships forged in the small-town flax mill were of a much different sort. Very few Canadians made anything from flax fibre without the help of businesses and other neighbours. This is not to say they did not grow, scutch, spin, and weave fibres, but very few families had the time or equipment to do all four.[39] The literature describing a transition to capitalism in colonial North America does not adequately explain flax, the reason being that before these mills appeared, there was little flax grown for subsistence, at least in Upper Canada. After the millers arrived, many families adopted home production and outwork and preferred renting their land and labour over farmer-led commodity production. In Quebec, flax was not a timeless standard on every farm, but rather a specialized commodity that appeared and declined depending on other developments in commercial farming. Early eighteenth-century flax production in Quebec was so small that Alice Lunn found that, "perhaps around 1707, linen had become so expensive that habitants had been forced to raise flax for their own use." However, even after receiving state subsidies on flax seed, they "soon became disgusted by the careful preparation which the material required." The expansion of textile production in the late eighteenth and early nineteenth century in rural Quebec was primarily due to imported and locally traded fibres.[40] Even in the flax-intensive areas of the UK, according to W.H.K. Turner, "flax was worked privately for family use and commercially for sale, and the one did not preempt the other."[41]

Historians often think of the Industrial Revolution in terms of the separation of producers and consumers through the creation of a mass consumer apparatus in the years between Gustavus Swift's innovations in meat-packing (c. 1877) and the First World War, and it occurred in the flax-paint commodity web in precisely the same period.[42] However, the change may be better described as a *disturbance* in the ecological sense. The mill systems of nineteenth-century industrial capitalism appeared around pioneer communities in Upper Canada and across the region at least two decades earlier – with dizzying speed, by any contemporary measure. When this production began in the middle of the nineteenth century, there was already considerable distance between the field and the end user.[43] In fact, a degree of industrial production made flax goods cheaper and more available to producers than they were before, and some elements of self-provisioning increased with access to industry – for instance, when families accepted partially finished fibre as payment for their flax crop.[44] The flax-paint economy strained and disturbed the relationship between producer and consumer – but a simple process of separation it was not.

The new producer–consumer relationships that emerge in the flax story are mainly about the transition from local exchange centred in the mill complex to global markets centred in the urban industrial complex. The rapid appearance of flax mills and their accompanying agricultural production in recently resettled areas suggests that the operating system for capitalism was already in place in the Upper Canadian countryside in the mid-nineteenth century. Capital immediately followed the railway through the western part of the province, where it flourished at the edges of the industrial world. The network of rural mills was typically river-based, but railways multiplied its terrestrial connections.[45] There, experienced immigrant millers applied their craft knowledge to pre-existing mill complexes and new mills, and they organized the land and labour necessary to furnish them with fibre. Industrial capital did not sever rural people's subsistence relationship with fibre; rather it offered farm operators one more choice for their annual land-use plan and provided the necessary infrastructure (seed, knowledge, and a market) for implementing it. Similarly, in nearby cities in this period, Robert Kristofferson argues, modern urban capitalism did not displace local craftworkers, but instead provided them with mobility, opportunity, and continuity.[46]

The transition that followed was a change not so much in how flax was acquired in eastern mill complexes but in how it was acquired from the West. The main reason for the shift was the increasing scale of production of a single product, linseed oil. What mattered to business was the efficient movement of products between distant places, a process that after 1877 was increasingly centralized in cities through the global grain-distribution system. To be clear, the most dramatic changes in flax were in geography and scale, not in the organization or modes of production. Horizontal and vertical integration occurred at all points in the period covered by this book, as did capitalization and industrial farming methods. What was new was, first, that a single branch of the output had taken over the industry, second, that inputs were as easily acquired from overseas and from other continents as they were locally, and third, that scientists played a more important role in shaping the raw material and regulating the product.

Commodity webs represent spatial as well as economic relationships, and flax illuminates the importance of place in the small towns and prairie fields at the edges of industrialization. John Riley argues that the Great Lakes region was to species what the "middle ground" was for First Nations. It was "an open and borderless crossroads."[47] Flax shows the same process at work for commodities; the region was a connection as much as it was a destination. Millers arrived from the United States, and flax spread quickly across the western part of Upper Canada, as it had in Ohio and New York. Flax producers expanded from there into Michigan and the Canadian prairies, and Midwest US flax seed poured back into Buffalo, Toronto, and Montreal mills. Place had a somewhat different importance in the Northern Great Plains and Prairies. There the prevailing settler myths oscillate between community-oriented ethnic minorities and profit-minded speculators who planned to farm wheat and get out. Merle

Massie focuses instead on the diversity of "prairie" experiences as people moved between grassland and parkland, participating in many commodity webs beyond the ubiquitous wheat story.[48] Further south, flax and other oilseeds complicate the idea of wheat monoculture and reveal the dynamic nature of farming on the Canadian-US border at the edge of the semi-arid prairie. Mapping these communities of producers at three scales – the farm level (1861–71), the eastern district (1861–91), and the western rural municipality (1890–1926) in a historical Geographic Information System (HGIS) – is a recently established way to move beyond regional generalizations and identify patterns of place. Additionally, account books from the original flax mill in Waterloo County (1856–65) allow us to tease out of some 5,000 daily transactions the relationships forged between countryside, mill, and consumers.[49]

Following on trends in world and Latin American history, specialty crops and other smaller commodity webs are beginning to emerge and form critical connections between Canada's most famous agricultural staples.[50] Some of the most useful texts on flax history range from the memoirs of industrialists to the latest scholarly books and articles. Whitney Eastman's *History of the Linseed Oil Industry in the United States* is a useful mix of statistics and historical overviews with personal reminiscences from his own career as a linseed oil miller.[51] Other texts include a well-researched history of local flax mills in Waterloo by Kathryn Lamb and studies of flax cloth in the United States by Kenneth Keller and Laurel Thatcher Ulrich.[52] Quebec historians David Thierry Ruddel and Sophie-Laurence Lamontagne and Fernand Harvey explore rural textile production in the late eighteenth and early nineteenth century, and Michel Boisvert examines the complex and often complementary relationships between sheep, fibre crop, and textile-producing regions of mid-nineteenth-century Quebec. Boisvert's thesis sets the stage for the next chapter of flax history in Quebec, the flax seed production that supplied Montreal's linseed oil manufacturers in the 1860s.[53]

Environmental historians want to read about a countryside with room for complexity and places that matter. Too often they find one of two extremes. On the one hand, rural people are depicted as self-sufficient characters living out a simple plotline in isolated prelapsarian communities. On the other, they are seen as capitalists in overalls, clearing trees and mining soil in order to convert – before moving on to new frontiers – featureless parcels of land into units of production. These units are often placeless, valued only by the percentage of forest removed and "improved," as well as by their ability to prepare staples for distant urban markets. No single commodity history can fully problematize this story, but by examining the actors inside and around the edges of the networks of exchange, many combined histories have and will.[54] This book attempts to portray the countryside for what it was and to understand rural people on their own terms. It describes the many facets of material goods stemming from a single plant in order to show the complexity of commodity production and the choices available to producers. This countryside was in perpetual transition. Its land was "discovered," measured, traded, and transformed irreversibly. Then, as population density increased and economies grew, there was another development: farmers specialized and pluri-occupational work declined in many of the most prosperous areas. Now the process of rural depopulation began, but this was more a matter about internal relocations than about net loss.[55] Farms consolidated and surplus populations moved to towns and cities. Today farms produce more food with less labour than ever in history. However, they require, and continue to require, more inputs. The fact that luxury commodities like paint could emerge from the soil at the same time as the food for these growing nations is another indication of the rapidly increasing productivity of agriculture.

A commodity history will take different shapes depending on whether it is told from the perspective of the products or the raw material. This book is organized according to the region and not the products of the web. There was only one flax plant, and although cultivation and industrial output changed, the plant did not. The first three chapters examine flax as it developed in the Great Lakes region, and the final two chapters focus exclusively on the northern plains. The two chapters in between examine the transitional period as seed displaced fibre in manufacturing, western farmers dwarfed the East in cultivated areas, and the urban industrial complex displaced the small mill in organization. Within each region we move through the fibre and oil webs, from the demand side first, beginning with the consumer goods and working back to the producers and promoters. The two basic materials in this book, fibre and oil, were in many ways the quintessential commodities of the First and Second Industrial Revolutions, respectively. To understand the plant, then, we need to understand why thousands of people would participate in a non-foodstuff agricultural commodity web like flax. What were the goods that consumers wanted? Who decided to manufacture them, and where? And who extracted the raw materials from the soils of a thinly populated country preoccupied with converting wildland to a granary? Another question posed in each section is why were promoters so enchanted by this plant, and why were they usually so unsuccessful in predicting and directing the shape of its commodity web?

Thus, the first two chapters of this book explain how southwestern Ontario millers acquired and transformed flax fibre for a variety of urban and local markets. Flax cultivation was tied to both local manufacturing and external sources of capital and knowledge, particularly from across the transnational Great Lakes region and New England cities. These factors were much more important than state support and other promotional efforts, but promoters were so persistent and so blindly committed to the fibre threads of the flax web that they merit a third chapter, one that investigates their failed and sometimes fraudulent attempts to shape the industry. Early fibre promoters were captivated by the promise that practical chemistry could transform flax into cotton and displace the dependence on Southern plantations. Later fibre promoters focused on the "labour problem," which some interpreted as a new racist view of First Nations labourers. Chapters 4 and 5 introduce the new demand for oil and the dramatic changes this brought to the geography and industrial organization of the commodity web. Chapter 5 concludes with a new breed of promoters born out of the success of the industry, particularly in the United States. These men and women were not promoting flax itself but rather its products and their regulation, all under the banner of new chemical standards. Industry hired marketers to advance their products' purity and performance, and consumer groups asked governments to test them. The final two chapters focus on the supply side of the flax-paint web. They establish, first, that farmers brought flax to the Northern Great Plains and Prairies as a first crop on new land, and then, that state scientists now helped shape flax production and understand its problems. Scientists were unable to keep the region's farmers in flax when other crops paid a better "rent" for the land. However, state and industry flax seed specialists did help form a new oilseed sector (dominated by canola but including flax seed and the occasional fibre promoter) that continues today.

CHAPTER ONE

The Edge of Industrialization

Finding a Northern Fibre

Like every story of work and material transformation in Canada, the history of the flax commodity web contains moments of tragedy and triumph. The Industrial Revolution shaped the biological, mechanical, and political landscape of the Great Lakes region and created unique opportunities for its residents. Where else could a Jamaican slave owner arrive via Charleston, start an agricultural society, and help convince a Unionist immigrant and local schoolteacher to start a flax mill that eventually hired escaped slaves, among other men and children? It happened in St Thomas, Upper Canada (later Ontario),[1] when a mill was established in 1864, but similar stories were told throughout the Great Lakes region. These hardly constituted a flax belt, for the fibre had never been grown or milled on a large scale anywhere in Canada. Yet it was immediately brought into production in this pioneer area through a transnational network of industrial knowledge, capital, railway transportation, and local labour. The crop was grown by both farmers and the mill itself. Its fibre was processed by both housewives and mill workers. Its consumption was encouraged by the fibre shortages of the American Civil War, and it was sold (duty-free) in the same textile sector that defined the conflict. Flax was, of course, not a simple triumph of industrialization; it was made possible by many tragic episodes as well. Flax mills were more likely to employ children than any other industry in Canada, and most of the back-breaking field work was performed in many periods by First Nations labourers with few viable alternatives (see

Chapter 2). And eventually, the increased consumption of flax seed for luxury commodities like paint and linoleum required an extended cultivation of flax on the most sensitive northern grasslands (see Chapter 6).

This chapter describes the large-scale introduction of the flax plant to Great Lakes agro-ecosystems as well as the framework necessary for its production. Flax's first life in this region was as a fibre cash crop, and its first significant appearance was in the mid-nineteenth century. Previously, farmers were extremely reluctant to grow flax because it had limited markets and was too difficult to process in the home without local infrastructure in place. When it did appear on Canadian farms, it was not as a result of state and expert promotions, although there were many. Most experts, officials, and commentators imagined a commodity they called a *flax staple* processed independently by farmers for export to Great Britain. What materialized in the Great Lakes region was something quite different, and the province's flax fibre web reveals the many threads of the fibre economy, the central role of rural mill complexes, and the effect of spinoff economies in the region's manufacturing sector.

A profile of the St Thomas flax mill suggests that industrial agribusiness was hardly foreign to the Upper Canadian countryside. This mill was located at the bridge on St George Street in the Kettle Creek valley. A nearby slaughterhouse gave the tiny neighbourhood the moniker "Bone Town." Built in 1864 by some experienced flax millers from Waterloo County, the Perine brothers, and managed by a Waterloo schoolteacher, the mill was the first to appear in this part of Upper Canada. A rare photograph of a busy afternoon in its first full season provides a vivid illustration of how, and where, the flax industry worked (Figure 1.1). The photograph shows an explosion of activity late in the summer of what was likely 1865.[2] Wagons are heaped, but sensibly, nowhere close to the size of the "brag loads" that appeared in photographs from other parts of Ontario. The mill and surroundings are a hive of workers, with at least five wagon teams visible along the road, upwards of a dozen hands at the mill and a dozen more in the yards behind the mill. Those yards were used for storing the raw material, and this image is a rare glimpse of the barn-sized stacks of flax straw, three adjacent to the mill race, three more along St George Street, and another half-formed in the field to the north. Several workers scamper over one of the piles, using ladders and manual labour to pile the curiously shaped stacks. The stacks are designed like thatched cottages, shedding the rain and snow and protecting all of the fibre beneath the top layer through

Figure 1.1   Perine and Young flax mill, St Thomas, c. 1865

the winter. This method was imported from New York where one author remarked that it was also recommended to cover the stack with canvas or boards.[3] Another landscape photograph of the valley in the 1870s and two panoramas of the city show that these enormous piles of fibre were semi-permanent "structures" on the outskirts of St Thomas.[4]

The landscape of the Kettle Creek valley was ideal for this industry. The valley forms a small (520 square kilometres) watershed that drains mainly agricultural land into Lake Erie to the south. Its circumfluent course defines the western edge of St Thomas, Elgin County's seat and largest city. When the mill was built, St Thomas was growing rapidly from a crossroad village to a newly incorporated town with 1,631 residents. Like many towns in this part of Upper Canada, St Thomas had been connected to a railway since 1856, and much of its growth in the following decades was encouraged by its role as a major junction.[5] The surrounding farmland was some of the best in Canada, and in addition to providing the mill seat, the river valley's slope provided perfect conditions for dedicated flax fields. At least six fields were laid out by fences running up and down the hill, protecting the crops from rogue livestock and allowing pullers to start at the bottom of the hill and move upward as they progressed.[6] Other views of the valley show that this was a common layout along the town's river-front lots, probably because it provided any fenced livestock the best access to water. The south-facing slope exposed the crop to maximum sunlight, and the mill at the bottom of the hill seemed to welcome the cumbersome loads of fibre in the easiest way possible.

The photo gives clues to the messy nature of local flax fibre processes such as pulling and retting, although much is obscured in the grainy image. Water and dew retting was "from the first to the last disagreeable, at every stage of the process" wrote an Irish essayist: "The manual labor is filthy in the extreme" and the steeped water kills fish and makes water unfit for cows.[7] The second field from the street shows less than a third of the crop still standing, suggesting that the harvest was almost complete and the date was late summer or early fall, before the leaves were dispatched from the hardwoods above the valley. The bare earth in the two fields closest to the road confirms that the crop had been pulled, not cut, and it appears that it had already been removed from the field at the time of the image. The darker section in the second field, downhill from the standing crop, was likely the newly pulled flax, retting in the dew and waiting to be hauled and stacked. The time of year suggests that the flax had not been fully retted. Retting took up to four weeks and a great deal of labour, so the fact that the stacks had already been built suggests that Perine and Young intended to ret the fibre later using the water method. It was illegal to ret flax in streams in Canada, and no doubt the location of these riverbank flax stacks violated the spirit if not the letter of the law. However, the strong pollution leeching into Kettle Creek from the abattoir and other industries in Bone Town may well have obscured the pollution from the flax mill. All of this riparian activity took place upstream from St Thomas's waterworks.

The St Thomas mill was representative of flax's role in the region's industrial, social, and environmental transformation. This was a place where former slave owners, Unionists, and escaped slaves worked together, not equally, and not exactly cheek by jowl, but together. The aforementioned slave owner was Benjamin Walker (c. 1817–1882), an English lawyer and plantation owner who had left Jamaica for Charleston after what the *Charleston Courier* called "heavy reverses of fortune, caused by the Emancipation Acts of the British Parliament." He moved again, with his wife Caroline Howard-Gibbon and four children, to Upper Canada sometime before the birth of their youngest child in 1859. In 1861 he was forty-four years old and worked as a barrister in St Thomas. Walker started the Elgin Flax Association in 1862, one of many wartime attempts to promote flax cultivation. As we will see in Chapter 3, virtually all fibre promotion efforts failed, and in all periods, including the American Civil War. Walker's effort, however, was a possible exception and may have been the most influential project of the lot. It may be coincidental, but a successful flax mill did open in St Thomas two years later, and unlike other promoters, Walker prescribed the kinds of flax processing that millers like the Perines actually used. Perine and Young were not part of the association, and they were likely more impressed by the town's railways than by its fibre experts. As railwaymen they would have visited St Thomas in the 1850s where they might have examined the mill prospects for themselves. If they did engage Walker in 1862, it must have been an interesting interaction between disparate political camps. The Perines were Unionists, Young was Presbyterian and presumably abolitionist, and Walker was almost certainly a Confederate. In the same year he started the Flax Association, he received news that his oldest son, Henry Pinckney Walker II, who had stayed behind at the Citadel Academy in Charleston, had been killed in action at the Battle of Secessionville.[8]

It is clear why Walker left Jamaica. Many plantation owners moved from the British West Indies to the southern United States after emancipation to operate in similar slave-based industries there.[9] It is less clear why he immigrated to Canada. Perhaps like other industrialists moving north across the border, he saw opportunity in the Upper Canadian fibre economy. Perhaps he just wanted a new beginning. Walker's four oldest children – Henry Pinckney, Maria Jane, Charles, and Fanny – were all born in Jamaica, and they were not Caroline's children.[10] Walker had lost his fifth child and likely his first wife in Jamaica in 1852. He then emigrated to Charleston where he had networks through her family. Her name was Maria Pinckney, and it was her family, not his, who owned a plantation and seven slaves in Jamaica.[11] It is not clear where Walker met Caroline, but she was English born and they were married in St Thomas in 1858.[12] Perhaps it was Caroline who led him away from Charleston and away from his slave-owning in-laws. Like many Upper Canadian families they moved several times within the new Dominion as well. By 1867 Walker practised law in Bothwell, with his surviving son working as clerk, and by 1871 he lived in East Toronto with Caroline and their youngest child.[13] He continued to promote flax fibre all through the civil war, sharing his knowledge with agricultural associations and the Bureau of Agriculture and seeking financial support from the latter for his efforts.[14] The Canadian government's efforts to encourage flax cultivation are explored in Chapter 3; the more significant actors in the commodity web were the millers and labourers who cultivated and transformed the plant into a range of consumer goods. The most salient aspect of the St Thomas story is the speed at which a mill could appear and then convert a river valley into a fibre mine. Walker understood part of the reason: the world was hungry for fibre. And the Perines and Young understood how and where to meet this demand.

Perhaps the St Thomas story could happen in any agricultural community experiencing economic growth and access to land, capital, and markets, but in the nineteenth century it took place most often on the edges – along the urban periphery and at the intersections between agricultural settlement and industrialization. These mills often hid in plain sight on the edge of small towns, dominating the modest skyline before being demolished, repurposed, or redeveloped. They existed beyond the border of most urban maps. When they did appear on fire insurance plans, they were often marked "silent," and many were operated infrequently enough to avoid the census. The farmers and millers on these edges were too often seen as people constrained by urban forces, living in places trapped in time when really the people shaping flax's commodity webs were adaptable, market-responsive agents rarely captured by the census, let alone controlled by the state. These peripheries were also central places, connecting the city and the country, encouraging mobility, and providing a place for one of Canada's first maker cultures. A close examination of manuscript census data in conjunction with business account books reveals many activities, relationships, and even whole commodities that escaped the notice of general statistics. The complex ways people manipulated industrial commodities in these peri-urban spaces demonstrates the critical role of the rural mill complex. A mill complex was any multi-structure site that processed multiple commodities. A single proprietor could own several complementary businesses on one parcel of land, or in the case of Bone Town, St Thomas, a range of owners might cluster operations in a single neighbourhood to take advantage of local factors such as labour, markets, water, energy supplies, and one another (Figure 1.2). Mill complexes shaped commodity webs both as the nodes where materials were transformed and the intersections where people moved from one commodity web to another.

Figure 1.2   Bowman's flax mill in a mill complex, Floradale, ON

Craftwork and manufacturing have been well-studied topics in US history, especially in textiles, but Canadian manufacturing was more or less ignored by historians, who focused on staple exports and commerce.[15] An early generation of Canadian historians believed, like K.W. Taylor, that "the history of modern manufacturing in Canada dates from 1879."[16] However, recent work shows that manufacturing was a very significant part of mid- and late-nineteenth-century Canadian economic development, including primary processing for the rural economy and what Ian Drummond has called "the production of relatively simple consumer goods."[17] As a quintessentially rural industry, flax has been overlooked in a literature that favours urban manufacturing.[18] The small flax fibre and cordage makers studied here have been more or less invisible, although recently environmental historians like Sterling Evans have examined similar processes in hemp and jute ropeworks and in sisal twine.[19] Businesses like the St Thomas flax mill illustrate the speed at which commodities and capital moved, appearing and disappearing, along the edges of industrialization. These mills supplied intermediate fibre goods and a variety of cordage and rough textile products to rural families, local industries, and urban manufacturers in Canada and the United States. Since mill complexes were initially so central to the flax story, they appeared along watersheds, particularly the Grand River watershed, but also on the Credit River, Kettle Creek, and eventually the Maitland River. By the 1870s there was no flax on the Credit and the flax fibre industry developed primarily in the small farming communities between Waterloo County and Lake Huron, roughly centred on Stratford.

Following an American pattern of flax production, three brothers, Moses, Joseph, and William Perine, immigrated to Waterloo County from Troy, New York. They initially built three flax mills with the intention of selling the processed fibre to textile and cordage mills in Massachusetts and New York State. Their timing coincided closely with the 1854 Reciprocity Treaty (the relatively liberal trade regime between Canada and the United States), but although Reciprocity eliminated the tariff on flax imported from Canada, other factors drew the Perines north of the border and helped them establish the flax industry there. The number of Upper Canadian mills increased from three to thirty-three during the American Civil War, when contemporary flax enthusiasts (and later historians) claimed that Canadian flax was one substitute for US cotton in the textile industry.[20] Certainly, production in some regions increased during the war, and Irish farmers briefly doubled their acreage.[21] Yet Canadian flax millers also invested heavily in businesses that prepared flax goods prohibited by US tariffs and only suitable for local markets. The Perines' Doon mill on a small tributary of the Grand River soon became a major twine and cordage manufacturing facility, and the brothers quickly established another much-larger textile mill on the Credit River, near Streetsville, with Toronto millers Gooderham & Worts. Ontario's flax industry continued to thrive beyond the abrogation of Reciprocity in 1866 and despite a steadily increasing US tariff on unmanufactured flax. A better explanation for the Perines' success in flax lies in the transnational entrepreneurial community forming its networks of trade and industrial knowledge and the growing manufacturing sector in Canada that provided the Perines a local market for their products.

The account books of merchant-millers help explain the demand for flax products and give us a picture of the commodity from the mill floor. A selection of the Perines' account books survives, but these are the only records they left. Moses Billings's Perines' daybooks and ledgers provide sufficient material on their linen and flax operations to determine industrial output, labour costs (and sometimes roles), and, to a certain extent, their relationships with farmers. Several historians have recently explored aspects of rural business activities referred to in account books, though most examined purchasing patterns using a much larger range of products.[22] The result is a picture of the production and consumption of a commodity that, while almost invisible in the historiography, was as common in the mid-nineteenth century as a grain sack, a bed cord, or a wooden ship.

The story of workers immigrating to the mill towns of New York and New England is familiar to many Canadians, but the mid-nineteenth-century flax web in the Great Lakes region shows the importance of the Americans who contributed to Canadian rural and industrial development.[23] In the mid-nineteenth century the expansion of railways in Upper Canada – especially the completion of the Great Western in 1854 – facilitated connections between that province and New York State.[24] The Perines were contractors and had worked on the Hudson, Cleveland & Pittsburgh and Troy & Boston railways with another brother, Nicholas, before moving on to Canadian lines. They first came to Canada to work on the Great Western, but by 1853 the young men had decided to invest in flax manufacturing.[25] By the end of the decade, the three brothers and their wives had young families and were situated in three small towns in Waterloo County. Each owned and operated a flax mill. The eldest, Moses Billings (1815–1898), processed the most flax and operated a sawmill on Schneider's Creek just above the town of Doon; Joseph Southworth (1820–1880) was located in Baden on Mill Creek, a small tributary of the Nith River, and William Danforth (b. 1826) in Conestogo at the confluence of the Conestogo and Grand Rivers. Their accounts show that flax had been part of their Canadian business plan almost as soon as they arrived. One of the first transactions Moses Perine recorded in Waterloo County included "Planks to Build Flax Mill," in April 1854.[26] According to a later article in the *Canadian Journal of Fabrics*, they shipped the "first carload of scutched flax to the United States three months before the reciprocity treaty came into force, and got a refund of $60 on the shipment."[27]

The Perine brothers were apparently the first in their family to enter the flax business, so their motives for working with flax and moving to Waterloo are not immediately obvious. Born and raised on a farm, the Perines were not well-known capitalists or even mill owners before they immigrated to Canada. They were born in Jackson, New York, near the Vermont border, and their accounts reveal that they left their homestead farm in that vicinity. Their father, John, had been a blacksmith and farmer in nearby Greenwich until his death in 1848, so presumably the farm was his.[28] The brothers took a careful inventory of the stock, tools, livestock, and grain – all to be sold to the new owner. Flax or flax equipment do not appear on the list.[29] Their mother, Hannah Billings, was initially the most likely source of any family capital used to secure property and equipment in Canada. The first few pages of Moses's account book record entries for "Billings & Ingram," a grocery in Troy, New York, that was established by Henry Ingram of Marlboro, Vermont, in 1830. Ingram later went on to organize and preside over the National State Bank, one of the largest lenders in Troy.[30] Moses Billings took the account book, together with the knowledge of double-entry accounting needed to fill it, with him to Canada. The most important export from Washington County to the Upper Canadian fibre economy, however, was surely a few bags of flax seed and a deep craft knowledge of the flax commodity web. The Perines almost certainly learned everything they knew about flax from the region around their home farm, where flax was cultivated extensively and where several scutching mills were in operation (Figure 1.3). Washington County and neighbouring Rensselaer County made up "the two greatest flax-raising counties in New York," according to one manual.[31] Large amounts of flax had been cultivated and milled there since the 1830s, and now local entrepreneurs sought to replicate the business model in new northern farm communities.

The Perines' knowledge of cross-border flax milling, flax markets, and shipping was undoubtedly the secret to their success. Their mother, Hannah Billings, was an important figure in the business and in her community until her death in Waterloo in 1875.[32] For several years she remained in New York State to liaise between the flax millers and their clients along the Erie Canal. In 1862 the accounts recorded that she was paid cash "at the falls" in November and then again the following April.[33] Eventually Hannah moved to Waterloo, but the New York family connections were maintained by other relatives who remained. William Danforth returned to the United States in 1868 and continued on in both the construction and flax trades in San Mateo, California. His brother Nicholas had also moved to California, just across the bay in Fruitvale.[34] The family was initially much better connected across the border than they were within it. The Perines were a Huguenot family by ancestry, and their slightly awkward integration into Waterloo's established Mennonite community was evidenced by the poor spelling of customer names in the accounts and, more seriously, by two Mennonites who cryptically recalled "many adversities in our business with the Perrines [sic]."[35]

The flax business grew slowly at first, but as prices improved during the American Civil War, the Perines were well positioned for expansion. In 1860 the Perines owned the only flax mills in Upper Canada, with the

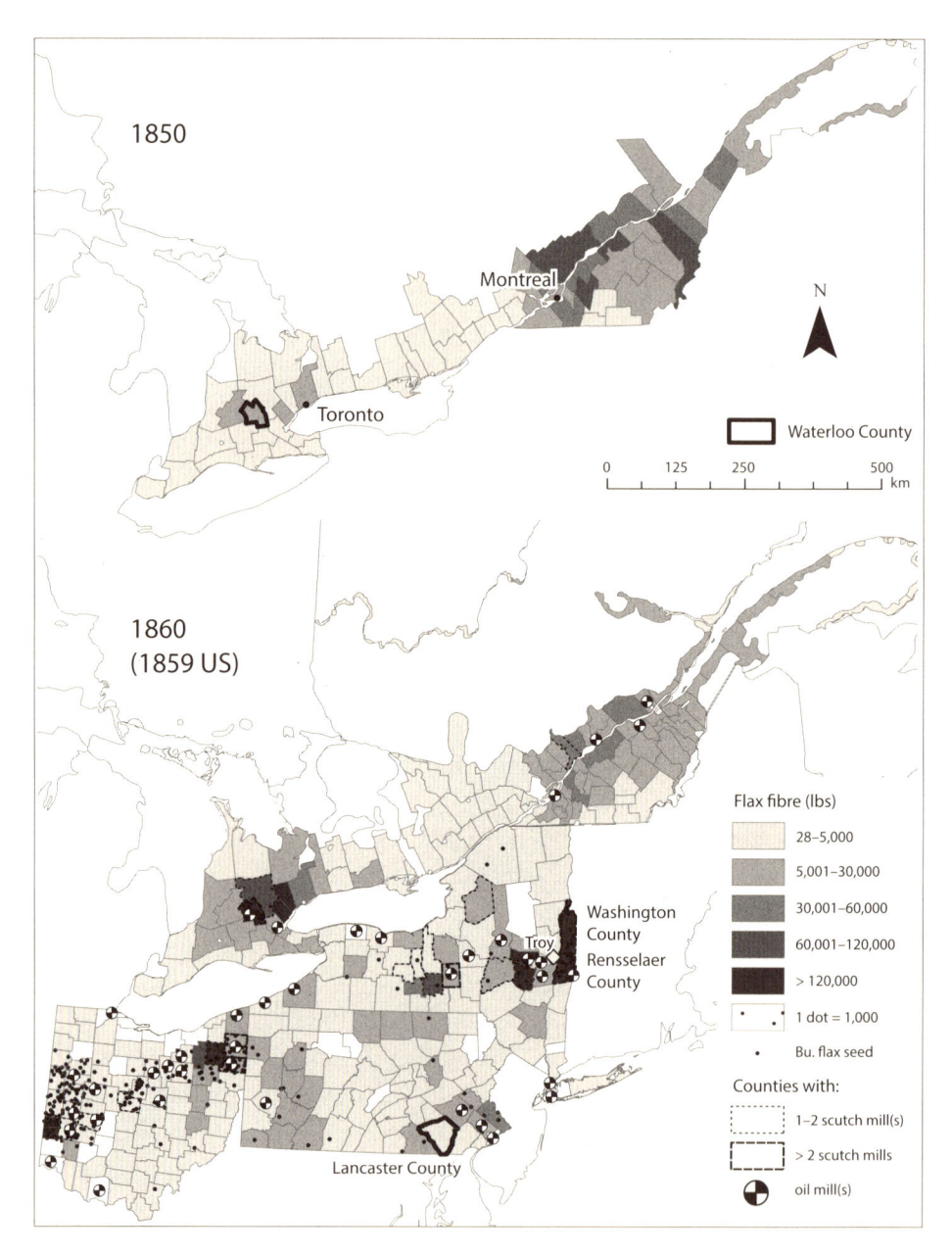

Figure 1.3 Flax output and industry in the Great Lakes region, 1850, 1859/60

Source: Canada, *Census of the Canadas, 1851–52*, Vol. 2 (Quebec: Lovell and Lamoureux 1855), 64–5, 164–5; Canada, *Census of the Canadas, 1860–61*, Vol. 2 (Quebec: S.B. Foote 1863), 218–19; United States, Department of the Interior, *Eighth Census, 1860* (Washington, DC: Government Printing Office [hereafter GPO] 1864). Canadian spatial boundary files for Figures 1.3 and 1.5 are used courtesy of Georeferenced Databases for Assessing Historical Data, University of Toronto.

exception of one in Halton, but five years later they owned seven of the thirty-three flax mills in Mitchell's directory.[36] The Perines often partnered with other businessmen when building new flax operations, as they did with Alexander Young in St Thomas.[37] Young's account books show that he was in regular communication with the Perines and often asked for their advice on flax samples, prices, and what quantities were needed.[38] Young had worked at several teaching positions throughout Waterloo County, and he had been a principal of the Central School in Berlin (Kitchener) when the Perines convinced him to manage the mill in St Thomas. He and his pregnant wife, Anna Eliza Keachie, decided to make the move, leaving his $600 per annum salary and their home for a stint in Bone Town.[39] Their youngest son was born that November, just as the mill pond was filling and the first flax was being prepared for production.[40] They were apparently so pleased with the opportunity that they named him William Perine Young.

The business occupied and transformed peri-urban river valleys, and although they may seem peripheral, these maker spaces helped define Upper Canadian industrial geographies. The Perines evidently tried to carve out their own industrial territory, as most of the new mills to the west and north of Doon opened in partnership with them. The only other Upper Canadian flax mill in 1861 was Colonel Mitchell's in Norval, on the Credit River. Mitchell briefly produced large amounts of fibre, but his business was not a significant threat to the Perines and may explain how they came to partner with Gooderham & Worts further down the Credit.[41] However, some competitors with significant capital were beginning to appear near Doon, such as Hunt, Elliott & Company in Preston and John Tilt's mill in New Dundee.[42] The Perine mill network extended toward pioneer territory. A more distant satellite mill was built in 1864 by William Hendry in Neustadt, a town established only seven years earlier in Grey County.[43] William was a relative of Charles Hendry, an established miller in Conestogo and another one of the Perines' partners in Maryborough, Wellington County.[44] The network of industrial knowledge hinged directly on the Perines in Waterloo County, but just as a rural mill complex combined multiple outputs from one source of energy, knowledge, and capital, the flax fibre web connected many actors involved in myriad industries across a concentrated region.

The Perines moved to Canada and found themselves in a culture fixated on "progress." According to Gerald Tulchinsky, that "meant, above all, the growth of industry."[45] Driven by demand and encouraged by brief

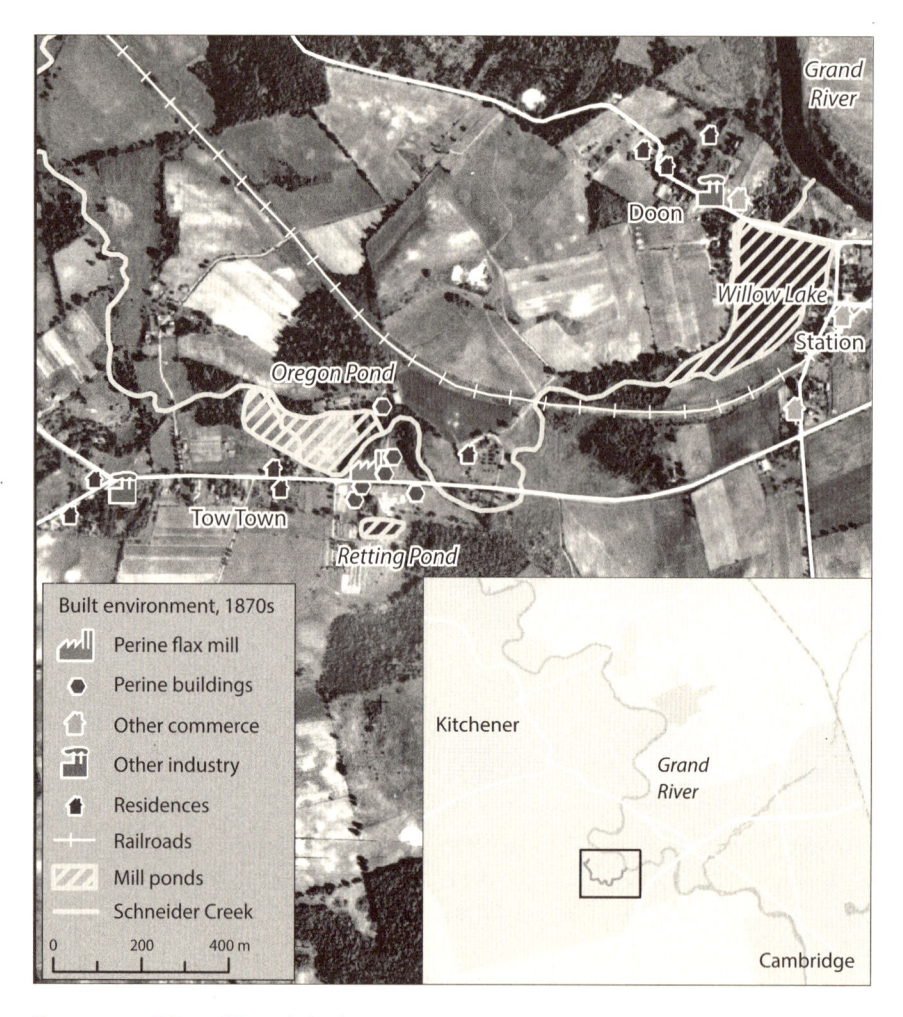

Figure 1.4   Map of Doon's built environment, ON, 1870s, showing the Perine house, mill complex, and mill ponds

upheavals in the region's political economy, the flax business depended on a range of factors for success. The Perines knew the land, and they knew the landscape of supply and demand. They brought a knowledge of the plant and the commodities it produced, a connection to strong markets for rough fibre, and an ability to mobilize labour and organize farm production on a few hundred acres of land near each flax mill. They invested in the products they knew from the flax farms around Troy, and they produced the kinds of intermediate goods needed in New York State's and

New England's manufacturing sector: paper products, rough textiles, and scutched flax or line. They also drew on their experience in transportation and other construction projects and serviced the local economy with an expanding spectrum of fibre and lumber products. Moses Perine's mill in Doon was part of a complex that included a retail store and a steady business in lumber (Figure 1.4).

Flax and lumber might appear as unlikely bedfellows, but repurposed sawmills were an excellent habitat for the primary processing stages of the flax fibre web. As the Perines and other millers started new operations in the 1860s, they gravitated toward the western Upper Canadian districts. By 1891 most new mills were constructed in Huron, Middlesex, and Lambton Counties along Lake Huron (Figure 1.5). As the forest receded and timber stocks declined, retired sawmills and their equipment became available for flax millers.[46] The large rotary power and dry storage available at these sites were essential to primary flax processing. The whirling heart of a sawmill was a perfect transplant for a scutching machine. Benjamin Walker's Elgin Flax Association imported two mechanical flax scutchers and ran daily demonstrations, one of them in a planing mill in St Thomas in 1862.[47] We know that many flax mills, including the Conestogo and Drayton properties, were formerly lumber mills, and that some flax mills, like the Livingston mill in Wellesley, Waterloo County, were later converted to planing mills. Another early flax miller, James McGee, converted his unsuccessful scutch mill in Weston to a sawmill.[48] In Huron County, the Crediton flax mill was owned by sawmiller and councilman Werner Schnarr.[49] In the adjacent Hay Township, the Kalbfleisch family built sawmill and planing mills in the 1860s with practically the same design as flax mills operated by that family and others in the 1870s. The flax mill in Hensall was powered by a steam engine purchased from John and Henry Kalbfleisch's sawmill,[50] and in 1910 the town of Zurich had two flax interests, Fred Kalbfleisch's "planing & flax mills" and the Zurich Flax Company.[51] As more steam-powered mills opened in the late nineteenth century, millers found that they could obtain some of their fuel from local biomass by-products. When flax millers could not find a market for the *shives* (waste material) produced from scutching, they could use them to help fuel the mill engines. Sometimes the shives provided a significant amount of fuel, and the flax mills in Wellesley, Crediton, and Dashwood burned "refuse" as their primary fuel.[52] Clearly, the by-products of lumber mills were also a readily available form of biomass for local steam engines.

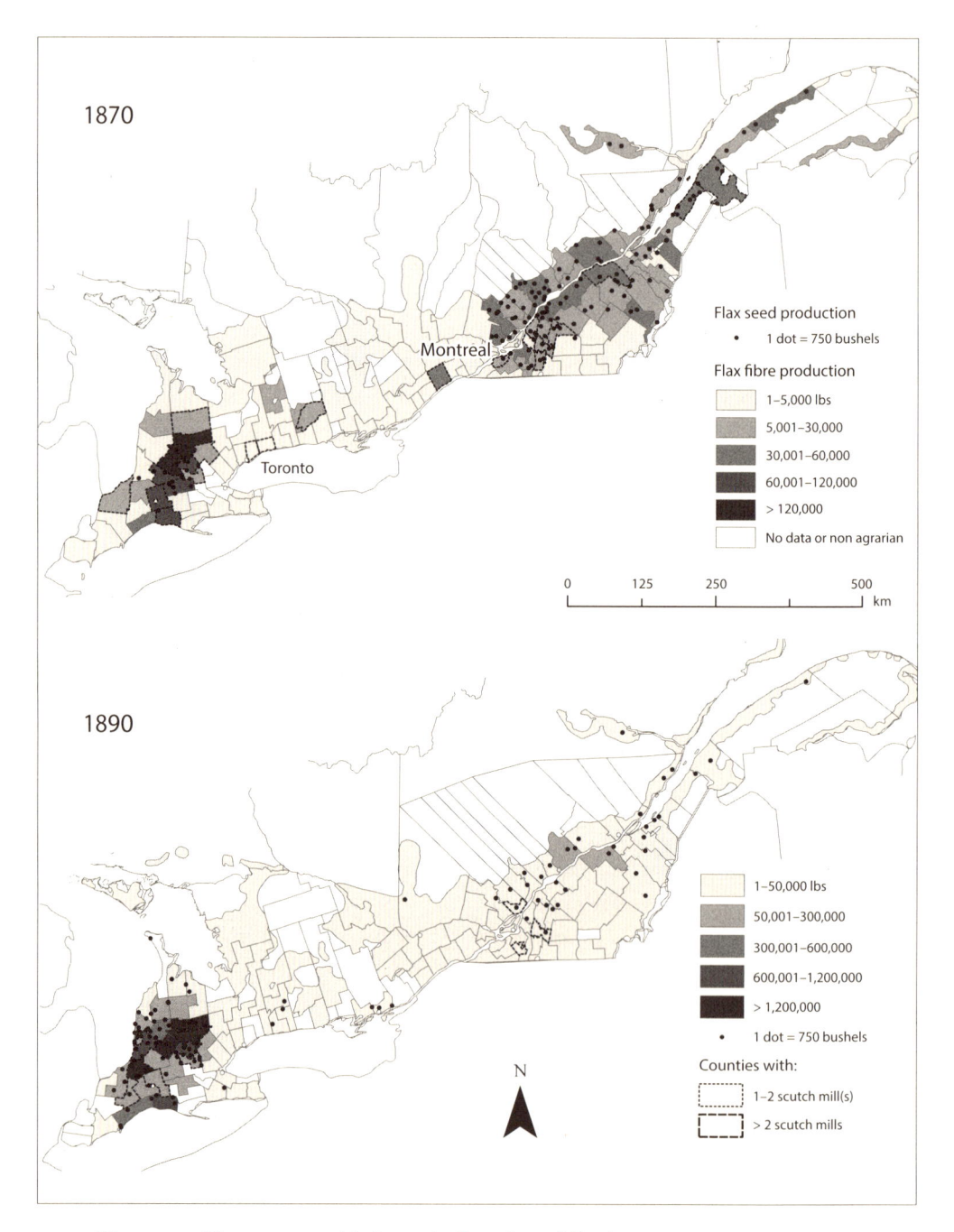

Figure 1.5 Flax output and industry in Ontario and Quebec, 1870, 1890

Source: Canada, *Census of Canada, 1870–71*, Vol. 3 (Ottawa: I.B. Taylor 1875), Table 51, 431; Canada, *Census of Canada, 1890–91*, Vols 3 and 4 (Ottawa: S.E. Dawson 1894, 1897), Table 1, 298.

Figure 1.6   One of the three central Perine mills, Conestogo, ON

In the early years, customers often bought both flax and lumber products from the Perines in a single transaction, and repeat lumber customers were among those who bought bushels of flax seed. This suggests that the mixed marketplace of the mill complex introduced the business to a wider spectrum of customers. This was how the Perines identified prospective producers and customers, and it was likely how farmers came to trust the Perines and grant them access to their land. As the flax fibre industry expanded in the nineteenth century, its success was founded on networks of trust based on everyday exchanges of goods and labour. Mill complexes were an integral part of this predominantly rural society in the mid-nineteenth century.

Flax mills combined a variety of the stages in the commodity web, depending on the labour, equipment, and other resources available to each business. The *Grey County Business Directory* painted one of the clearest pictures of how nineteenth-century flax mills operated. "The mill is frame," the directory explained of the Hendry and Perine mill, and was divided into two parts. One side was open and used for storing the flax straw, while the "other half contains the scutching machinery, which is very simple." The mill's machinery included a mechanical brake that pulled the fibre multiple times through a "heavy sharp-ribbed iron roller" before it was ready to be scutched, a mechanical process that separated the flax fibre from the waste. The fibre was shipped by rail and wagon team to the Perines' main mill in Doon, where it was spun, woven, and bleached, and the waste material could be sent in bales to be cleaned and processed for use in oakum.[53] According to the directory, the Neustadt mill dressed the flax by retting and scutching the fibres. Since flax was processed to different degrees at each of the mills, the Perines were responsible either for sharing among the mills or for providing the necessary machinery to them. The layout of most Upper Canadian flax mills was relatively standard, and the directory's overview would have described the Conestogo mill in Figure 1.6 quite accurately.

Expansion

In the early 1860s, the Perines ventured into industrial textiles and cordage to supplement their fibre-processing operations. One writer stated, in 1863, that the Perines had just begun to add bags and toweling to their line of cords and twines and were planning to incorporate more and finer textiles.[54] Understanding more of the way mills worked and what was required to manage them helps explain why this business was so soon consolidated within the operations of one mill. Between August and November of 1863, the Perines spent over $7,200 on spinning and scutching equipment and "one sett of long line spinning machinery ... mounted in perfect order."[55] In addition to acquiring machinery, the Perines hired craftsmen throughout the summer of 1863 to construct a boarding house in Doon.[56] The capital for such purchases was secured in part by the Perines' relationships with US customers and perhaps through the Billings & Ingram Company in Troy. Much of the new equipment came from the mill's leading customer in Boston, M.S. Chambers, and was credited to his flax account.

The improvements at the Perines' Doon mill involved a combination of local knowledge and materials and imported textile machinery (Figure 1.7). Local woodworkers, bricklayers, and labourers were critical for building and maintaining the Perines' facilities over the years, and the business occasionally hired nearby millwrights and metalworkers, such as Beck & Clair in Waterloo, to fashion manual flax-breaking equipment. Mair, Inglis & Company of Guelph, Ontario, constructed flax-scutching machinery (and later appliances), specifically machinery for "making coarse tow into paper stock."[57] But eventually the Perines expanded at such a

Figure 1.7 Doon Twines, Doon, ON – rear of visible building with drying racks on the right and a tennis court in the foreground

rate that they began to buy all of their machinery elsewhere. For instance, although a brake could be bought in Montreal for $300, duty-free, the Perines spared little expense when buying equipment, never paying less than $400 for a brake and always buying from US suppliers despite the duty.[58] The Perine brakes may have been portable, for on at least one subsequent occasion a brake was taken from one mill to another.[59] The mills' industrial activity fluctuated periodically, so it is likely that some of the mills shared equipment on a seasonal schedule. Many Canadian mills installed machinery made in the province, although woollen millers in the Humber valley at least had always turned to the United States for their technologically advanced textile machines.[60] Before the 1860s, the decade

Table 1.1 Value ($) of Perine flax outputs by country, 1857–1865 (part)

Year	Country	Flax seed	Flax fibre	Textiles	Grand total	% of total
1857	Canada	38	625	1	664	9%
	United States	26	5,702		5,728	75%
	Unknown	82	1,171		1,253	16%
1857 Total		146	7,498	1	7,645	
1858	Canada	68	313	44	425	11%
	United States	459	939		1,398	36%
	Unknown	85	1,960	1	2,046	53%
1858 Total		611	3,213	45	3,869	
1859	Canada	116	646	91	853	34%
	United States		1,013		1,013	40%
	Unknown	266	376		642	26%
1859 Total		381	2,035	91	2,508	
1860	Canada	12	577	114	703	9%
	United States		6,747		6,747	89%
	Unknown	23	63	56	142	2%
1860 Total		34	7,387	170	7,592	
1861	Canada	285	792	44	1,121	5%
	United States		19,327		19,327	93%
	Unknown	115	114	41	270	1%
1861 Total		400	20,232	85	20,717	
1862	Canada	910	1,362	177	2,449	20%
	United States	1,986	7,234		9,220	77%
	Unknown	62	146	163	372	3%
1862 Total		2,958	8,743	340	12,041	

/continued

when flax millers acquired a great deal of processing equipment, Canadian implement makers had had no reason to master the US and British technologies required to make large-scale flax-manufacturing equipment. But rural industry fit well within the Canadian manufacturing sector, and other companies quickly moved in to meet the demand for flax machinery and make other mill improvements.

The primary flax output from the scutch mills was fibre for coarse cloth and cordage – intermediate products that were inputs in many local and US businesses (Table 1.1). The largest industrial output in the early years was unmanufactured or simply scutched flax sold to one major US client, Smith, Dove & Company, in Andover, who had manufactured flax with machinery since the 1830s. Their main product was a strong linen thread

Table 1.1 (continued)

Year	Country	Flax seed	Flax fibre	Textiles	Grand total	% of total
1863	Canada	564	1,746	487	2,797	17%
	United States		13,241		13,241	81%
	Unknown	42	201	156	398	2%
1863 Total		605	15,188	643	16,437	
1864	Canada	64	8,519	6,917	15,500	46%
	United States		15,437		15,437	45%
	Unknown	67	598	2,385	3,050	9%
1864 Total		131	24,554	9,302	33,988	
Part 1865	Canada	454	4,795	3,639	8,888	51%
	United States		7,129		7,129	41%
	Unknown	129	112	1,094	1,335	8%
Part 1865 Total		583	12,037	4,732	17,352	
Grand Total		5,850	100,888	15,409	122,148	

Note: The last source (in this note) is the last Perine account book extant, and its coverage of 1865 is only partial. I estimate that it covers most transactions into May or June, but since ledgers are arranged by customer accounts, those with the most transactions were carried forward to the next available page. Around June 1865, those pages would have begun to appear in a ledger that is now missing.

Source: DHC, "Daybook," Perine brothers, 2006.023.024.1; "Daybook," Perine brothers, 2006.023.024.2; and "Ledger," Perine brothers, 2006.023.024.6.

for use in carpet, sail, and shoemaking. Their flax was mostly imported from Belgium.[61] They were a major buyer of long line flax (included in "Flax Fibre" in Table 1.1), a product that brought the Perines an average of $5,600 per year from 1857 to 1864 (with sharp year-to-year fluctuations). After 1861 the Perines cut ties with Smith Dove & Co, and 1862 became a pivotal year, as long line flax sales began to be replaced by more-domestic processing and consumption, and by exports of tow. Tow was a shorter fibre and a by-product of scutching, and after January 1864, sales of long fibre (line) practically disappeared and the shorter, coarser fibres took its place. By then, the central Perine mill in Doon was a large-scale cordage, bag, and coarse cloth operation.

The semi-processed products the brothers had come to Canada to supply were still important, but the Perines quickly developed the capacity to add value within their own mills. The export of unmanufactured flax fibre and seed to the United States was important in all years, but these exported fibres took several forms and the processing became more sophisticated in the 1860s. As the Perines began to hackle a greater share of their flax fibres for use in cordage, they exported less long-fibre flax and produced more tow by-products. Tow itself was sold in coarse and fine varieties. The more commonly sold variety in the United States was fine tow, which was used for twine and bagging. Its largest buyers were in Boston, especially manufacturers like Chambers. The most regular buyer of coarse tow was a paper mill in Niagara Falls. New flax fibres and recycled linen rags were important and apparently constituted scarce raw materials for the mid-nineteenth-century paper industry.[62] Manufacturers in Buffalo and Troy were known as early movers in the paper industry, incorporating rag and wood pulp technology in the production of that commodity in the 1850s.[63] The sales of tow grew steadily in Canada and the United States in the 1860s and rapidly in 1864, and like the longer scutched flax, it had only one or two destinations in the early years of production.

After 1861 the distribution area included cities such as New York and Pittsburgh. This was how the radial strands of the flax fibre web began to expand beyond and compete with other commodity suppliers in the Great Lakes region. Although semi-processed exports were critical to the business at first, the value of domestic fibre sales jumped fourfold in 1864 and appeared to be rising further in the early months of 1865. If war and Reciprocity drove the growth of exports, then something else accounted for the expansion in Canadian fibre markets. The variety of outputs offers a partial explanation. In the Canadian market, the most valuable output was usually coarse tow, purchased by a variety of upholsterers, cabinetmakers, textile mills, hardware stores, and paper manufacturers. That these sales grew shows an expanding place for flax in the local economy that has little to do with the civil war and Reciprocity.

An increasing number of cabinetmakers and other upholsterers used Perine tow in their craft. Of those clients who could be identified as cabinet- or carriage-makers through the 1871 industrial census and local business directories, four customers paid over $1,000 each for Perine tow from 1860 to 1865. Dozens more bought small amounts of tow for use in their upholstery. According to one history of upholstery,[64] tow was used, along with horsehair and sometimes grass, as batting. Of the more than 500 cabinetmakers in the 1871 industrial census, only 8 reported the use of tow as a raw material, but it was not common to report these secondary materials and only a dozen reported any form of cloth or horsehair used in upholstery making. One historian of Jacques and Hay, Toronto furniture manufacturers and Perine customers, mentioned that they used horsehair for upholstery in this period.[65] However, this aspect of the accounts suggests that local flax became a new, but perhaps temporary, part of the upholsterers' business strategy. The data from Moses Perine's accounts show an incomplete picture, and as more mills appeared in southwestern Ontario in the 1860s and 1870s, similar networks of intermediate products became available to local trades and urban industries.

The transition from fine fibre to tow was a shift in the commodity web that showed adaptation to local agro-ecosystems and the difficulty of procuring good Canadian flax. Apart from the testimonies of the usual flax boosters, most indicated that the flax grown in the Great Lakes region was unsuitable for long line fibre. This was not about climate and soil conditions for growing flax, but about cultivation methods. Harvest labour was difficult to find at precisely the moment that flax should be pulled, and shortages meant the crop was often cut or left in the field until pullers arrived. The Perines' response to labour and material shortages was to alter the production process and, ultimately, the marketable product.

The evolution from long fibre to tow had to do with the parameters of the local flax supply and a growing market for coarse tow. The transition also represents the opportunity for the Perines to use their own long fibres in the linen mill and sell them in the form of finished products. Yet even these "finished products" were intended for use with other commodities: rope sold to harness makers, broom cord sold to broom makers, bed cord sold to bed makers, and textiles sold to those who made a variety of cloths and garments. Possibly the most stand-alone of all the products were grain bags, a commodity whose usefulness was determined by its precarious place in a supply chain that was then shifting from bag to bulk transport.[66]

Local Textile Markets

The introduction of spinning equipment and other improvements in 1864 was reflected in accounts that showed an increase in consumer goods for local markets. As the Perines began to manufacture bags, ropes, and twines, for the first time the majority of their industrial output was sold in the Canadian market. Cordage was bought by many dry goods, hardware, and grocery merchants from St Thomas to Toronto, but most of the sales were concentrated in the towns and villages of Waterloo County. In 1864 the largest customer for ropes and cords was John Rankin, a general merchant in Dundas who spent $876 on halters, cords, and machine ropes in September alone and continued to make cordage purchases for the rest of the year. Twines and lines were another important product, perhaps the one most clearly tied to the agricultural market. Twine sales were highest through the summer and fall seasons.

As flax fibre expanded in Canada, businesses began to focus on local markets and thus local marketing strategies came into play. The clearest sign of a move toward domestic markets was the mill that opened in partnership between the Perine brothers and Gooderham & Worts, in Streetsville, Peel County. The Streetsville Linen Manufacturing Company was opened in 1865. It was comprised of a scutch mill on one side of the river, with outdoor vats capable of holding 25 tons of flax for retting, and a five-storey linen mill on the other side of the river. In 1867 Gooderham & Worts announced they had 100,000 bags made from "pure Canadian flax" for sale. Other businesses also offered goods made locally from flax fibre, including Brown & Company's flax mill in Stratford, which had "about 300 yards of bagging ready for market."[67] It is significant that the Perines and Gooderham & Worts advertised consumer goods made from "pure Canadian" fibre. This was more than just nostalgia or a nod to the industrial capability of the Dominion on the occasion of its Confederation. Other Canadians praised the increase in flax production in this period, hoping it would replace inferior imports such as bags made of jute or so called "sea weed" from India.[68] Beatrice Craig has shown that New Brunswick merchants sold homespun linen and other cloth, advertising "Genuine Domestic Linen … hand dressed and manufactured from flax in this province, entirely free from the process of machinery and chemicals."[69] The superiority of locally produced flax and flax products is implicit in the writings of many promoters, but these advertisements specify both the advantage of buying Canadian and the suspicion surrounding mechanical and chemical processes. As Chapter 5 explains, this suspicion mounted in the following decades as consumers and reformers began to promote greater product purity in flax and other manufactured goods.

As the Perine business grew, its customer base became much more complex in Canada and slightly less diverse in the United States. Both the local distribution area and the business clientele increased. The analysis of the geography of their sales is based on those clients whose location could be identified – these were less than a third of the 535 customers listed, but they represented over 90 per cent of the value of textile sales. Many of the remaining small customers were no doubt local to the Waterloo region, and one can be confident that most lived in Canada. Expansion in the first half of the 1860s was about increasing their products and broadening their

Table 1.2 Value ($) of Perine flax cordage/textile outputs, 1857–1865 (part)

Country	From	1857–63	1864	Part 1865	Total
Canada	Dundas	0	2,430	870	3,300
	Hamilton	71	1,204	1,479	2,755
	Conestoga	13	777	167	957
	Baden	0	181	370	551
	Berlin (Kitchener)	223	162	73	458
	Galt	143	194	41	378
	Toronto	0	171	187	359
	Hespeler	43	226	50	319
	Maryborough	34	267	15	316
	Doon	1	314	0	315
	Montreal	0	89	225	315
	Streetsville	0	206	75	281
	St Thomas	18	232	9	259
	Guelph	162	39	19	220
	Waterloo	93	74	19	186
	Malahide, Elgin Co.	20	135		155
	Preston	77	37	16	130
	Neustadt	0	87	22	109
	Mitchell	0	69		69
	Norval	31			31
	Bridgeport	0	20		20
	Blenheim, Oxford Co.	18			18
	Freeport	4		1	5
	Waterloo Co.	2	2		3
	Plattsville	2			2
	Roseville	2			2
	London	1			1
Canada Total		958	6,917	3,639	11,514
Unknown		417	2,385	1,094	3,895
Total		1,374	9,302	4,732	15,409

Note: All Canadian locations were in Upper Canada, with the exception of Montreal. Three customer accounts are only tentatively identified – the Montreal, Mitchell, and Guelph customers; transactions in these accounts amounted to less than $400.

Source: DHC, "Daybook," Perine brothers, 2006.023.024.1, "Daybook," Perine brothers, 2006.023.024.2, and "Ledger," Perine brothers, 2006.023.024.6.

customer base. The number of textile buyers increased by over five times, to 175, by 1865.

The buyers of flax fibre also increased over that period, from thirty-one to seventy-four. Sales in the United States were made in between three and six cities each year. The Canadian buyers of Perine fibre flax lived in about eleven different places in 1860 and seventeen by 1864, as the business's coverage expanded west of Toronto. Cordage was purchased at first by customers from only eight places, mostly near the Perine mill, but from twenty locations by 1864. At this point the largest shipments of textiles and cordage went to Hamilton and Dundas, followed by the Perines' own stores in Baden and Conestogo. The next-largest customers lived in the nearby cities of Berlin (Kitchener) and Galt (now Cambridge) and in merchant houses in Toronto and Streetsville. The Streetsville account was not Gooderham & Worts, but actually represented a new relationship with the Barber Brothers, whose woollen mill was completely rebuilt after a fire in 1862.[70] They were the third-largest tow consumer in 1864, ranking behind two Boston clients and their textile and cordage purchases here included warp, crash, and yarn, as well as twines and ropes by the bag. Rural periodicals imagined that Elliott and Hunt, the Hespeler mills, the Barber Brothers, and most other wool manufacturers west of Toronto were incorporating flax into their production, but if they did, it was only on a small scale and for brief periods.[71] The accounts show that at least the Barbers experimented with flax in their products as part of their rebound strategy, and for the Perines that meant a new market for both fibre and manufactured goods.

Trade policy was another factor that shaped the type of industrial output pursued by millers. For most of the nineteenth century, flax could enter the United States under relatively small tariffs. The first consistent tariff on imported flax was a 15 per cent *ad valorem* tax in the 1840s and 1850s, from which the British North American colonies were exempt under the Reciprocity Treaty, beginning in 1854. However, the United States briefly eliminated the flax tariff for all countries between 1857 and 1861, the year the Morrill Tariff came into effect. In 1868, two years after the agreement had ended, the *Canada Farmer* suggested ways to make flax exports to the United States more profitable. Hitherto, most flax sent to the United States was only scutched, but the duties "on the other side ... make it very profitable to carry the process of preparing flax in this country a step further and hackle it."[72] Increasingly in the 1860s the Perines processed hackled flax, whereas at the outset they sold mostly scutched fibre. The US duty on scutched flax was $15 per ton and was the same on hackled, or combed, flax. But tow, a by-product of hackling, was only taxed at $5 per ton in 1868. This meant that shipping an amount of scutched flax in its hackled form would average out to $10 duty per ton. Flax was usually baled and shipped by rail car (Figure 1.8), and in some cases millers learned to

Figure 1.8 Scutched flax bales in a rail car

disguise bales of the long-unprocessed fibre by surrounding it with hackled flax and thereby paying only the lower duty.

The end of Reciprocity had little effect on the production of flax goods for local markets. These were growing steadily through the time covered by the Perine accounts. However, several businesses adapted their production of semi-processed fibres in Canada (line and tow) to mitigate the effects of the US tariff. The A.H. Hart Company, of Troy, established a flax-hackling branch in Stratford, Ontario, in 1868, and according to the *Canada Farmer*, they were buying all the scutched flax they could find.[73] The Perines recorded a little business with Hart & Co., of Troy, in 1862, although it was for sawmill equipment. Doon itself became known as "Tow Town" in the late nineteenth century, which suggests that hackled flax remained an important product of the mills. There are insufficient business records to determine if minor trade advantages encouraged Canadians to export more hackled flax after Reciprocity and the steady rise of fibre tariffs, but the transnational connections formed through these relationships had lasting effects in other ways. When Hart first arrived in Ontario, he was welcomed by the flax promoter John A. Donaldson. Hart's experience reveals the continued importance of the larger Great Lakes region to the fibre industry. As we will see in Chapter 5, another major Ontario fibre miller went on to own the Hart Flax Spinning Mills and then merged it with the Linen Thread Company, one of the largest thread and twine companies in

North America.[74] All of these new businesses developed on the heels of a major readjustment in the flax industry in the mid-1860s.

Through the Perines' growing rural network and Gooderham & Worts's established market in Toronto, the Streetsville factory seemed well situated to supply both the farm hinterland and the growing Upper Canadian cities with linen goods. Gooderham & Worts had been purchasing and constructing mill complexes from Norval down to Streetsville along the Credit River, but their flax mill was most likely located north of the village of Streetsville, on the site of their Alpha Knitting Mills.[75] It was fully integrated vertically, operating everything from the flax production and scutching on one side of the Credit River to the linen bags, rope, and cordage for wholesale on the other. A mill's role in the cultivation of flax is explored in Chapter 2. Streetsville was St Thomas and Doon writ large. It was built in 1865 at a cost of over $100,000, with a company known for its pecuniary strength and "unlimited" credit rating.[76] In its first winter (1865–66) between 70 and 100 workers processed 900 tons of flax, the product of about 300 acres. This was apparently only a quarter of the mill's capacity. The linen mill spun and wove long line fibre into double-webbed linen for seamless bags, half a yard each in length. The bags were cut and hemmed on site and baled at 100 bags per bale. The mill was capable of producing up to 2,000 bales per week, worth $40 per bale. It also produced counter twine from the line and cordage from the highest-quality tow. The "refuse tow" was actually a by-product sold to paper manufacturers.[77]

Readjustment and Consolidation

A readjustment in the linen-manufacturing end of the commodity web began in 1866, and the Streetsville factory's production of "pure Canadian flax" soon ground to a halt. The scutch mill was destroyed by fire in August 1866, and the blaze ruined a great deal of the raw material. The *Journal of the Board of Arts and Manufactures for Upper Canada* assumed the factory would be quickly rebuilt, since "everything the members of this firm engage in is pursued with energy."[78] The Perines' records stop in 1865, but we know they struggled like the rest of the flax industry in the late 1860s. They experienced several other mill fires, depressed postwar fibre prices, and higher tariffs on fibre shipped to the United States. The fires were enough to interrupt the raw material supply throughout the Perine network and to significantly slow production at the Streetsville mill. Gooderham & Worts maintained ownership of the property, but another fire destroyed the linen mill in January 1868, including large stores of flax in a storehouse and outside the mill. The millers suspected arson.[79] Linen production never resumed, and ten years later, on an early spring day, the "ice in the River Credit suddenly broke up" and carried away what remained of the mill's bridge and waterworks.[80]

Almost as quickly as the Perines had expanded into several new scutching mills, they sold the new acquisitions and consolidated their interests in Waterloo County. The brothers sold the Baden mill to James and John Livingston and sent Joseph Perine to Conestogo and William D. back home to the United States in 1868. They left the partnership with Alexander Young in St Thomas in 1868, and by the following year, J. & J. Livingston owned that mill as well. Young returned to schoolteaching in Berlin (Kitchener), and despite positive contemporary accounts and credit ratings, his obituary called the whole operation a failure.[81] Fire burned down the Neustadt mill sometime in its first few years, destroying the property, machinery, and the entire year's crop, which sat processed and ready to ship.[82] The mill was owned solely by William Hendry in 1869 and by Mrs Washburn Hendry (presumably William's wife, Sarah) in 1871.[83] The Perines kept the much-improved Doon mill, but they divested from Streetsville, St Thomas, and all of the other co-owned mills.

Although the Perine business grew and transformed dramatically in the early 1860s, it was reorganized just as quickly in the last half of the decade. It went from supporting three families, several partners, up to a dozen mills, and two or three hundred permanent mill workers down to a single mill and only two of the brothers, Moses and Joseph. It is possible that the Perines had expanded too far, that the combination of decreased demand after the war and a tragedy at the Streetsville mill meant that they almost lost their linen shirts. Ultimately, Moses Perine found a niche industry that operated in twine and cordage more than in other linen products. With the growth of the binder twine market in the late nineteenth century, the Perines would have abandoned flax for fibres that worked better in binders, deterred twine-eating insects, and presumably cost less to produce.[84] Their expertise in flax production became less important than their knowledge of the Canadian market for fibre goods.

Rural Factory Labour

To produce these consumer goods, the millers required access to a steady labour force. In 1861 Moses Perine and his wife and daughter boarded ten male labourers and two women, who were likely also employed in the flax mill. It is clear from the Perines' experience with outworkers and from several references to other boarding families in Moses's accounts that the mill required additional labourers to produce their fibre, cordage, and rough textiles.[85] Outwork weavers were a regular part of rural textile production in the nineteenth century in the United States, but much less so in Canada.[86] The Perines appear to have employed them in the nascent stages of their production. Ernst Firey, for example, the Perines' most adroit cordage maker, was charged for his raw materials and credited for his finished goods. Those cords and twines were then sold to customers from the Perines' store. Firey's production was highest in the fall of 1860 and the winter of 1861, when he was credited $141 for bed cords, twine, and something known to the mill as bail rope. In the following years, his credits included time worked for the Perines, mostly in the first few months of the year, and in two cases in January 1862, he was credited for "spinning broom twine from tow."[87] He continued to work for the Perines until 1863, at which point the mill's cordage operations expanded and probably made outwork less necessary. By 1864, most of the weavers employed worked in the mill, but at least one, George Ellendorf, was debited for yarn and credited mostly for "yarn returned in cloth."[88]

Like the Perines, late-eighteenth-century flax and hemp millers in Virginia hired skilled spinners from Europe. Indeed, some of Charlotte Erickson's subjects in *Invisible Immigrants* were skilled flax workers who had immigrated to work in the mills in Rensselaer County, New York. One of these mills was the Dunbarton Flax Mill in Greenwich, which differed from the Perines' migration because it was a branch of a pre-existing company in Gilford, Northern Ireland. The mill invested heavily in the Washington and Rensselaer region and sponsored the emigration of skilled workers to the Greenwich plant. The company clearly had great faith in the flax-production capabilities of the region, but due to the lower-quality local fibre, it seems that they too focused on rough textiles and cordage.[89] One of the Perines' immigrant flax workers was Robert Printer, an Irish Protestant "flax dresser" who lived next door to Moses.[90] But most of the 1861 labourers were born in Canada, and although some external inputs were necessary, a large and local male and female labour force was important to the industry's, and especially the Perine mills', operations. Initially ten of their twelve boarders were male, suggesting that female workers boarded elsewhere or were not employed as often as men. Eventually, women became a much more important part of the Perines' workforce.

The following chapter will show that the Perines employed more women than any other business in the cordage and flax fibre industries (see Table 2.3).[91] Recent immigrants and temporary workers from Europe remained significant as late as 1911, when the lodgings of immigrants and others employed by the Perines surrounded the Doon mill. Of the sixty fibre workers closest to the mill thirty-two were born in Europe (England, Scotland, and Germany).[92] Moses Perine's business began as a small mill complex that employed outworkers, processed diverse fibre products, and furnished linen to Mennonites for home manufacturing, but by the end of the century, it was a large rural factory with skilled labourers and a single line of cordage products made from a global supply-shed of raw plant fibres. This will be discussed further in Chapter 2.

Cordage Moves to Town

The Perines experienced several transitions in the flax fibre industry, and by the late nineteenth century they had produced a large and resilient family business. From the 1860s until it moved to the county's urban centre, the Perines' Doon mill was the most important industry in Tow Town and during the 1880s in all of Waterloo Township. Soon after 1861 the Perines could be counted among the leading one hundred rural property-holders of Waterloo Township, and by 1881 they were seventh in the township with property worth $9,000. At the turn of the century their mills were worth $31,200 – the most valuable property in the township.[93] They employed 190 workers in the mill and continued to increase those numbers during harvest time to almost 400 employees.[94] The Doon location remained the business headquarters well after Moses's death in 1898. Moses's son, Edward Graham, married Helen Margaret Hepburn in nearby Paris and continued working in the family business throughout the 1890s. Edward eventually took over management of the cordage mill in this period, operating it in Doon until his untimely death in 1911 at the age of forty-five.[95]

With the deaths of these proprietors, Hartman Krug became the majority shareholder and owner in 1912. He renamed the business "Doon Twines" and promptly moved it out of Doon. He built a new factory in Berlin (Kitchener) at the end of the First World War in order to gain larger facilities and better access to transportation. The business operated in that location for almost another century, changing its name again to Canada Cordage Inc. and finally closing in 2007.[96] As it grew, it increasingly relied on other fibres for its cordage products. The business began to import sisal

Figure 1.9   Doon Twines Ltd products

and jute, but it specialized in the latter. During the Second World War, regulators prohibited the use of sisal in twine, and Doon Twines became the only mill in Canada with the capacity to produce cordage from jute. It seized the opportunity and used its new business to expand operations and diversify its products. By the 1960s, the company had added a line of synthetic fibres, and its use of flax and other natural fibres decreased again.[97]

The storefront display of Doon Twines products in Figure 1.9 shows the importance of the range of goods produced by the company in the mid-twentieth century. By then, they were producing only cordage. The fibres in the background were brought to the mill from global markets. Flax is the fibre hanging on the display backdrop, second from the left, between hemp and sisal. The thirty or more products in the display identified the fibres they were made from, and some of the uses such as the "clothes lines" in the middle of the case. To local customers it was important to know that these mills provided wealth and opportunity for local residents and that some of the consumer goods they purchased there were "Made in Kitchener."

At first glance, the Perines' business fits neatly into the traditional framework for understanding the Canadian economy in this period. Deeply connected with the US business world, it exemplifies Canadian business developments that benefited from the capital, entrepreneurial talent, and industrial knowledge of immigrants from south of the border. Nascent producers of agricultural raw materials were encouraged by free trade until the end of Reciprocity and the start of Confederation and eventually by a policy of protection, at which point the Perine mills were able to manufacture goods for the domestic market. While seeming to fit this pattern, the mills produced natural and intermediate products for US customers until 1866 when political restructuring encouraged the Perines to focus entirely on manufactured twine – their primary output according to the 1871 census. However, the Perines' decision to immigrate to Canada was made *before* Reciprocity, and their accounts show that an increasingly large proportion of their industrial output was sold to Canadian customers without the help of protection. The picture of flax and flax products uncovered in the Perine accounts included the contributions of large, urban textile mills that used highly processed flax as well as local saddlers and furniture makers who used intermediate products. In rural Upper Canada we see small flax manufacturers, specialized outworkers, and ordinary farmers all contributing to the commodity's web, and each performed a variety of interrelated roles.

In other words, the accounts show that not only was manufacturing occurring in pre-Confederation Canada, but the networks created through the flax fibre web were regional and rooted in place. The Perine mills show that local markets for flax products were growing in a variety of rural and small-town places, and the southwestern Ontario flax industry was ideally situated in an expanding network of infrastructure and industry. They benefited from the transportation and communication network that they helped build as contractors and that they established through family members. This facilitated trade with urban centres in Canada and the United States. The mill produced a variety of industrial outputs that were marketable in local stores, Canadian textile, cordage, and upholstery mills, and large US factories. Rather than being dependent on political events beyond their control, flax producers in this region were part of an integrated system, a network of internal resources and outside contributions. North American Reciprocity and the American Civil War played certain roles in the establishment of a small Canadian flax industry, but even more important was the framework of industrial production that built on several decades of social and economic growth.

The flax fibre industry that developed in Upper Canada and upstate New York was not the only flax business on the continent, but it was certainly the largest and most stable system. Flax also appeared in Lower Canada and the Maritimes, especially among French Canadians, but in much smaller amounts and with less mechanization. Several Nova Scotians attempted to open flax mills at various points, but these were short-lived.[98] There is no single reason for this, and at least two phenomena brought millers like the Perines to Upper Canada and helped establish the flax industry there. The first phenomenon was on the demand side: Upper Canada had a large and growing manufacturing sector in rural and urban areas within close proximity to the flax mills. Second, the province also had supply-side solutions in the rich farmland and watersheds of southwestern Ontario that gave flax millers the ability to open a network of mills and convert places to products and landscapes to linen thread. The next chapter explores this supply side more closely, asking what enabled millers to grow flax and what possessed families and First Nations labourers to harvest it and process it in the field.

This glimpse of Canadian industrialization in the countryside shows, first, that what worked in the upper Great Lakes region worked in Upper Canada and that material transformation continued across both parts of this region. Second, despite the strictures of a thin population base, labour-intensive industries with diverse outputs were established in rural areas. The pattern of establishment hinged on how what Philip Scranton calls the "matrix of accumulation" was organized to provide a base for production and distribution.[99] The matrix included a range of material, social, and external factors that influenced businesses. Westward expansion of flax milling in Ontario was about following the disappearing forest, repurposing existing mill equipment, and establishing relationships with farmers who needed income but were otherwise reluctant to try such risky and labour-intensive commodity webs. The region's developing markets, transportation grid, and vibrant network of mill complexes allowed flax millers to bring their expertise and client base to new sources of farmland and labour. The small-town mill complex was not only a place where material from a local catchment area was transformed for the market, but, as we will see in the next chapter, a place where people could build networks of trust – where they could produce and consume in multiple, interrelated commodity webs.

CHAPTER TWO

Everyday Exchanges

*Growing and Harvesting
Flax in Ontario*

In 1858, the same year plantation owner Benjamin Walker left his son in Charleston and emigrated to St Thomas with his new wife, a young black man was running in the same direction. Eighteen-year-old Nathan Ryan was a slave to Henry Davis of Mount Aviv, Maryland, when he escaped his master's bloodhounds, forded rivers, and found his way to Upper Canada with the assistance of the Underground Railway. Ryan soon arrived in St Thomas, where he would work in the Bone Town flax mill until it closed in 1894.[1] The briefest of biographies suggests he arrived early in the civil war and found work with the Perines and Alexander Young, hauling and eventually milling the fibre that was made so valuable by the dismantling of slavery in his previous home. There were no African-Canadians in St Thomas according to the 1861 census, but the twenty-one-year-old may have been living in the mill vicinity even then. Ryan's first photograph was taken in the Kettle Creek valley, where he almost certainly would have been among the workers hauling flax to the mill in 1865 (see Chapter 1, Figure 1.1).

As Walker sang the crop's praises to local newspapers, others were getting ready to mobilize both the productivity of the Kettle Creek agro-ecosystem and the labour of hundreds of rural and small-town workers. These were people living and working in what historical geographers call socially and spatially marginal places. Unskilled field work and most of the labour in the mills were not specialized trades like the flax weaving in

Chapter 1. However, we know that beyond racial and class oppression, relatively powerless people like Nathan Ryan often used the margins as "sites of creativity and power," places that connected people as much as they separated them.[2] For almost a century, some Ontarians, such as Joseph Waddelove and dozens of other Munsee First Nations, became quite central in these spaces, arriving every summer to harvest flax at competitive wages and terms. To the Munsee these were not marginal spaces at all, but rather familiar lands to which they returned over the nineteenth century to hunt and gather wildland resources for baskets and other goods. As pioneering deforested much of these traditional wildlands in southwestern Ontario, hundreds of First Nations people returned to work and camp in the fields or in specially provided lodgings at mill sites. This itinerant wage labour became a long and mostly unremarkable rural tradition, until the confluence of a declining industry and confining federal policies dissuaded off-reserve labour and replaced it with postwar migrant-labour programs. Consistent with this shift, racism in the flax industry actually increased in the twentieth century, a situation reflected most clearly in the stance of flax promoters and the murder of Adam Seneca, an Anishinaabe teen from Chippewa of the Thames who was brutally killed after a day of flax pulling near Stratford.

The most striking aspect of the nineteenth-century fibre web was the degree of land and labour management the rural mill complex required to produce it. Only part of the flax used in local manufacturing came from crops grown independently by farmers; millers like the Perine brothers acquired the rest by developing complex arrangements with producers and labourers in a system called *flax factorship*. Under flax factorship, a miller either bought flax from producers in its standing state or rented land from farmers to grow it with few or no inputs from the landowner. This form of industrial agriculture appeared in Upper Canada in the mid-nineteenth century, a period not often associated with rural industrialization. Kenneth Sylvester notes that the hallmarks of industrial agriculture – corporate control of production, wage labour, and the absence of small producers – were uncommon in Canada, at least in the West. Colin Duncan argues that the inputs for most forms of industrial agriculture are derived mainly from off the farm – a process that increases ecological vulnerabilities on the land – and Deborah Fitzgerald has pointed to the First World War as a turning point in the advance of this form of agriculture.[3] The sudden and persistent success of flax factorship in New York and Upper Canada suggests that industrial forms had thrived in the Great Lakes region at least three quarters of a century earlier. This chapter explores the supply-side solutions to the new fibre markets presented by the region's rural manufacturing sector. Farmers rented their land and contributed their labour to participate in the fibre web because flax factorship was flexible and flax overlapped with the other commodity webs in which they operated.

The Nineteenth-Century Geography of Flax

European flax cultivation was specialized, and a single farm family rarely performed more than a few stages of production.[4] In North America, farmers performed more stages, but they rarely approached self-sufficiency.[5] Independent flax producers usually fell short of the quantities needed by their own families, let alone the market. One explanation for the widespread reluctance to grow flax in Canada is provided in the diary of John Tidey. This document chronicles the young family's daily chores and accomplishments in Brock District, Upper Canada. He describes his home as "a hive of bees," with incessant clearing, planting, chopping, carrying, repairing, cleaning and cooking. His wife was involved in all but the heaviest lifting, "and a stranger that had heard her would believe that there was nothing left for anyone else to do." However, when it came to clothing he complained that he had no way to buy any and that his wife, who had access to a hired girl, plenty of flax tools, and her own skills in spinning and weaving, did not produce enough clothes to improve her "shamefully slovenly personal appearance."[6] Tidey wrote about seeing homespun clothes on his neighbours and clearly they would have been valuable to Canadian families. Yet, even when pioneer families possessed the skills and equipment for flax production, the time they could spend on farm making was usually more valuable than the time it took to make linen. Flax was difficult, and North American pioneers found that other crops simply paid a better rent for the land. However, flax was familiar, and when certain conditions were in place, most critically a strong market price but sometimes a farm or household need, farmers found they could pick it up with little adjustment to their normal routines.

In the early nineteenth century, flax was at best a secondary crop in British North America, and when it was grown by early Ontarian pioneers, it was limited to small areas. In the 1840s, according to J. David Wood, flax concentrations appeared in the earliest resettled areas of Upper Canada, including Niagara, near and north of Toronto, and to the east.[7] However, a concentration might be overly generous. The four districts produced less than 24,000 pounds of flax in total, and although this represented over half of the province's output of flax, it was the product of about ninety acres. In 1850 provincial production had increased to a little less than 60,000 pounds, and four counties, Perth, York, Halton, and Waterloo Counties, raised almost half of that amount. The most striking feature of these data for Ontario is that most other counties, including pioneer areas such as Huron and Grey, produced less than 800 pounds, or about three acres, of flax (Figure 1.3).

There were also regional specializations in fibre production and linen and cloth production in other parts of the country.[8] In Lower Canada (Quebec) a much larger concentration of flax cultivation emerged in the Montreal Plain and the Centre-du-Québec regions at mid-century in response to a demand for locally produced linen. In New Brunswick, as Beatrice Craig and Judith Rygiel have pointed out, there were concentrations of subsistence and market cloth production in various regions, but here as well the scale of flax production was extremely small.[9] For instance, linen shirts were popular with farmers in Madawaska, New Brunswick, and many people produced cloth of various kinds in the parish, but the 1861 and 1871 censuses show only between 2,500 and 4,500 pounds of flax, or the product of eight to eighteen acres.[10] In a county with over 30,000 acres of cleared land for agriculture, it appears that very few farmers were producing any flax at all. It is possible that linen thread was being imported for manufacturing linen, but far more likely that local weavers had switched to imported cotton for their household production of light textiles. The early paucity of flax farms supports Craig's argument that market relationships were critical to "homespun" linen production in New Brunswick and eastern Quebec.

In the 1850s the landscape of Canadian flax farming shifted. Lower Canadian flax production all but disappeared, and Upper Canadians began producing large amounts of flax for new domestic and US markets created by the addition of scutching mills to small-town mill complexes. The transitions were rapid and dramatic. Upper Canada's flax crop expanded twenty-fold by 1861, and in Waterloo and Halton Counties it increased almost sixty times. These were the first major concentrations of flax fibre in the country. Neighbouring counties, such as Peel and Wellington, each produced less than half Halton's amount, and the remaining counties averaged just over 5,600 pounds each. Production then continued to rise incrementally in southwestern Ontario until the First World War, according to the decennial censuses, but as we have seen, flax milling experienced a dramatic expansion and decline during the American Civil War.

Considering the value of flax fibre, most counties could attribute almost no agricultural income to flax. However, in Halton the 380,000 pounds of finished fibre alone (not including by-products) was worth over $41,800. That was equivalent to about a fifth of the value of the county's hay crop or a quarter of its potato harvest.[11] By comparison to major crops like wheat, flax was still very small. In Waterloo, where all enumerators were instructed to record the quantity and value of flax seed in the 1861 census, farmers grew over 369,000 pounds of fibre; combined with the flax seed, this was worth over $55,600. The first major increase in flax in Waterloo and Halton was substantial, but despite its new profile it was still worth a modest amount and looked like a rather ordinary contribution to the gamut of agricultural commodities. What makes Canadian flax so intriguing is the fact that, like US flax, it appeared so suddenly in two very specific areas. This geography is key to understanding the role of flax in the local economy. In each place, flax initially passed through the hands of only one business, the Perines in Waterloo and Colonel Mitchell in Norval, Halton County. The single stream turned into a small flood of new producers, although all were within a concentrated area and all had widely varying levels of success. Flax was not a commodity evenly represented across the province, but in Wilmot, Woolwich, and Waterloo South and North, the townships where the Perine flax mills were found, over 28 per cent of farmers grew it (Table 2.1).

If account books are the best way to examine the millers and makers in the flax fibre commodity web, then the production of the raw material is best viewed through the manuscript censuses. The census records flax production from the farmer's perspective and supplies data on individual flax producers that complement the business records. By examining the agricultural schedules of the 1861 and 1871 censuses in the Grand River and Kettle Creek watersheds, we can see flax on a farm-by-farm basis and imagine the relationships that connected farms to mills. The agricultural schedules of the 1861 manuscript census reveal that 566 farms reported growing either flax or flax seed in Waterloo County (Table 2.1).

The seventy-six flax farmers in Waterloo South were the largest producers, and they lived closest to the original Perine mill in Doon. There

Table 2.1 Flax farmers in Waterloo County, 1861

	Waterloo S	Waterloo N	Woolwich	Wilmot	Wellesley	Total
No. of farmers	330	257	408	539		
No. reporting flax or flax seed	76	85	132	165	108	566
No. reporting flax	69	84	123	165	78	519
Volume of flax, lbs	110,994	27,842	59,874	93,979	25,868	318,557
Median	900	100	200	200	100	
Mean	1,609	331	487	584	332	
No. reporting flax seed	73	72	109	152	98	504
Volume of flax seed, bu.	3,037	1,325	2,110	5,826	1,805	14,103
Median	30	6	10	12	13	
Mean	42	18	19	38	18	
Value of flax seed	$3,645	$1,610	$2,497	$5,143	$2,133	$15,028
Median	$37.00	$7.50	$11.25	$13.00	$15.00	
Mean	$49.93	$22.37	$22.90	$34.06	$21.99	

Source: Canada, *Census of Canada, 1860–61*, Agricultural Schedule for Waterloo County, Canada West, Reels C-1077–C-1080.

were hardly any small producers in this township, and their confidence in the crop likely reflected confidence in the millers. Their proximity to the Perine mills afforded them easier access to the industrialists' expertise and resources. At first glance, it might seem there was a continuous flax belt around the town of Berlin (Kitchener) in 1861, but even in what appears to be a homogeneous zone of commodity production, the mapped data show subtle patterns in land use. For example, in the westernmost townships, Wellesley and Wilmot, most of the flax farmers were near or northwest of the Baden mill, and production decreased as proximity to the town of Berlin increased (Figure 2.1). In these two townships, the ratio of bushels of seed to pounds of fibre was almost double that of the ratio in the east. These farmers had quickly adapted their cultivation practices to produce a crop that favoured seed for linseed oil or sowing rather than fibre for making linen.[12] The US census data for 1860 similarly show that several counties in Ohio produced mainly seed and very little fibre (Figure 1.3). The rural mill complex offered a range of markets for flax, and even in the early stages of the mid-nineteenth-century flax fibre expansion, farmers were responding to local variations. Seed production was already displacing the value of fibre.

Figure 2.1 Flax in Waterloo County, 1861

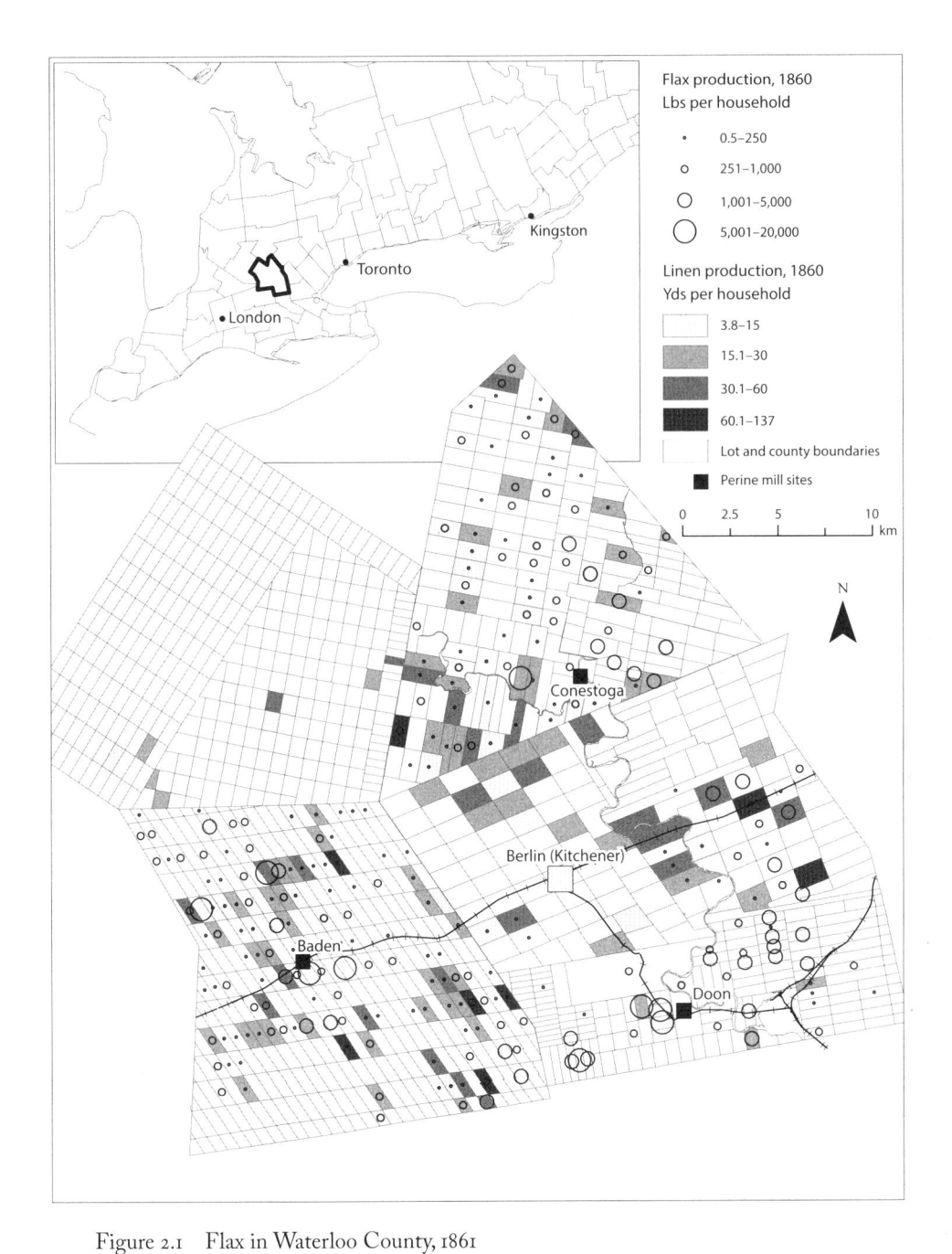

Sources: Canada, *Census of Canada, 1860–61*, Agricultural Schedule for Waterloo County, Canada West, Reels C-1077–C-1080.

The mid-nineteenth-century censuses also reveal that independent farmers were only part of the fibre supply chain and that other sources were required to satisfy mill demand. Flax millers, from Streetsville to St Thomas, were also engaged in flax factorship, a mill-directed form of production. In some cases, millers acquired over ten times as much fibre through this process as they did through farm deliveries. The census and other statistical data missed the surprising reality that far more people were involved in producing the flax fibre web than generally thought. We see some of the fingerprints of factorship in the census, such as B.J. Tiffs, a fifty-year-old man in Waterloo County whose occupation was listed in 1861 as a "Flax Raiser." Strangely, according to the agricultural census of the same year, Tiffs grew no flax. Instead, the crops he cultivated likely belonged to the local flax millers. He was a labourer in an industrial farming system that produced much higher volumes than those recorded by individual farmers in the census. Waterloo County farmers reported 369,000 pounds of flax in 1860, but on the manufacturing schedule the three Perine mills reported using 4.8 million pounds of flax for that year.[13] Colonel Mitchell's mill in Halton County reported using 1 million pounds, although farm production there was reported as less than 40 per cent of that. Similarly, in 1870 Ontario farmers reported growing 1.2 million pounds of flax, but all the Ontario mills together reported producing flax for a total of 26.8 million pounds of raw flax.[14] As the discrepancy suggests, there are many problems with the flax data in the Canadian censuses, so much so that M.C. Urquhart called the historical data on flax fibre "the least satisfactory series of the whole lot."[15]

Flax Factorship

Some flax promoters and most contemporary flax manuals explain how this mill-directed production yielded the large quantity of flax fibre used by millers. However, as we will see in Chapter 3, most promoters misunderstood the entire system. Benjamin Walker was one promoter who got it right. His Elgin Flax Association attempted to "call into existence a class of Flaxmen the same as the 'linier' of France."[16] Alexander Kirkwood also referred to the miller in charge of French and Belgian flax factorship as the "linier, or 'flax man.'"[17] In 1852, E.F. Deman's industrial manual for flax scutchers, *The Flax Industry*, explained what had become a standard primary processing system in Flanders – flax factorship. Deman explained how to encourage flax cultivation among local farmers and see if "a sufficient quantity is likely to be grown in the immediate neighbourhood – say ten miles round." One mill performing all of the field work and primary processing could probably manage up to 500 acres, although many had smaller sources of raw material.[18] Kirkwood and other Canadian promoters said little about the industry's organization in North America, but the flax production system that emerged in Waterloo County and other parts of the eastern Great Lakes region was very similar to what Deman described. Money and flax changed hands in several ways under this system. According to a farmer in Washington County, New York, some farmers sold the crop standing or before pulling; some pulled, stored, and then sold; and others retted and rippled the crop before selling semi-processed fibre.[19] Deman noted that Flemish flax millers sometimes bought the crop a few days before harvest, keeping the straw and leaving the seed for the farmer. At the point of purchase, the "farmer has no more to do with the management of the crop" except to rent the miller "his barns to take off the seed, his horses and wagons to remove his flax, and the use of his meadows and waters" for retting.[20] The only way to guarantee a steady supply of raw material, according to Deman, was to provide a way for the farmer to "sell his flax in a green state to factors or merchants ... the farmer being too independent to give his attention to so many *after processes*."[21]

Flax factorship was not easily captured by official statistics, and the problematic census categories reflect an evolving sense of what flax was, what it could be used for, and why it was important to record it at all. From botanists to general encyclopaedists, experts and promoters often disagreed on the function and classification of the plant; they called it both a fibre crop and an oil crop, and in 1912 it was simply grouped along with tropical commodities as a "plant of great commercial value."[22] Promoters were sufficiently confused about the nature of the plant that they even trusted charlatan alchemists who argued that flax could be converted to cotton through chemical transmutation. Thus it was perhaps unsurprising when the official labels used to identify scutching mills and other fibre establishments changed periodically.[23] The reporting of such establishments was also suspect. None appear in areas with moderate to high fibre production like the northeast coast of New Brunswick or the Quebec districts north of the St Lawrence. In the 1871 Census of Prince Edward Island, enumerators listed several scutching industries where there was clearly no flax to scutch.[24] Hemp and flax were initially recorded together in Canada and the United States, because both had similar uses. In 1861 enumerators in Waterloo County very infrequently noted "Hemp" in the margins to differentiate it from flax, which had now become vernacular. As well, the use of flax seed for linseed oil grew later in the century, and so by 1871 the census began to record bushels of flax seed. In fact, Waterloo County census officials deemed flax seed an important enough part of the crop in 1860 to declare it a new handwritten category on every schedule in the county.

By 1871 the first data on Canadian flax seed showed that Ontario and Quebec had concentrations of raw material for the linseed oil industry similar to those south of the border (see Figure 1.3). It might appear that Quebec's flax fibre industry was stable (it supported about 1 million pounds of fibre per year), mostly without the help of mechanical scutching outfits like the Perines'. However, between 1850 and 1890, the nature of both the geography of production and the processing industry changed frequently. In fact, flax production in Quebec closely resembled Ontario's. Both provinces developed fibre and oil industries, and Quebec was the country's oil centre until at least 1871. Quebec scutching mills appeared slightly later than Ontario's and processed much smaller quantities of fibre, but Quebec farmers were responding to three different markets – industrial fibre scutching, the local and home-manufactured linen industry, and a linseed oil market that resembled Ohio's – with large mills in Quebec County and Montreal and specialized areas that produced seed and abandoned the fibre. Just as the wool and flax industries in Ontario were divided east from west, Quebec's flax and linen industries were mostly in the centre and east of the province, whereas wool was concentrated in the west.[25]

In southwestern Ontario, the provincial flax fibre industry had changed little in terms of its scale and production in concentrated areas in the 1860s (see Figure 1.5), but the centre of production had shifted west and would continue to move in later decades. Flax in the Credit valley and points further east had practically disappeared; there were few flax producers in immediate proximity to the mills in Conestogo, Doon, and Baden; and the major producers were now in sporadic concentrations along a line stretching from Wellington North to Elgin East on the shores of Lake Erie.

Figure 2.2, which uses the last available Ontario agricultural schedules, presents a snapshot of independent flax producers and mill locations in Ontario. The crop continued to make sense mostly to farmers in highly concentrated areas, but the location of those areas was constantly moving. The major flax fibre growers of 1860 were now almost entirely uninterested

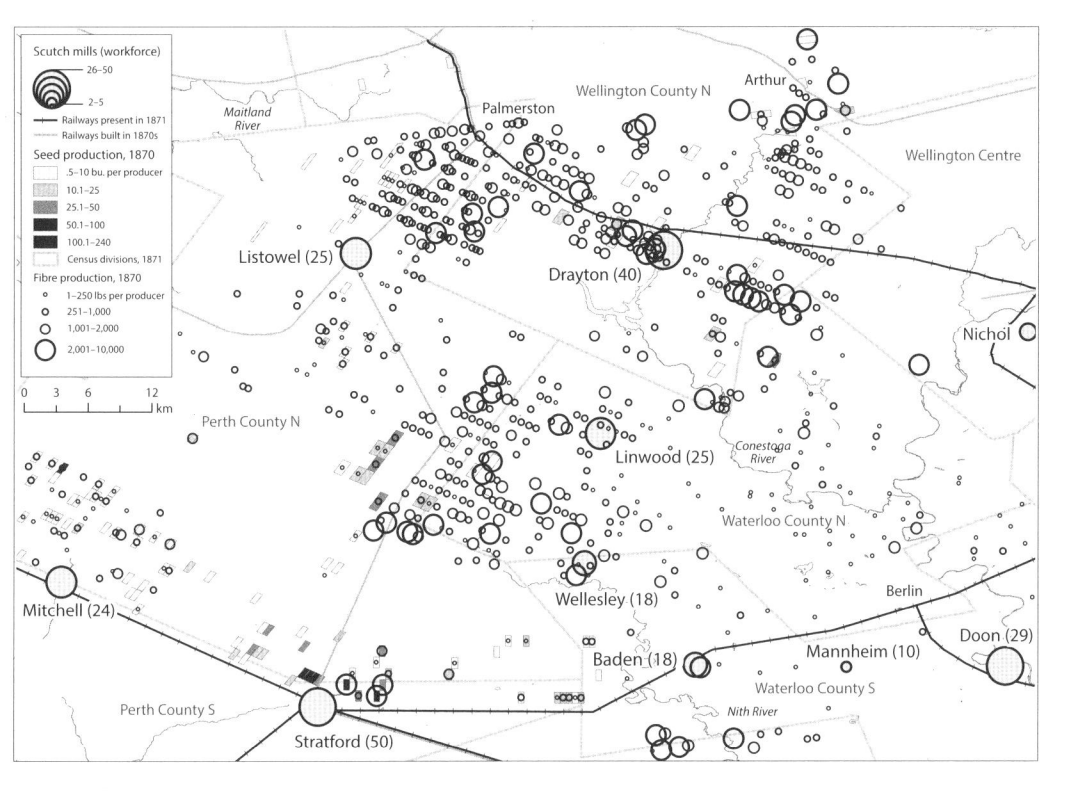

Figure 2.2   Flax fibre, seed, and scutch mills in southwestern Ontario, 1870–1871

Source: Canada, *Census of Canada, 1870–71,* LAC, Agricultural Schedule for Waterloo County, ON, Reels C-9943–C-9945, Mornington, Perth County, Reel C-9941.

in its cultivation. As we will see in later chapters, flax seed production tended to move west in the Midwest and Northern Great Plains, leaving some older areas for good. Ontario's fibre production probably moved west for different reasons. By 1870, the main market for raw flax was no longer in Doon, but stretched from the western edge of Waterloo County into Oxford and Perth Counties. This market was shaped in large part by railways. We saw how critical the railway was to the location of mills in the 1850s and 1860s, and now producers decided to plant flax based in part on their proximity to rail transportation. A clear line of flax cultivation was visible, stretching roughly from Fergus to Palmerston, cutting through Peel and Maryborough in Wellington County and Wallace Township in Perth County. The reason was clearly the presence of flax mills and the construction of the new Wellington, Grey & Bruce Railway. Charles Hendry

Table 2.2 Flax farmers in select townships, 1871

County Township	Waterloo Waterloo S	Waterloo N	Woolwich	Wilmot	Wellesley	Perth Mornington	Elgin East Yarmouth	Total
No. of farmers	675	n/a	890	n/a	935	634	1,081	
No. reporting flax or flax seed	11	41	57	41	112	144	41	447
No. reporting flax	7	25	36	28	107	140	41	384
Volume of flax, lbs	2,736	5,275	7,615	29,333	63,619	81,376	68,226	258,180
Median	100	35	63	650	450	500	1,400	
Mean	391	211	212	1,048	595	581	1,664	
No. reporting flax seed	10	35	50	37	110	15	3	260
Volume of flax seed, bu.	128	147	1,147	1,812	1,726	287	34	5,281
Median	3	2	5	20	12	18	12	
Mean	13	4	23	49	16	19	11	

Source: Canada, *Census of Canada, 1870–71*, LAC, Agricultural Schedule for Waterloo County, Reels C-9943–C-9945, Mornington, Perth County, Reel C-9941.

owned the mill in Maryborough, and the Livingstons operated the mill in Listowel. However, the railway did not quite reach either town. The tracks had only been laid to Alma by 1870. Evidently that was close enough for many flax farmers, and the promise of an extended line was incentive enough for others. Construction of the line to Palmerston was finished in 1871, and the rest of the line to Lake Huron by 1873.[26] This kind of analysis is only possible with Geographic Information Systems and farm-level data. Flax was significant in Waterloo, Wellington, and Perth Counties, and farm-level mapping best demonstrates how flax production was concentrated in the area where these three counties met.

As later chapters will show, Baden eventually became the processing centre for most of the province's flax seed. Table 2.2 shows that specialization in seed or fibre was already evident in 1871. In fact, the major flax fibre farmers in 1871 grew very little seed. Southwestern Ontario farmers maintained this increased flax production in 1881 and grew even more flax in 1891, the last year that the census recorded flax grown for fibre. By this point it was clear that Ontario's production could supply the raw materials for a steady flax fibre industry, and yet the centre of production moved continuously over several decades. Writing with years of experience in the

USDA (United States Department of Agriculture) fibre division, Charles Dodge noted the strength of the Canadian system, arguing that the only industry that would work in his country would be run on similar lines. What he wanted was a network of "small central factories established in flax-growing locations, where the farmers will be able to make contracts to grow certain quantities of the straw." The mills would be responsible for everything thereafter, thus obviating the need for "skill and technical knowledge" and start-up capital. The flax millers in Yale, Michigan, the only state with a functional flax industry to Dodge's knowledge, were organized and managed by a Canadian company.[27]

The Perines' Flax Factorship

Procuring raw materials and persuading farmers to grow flax were constant problems for the Perines. In March of 1856 the brothers placed a notice to farmers in the local English-language press. It was simply titled "Flax Wanted," and it promised to "pay cash for any quantity of well rotted and well handled flax straw, delivered at their mills."[28] Christian B. Snyder, a local Mennonite farmer, kept a diary that year in which he recorded that although he had not grown any flax in 1856, he had travelled in the fall "to Conestoga vilage [sic] to see Wm Perine."[29] His visit to the Perine flax mill likely concerned the ways the farmer could add flax to his already broad spectrum of agricultural outputs. It is not clear what Snyder discussed with William Perine, or whether he ever did business with the flax miller in the future. But many of Snyder's neighbours did, and by 1861 the census data showed that local farmers had indeed responded to the new market for flax fibre. The Waterloo County mills advertising for flax modified the way industry and agriculture met, and within a few years many local farmers had decided, through various methods, to incorporate flax cultivation into their farm strategies. They had also allowed millers to manage parts of their farms.

In the mid-1860s, articles in the *Canada Farmer* claimed that the Waterloo flax millers could offer unlimited seed to farmers and that the amount borrowed to sow could be returned from the harvest or deducted from the sale of the flax. This form of exchange was followed in many transactions in the Perines' account books. On 9 December 1861 Christian Furtney was credited with over 4,600 pounds of flax and 40 bushels of flax seed "less 4 bu. used for sowing."[30] Furtney was a much larger producer of flax and flax seed than the average flax farmer listed in the census, but what matters here is that he enjoyed one of the common manufacturer–farmer relationships visible in the Perine accounts.

The Perines' account with another farmer, David Hilborn, shows a different kind of relationship with local producers. In July 1861 Hilborn was credited with $72 for an amount of flax similar to the previous year's harvest. He supplied the Perines with almost 5,000 pounds of green, unprocessed flax and 26 bushels of flax. However, he was subsequently debited almost the total value of the flax he sold in order to pay for the 44 "bushels flax seed [he had] borrowed to sow in 1860."[31] After paying for the previous year's seed, he was left with less than $7. Under normal conditions, 44 bushels of seed typically yielded closer to 84,000 pounds of green flax, but Hilborn was only paid for 5,000.[32] Our view of the Perines' strategy for acquiring flax and Hilborn's strategy for making money is supported by other transactions. In March 1860 Hilborn was credited with a $33 "flax land note," and on three subsequent occasions that year he was credited a total of $32 for pulling flax. He thus earned income from flax by selling fibre and seed, renting some of his land, and working on the harvest of the Perines' own flax crop.[33] The several tons of product missing from the original 44 bushels of seed might have been lost in poor harvests, but a more likely explanation is that Hilborn redistributed the seed among neighbours and his own siblings. In 1861 the twenty-nine-year-old Hilborn and his wife, Sara Ferguson, had no children. They lived next to his parents, his younger brother Jacob and sister-in-law Caroline, and Jacob's three young children. Jacob was the only Hilborn to report flax in the 1861 census, but at least three other siblings farmed nearby and the census shows a concentration of flax producers to the north of the Hilborn properties (Figure 2.1).[34] Based on his activity in the Perine accounts, Hilborn more than likely worked in these flax fields as well.

David Hilborn was not the only flax farmer credited for both labour and land. In many other places, flax millers like the Perines and their St Thomas mill partner Alexander Young recorded relationships with flax farmers and field labourers that indicate they used the system of factorship. The Perines frequently credited people for "pulling flax."[35] Young's flax harvest involved hiring individuals to cut or pull flax from land he had leased and then spreading it in the fields.[36] Spreading allowed the dew to ret the fibres and prepare the plant for breaking and scutching. Perine employees harvested both flax owned by the business and that owned by individual farmers. One man was paid in 1862 "for pulling our own flax," and a farmer was debited that same year for seven hands to spread his flax and one boy to "take it up."[37]

Furthermore, some farmers worked with the crop until it reached the mill gate; some also took processed fibre back to their homes to spin. The Perines themselves indicated that home production was an important part of the industry, and they considered it a Canadian contribution to their knowledge of flax processing. In 1895 an elderly Moses Perine recalled that "the settlers' wives after the straw was brought in to be scutched, would take the dressed line [fibre] back and spin [and weave] it up into cloths, towelings, etc."[38] One outcome of this relationship appears to have been a concentration of home-manufactured linen, especially in Wilmot Township (Figure 2.1). Waterloo was not the only region where farmers spun some of their own crop. The mills of Brown & Co. in Stratford, according to the *Canada Farmer*, "had some of the [1864] crop spun by farmers."[39] This brief reference seems to indicate that the mill relied on home production as part of its rough textile manufacturing. Although the Stratford mill's farmers-cum-spinners appear to be outworkers, the Perines' farmer customers were not. As early as 1859 farmers bought small quantities of flax, which the Perines noted was "to spin,"[40] and many similar purchases were recorded without the telltale title. Most flax debits in the daybooks are for less than a hundred pounds, and even though the intended purpose of the outgoing fibre was rarely given, it is hard to imagine such meagre amounts being useful for anything other than small or home production. Although the individual amounts were small, the total fibre taken away to spin represented 19 per cent of the amount dressed in the mill. Many of the customers who bought small amounts of flax were also credited for flax harvested from their land. It appears that the exchange of semi-processed fibre for labour and land was a unique contribution by Canadian farmers.

Although there are still many gaps, the accounts show that the Perines were able to partly satisfy the voracious appetite for fibre in the nineteenth century by incorporating a local variation of the European and New York flax factorship system into Upper Canada. The practice helps explain the discrepancy between what was recorded in the Census of Agriculture and the raw material needed to meet the demands of the mills. Scraps of evidence are also visible in the census manuscripts. In the 1860 agricultural census for Wilmot Township, "Perine Brothers" appears on a lot adjacent to their mill in Baden, but their entry recorded none of the standard farm details. Rather, the census simply states the lot number and the "27 acres land rented." The yield columns predictably list only flax and flax seed. In one of their properties around Meadowvale, Gooderham & Worts cultivated a 32-acre field of flax "on the flats of the Credit." The field had been mainly sod, "unturned for 17 years," when the millers ploughed, harrowed, rolled, and sowed it to flax in the last week of April 1865. This may have been one of several flax fields owned by the mill, and together with Colonel Mitchell further up the Credit in Norval, these millers distributed enough seed for 1,200 acres of flax. This industrial flax site is now the Meadowvale Conservation Area, where hikers follow the river through forests and watch for "remnants of early settlement with mill and farm ruins still visible."[41] By 1871, millers produced much larger crops on rented land. One enumerator noted in the industrial schedules of the census that a large flax mill in Oxford County "raised 325 acres of flax in other townships on rented land of which the owner knows nothing of the production."[42]

The Perines' method of factorship was also practised elsewhere in North America. In 1863, for example, the *Canada Farmer* made note of a new flax mill in Janesville, Wisconsin, that contracted for 500 acres of flax and whose method of payment resembled the way the Perines bought flax. The mill "takes all the risk," the paper stated, by paying the farmer $20 per acre, supplying them with seed, and sending "one of their own men" to oversee the retting.[43] Unfortunately, there are no surviving examples or even details about contracts between flax millers and farmers, but two early proposals on both sides of the border give us a glimpse of what millers were willing to offer. In Galt, Upper Canada, according to W.H. Smith, the Forbes paper mill offered the farmer £20 per ton of good flax in 1850 or "double the sum per acre that he was accustomed to make by growing wheat."[44] In Fall River, Massachusetts, the owners of what became the American Linen Company presented flax samples and offered guaranteed prices for twelve months thereafter to farmers whose flax matched the quality of the samples. American Linen was one of the largest US linen producers in the late nineteenth century, but like Smith, Dove & Company in Andover, it relied mainly on imported fibres. There is no evidence that the price guarantees continued beyond 1854.[45] A similar method of contracting with farmers was practised earlier by the Boston Linen mill.[46]

Evidently, successful flax production required a very specific set of skills and labour to provide the raw material. Flax specialists argued that factorship was the only efficient system because of the skill required in processing flax. When the Flemish miller had procured a crop of flax, Deman explained, he turned his attention "to its after-management; and the gangs of labourers employed by him soon become expert under his skilful superintendence in their peculiar branch of business, from the right system of pulling to that of the last handling required, which never can be performed or generally understood by the farmer himself."[47] As we will see, flexible field labour became a critical element in a successful flax-growing community in the Great Lakes region.

Labour Gangs

Since its earliest incarnation in the 1850s, Canadian industrial flax production required two major ingredients: flexible work gangs and skilled managers and millers who knew how to process, handle, and schedule the raw materials properly. Small quantities of fibre were harvested in flax bees among neighbours, but flax millers usually had to organize production to supplement what farmers would not grow. The millers shared resources and responsibilities with farmers by renting their land and often their labour, and most of the labour involved in pulling flax by hand was organized and paid for by the mill. The labour organized by flax millers supports the argument, made by Jan de Vries, Daniel Vickers, and others, that productivity was high in the late nineteenth century and growing because of the work done by women and children.[48] In the flax fibre web, the key aspect of industrial farming was not mechanization but the knowledge of managers and the strength of markets.

Flax had to go through a long list of manual stages before it became scutched fibre or tow. The fibre industry presented special challenges because flax processing was intensive and if processors wanted the best possible product, they had to pull it by hand and it had to be done in a narrow window at maturity. At harvest, flax could either be pulled or cut, but reaping shortened the straw and made it less useful for making fibre. The manual processes required for harvesting flax fibre were unique in Canadian labour history. The remarkable aspect of flax production was not that it was so small or confined to one region, but because it was so large, considering the labour constraints in the Canadian economy.

Pulling an acre of flax required approximately 3.5 person-days – cotton picking took 6.5 days per acre – but processing the pulled flax required timely visits and additional, if less strenuous, work.[49] Such a labour-intensive industry was possible only because there was demand for the crop's two products, fibre and seed, and farmers and millers were able to adapt to markets for both products, sometimes in a single growing season. Flax millers employed large gangs of flax pullers to harvest the mill's crop every year in late July and August – sometimes drawing on the women, children, and seasonally employed men in nearby towns and often hiring gangs from distant mills or First Nations reserves. First Nations labourers earned a reputation as Canada's flax pullers and were involved in the Ontario harvests for at least half a century.[50]

Scarcely any memoirs have survived that recalled waged work in the flax harvest, although this form of hired work was exactly how the overwhelming majority of the crop was produced. For the most part, harvesting and processing flax fibre was not reciprocal work and not something that all farmers identified with. Millers "reciprocated" with cash or credit, and production was so concentrated, time sensitive, and mill centred that family-organized bees were inadequate for harvesting large amounts of fibre.[51] Community was still important to flax labourers. Men, women, and children – often relatives and neighbours of one another – harvested the crop together. Millers sometimes provided transportation and housing, they occasionally commissioned photographs of the workers, and one miller in Tavistock even collected or wrote flax-pulling songs in German for his workers. But there was little a miller could do to mask the back-breaking nature of flax pulling. When people experienced community in the flax harvest, they experienced it on the millers' terms, and when people recorded their few memories of working in flax mills or flax fields, they were more often memories of wages, not merriment.

An alternate form of flax work was the hired or reciprocal labour organized by farm operators. This would have represented a distant secondary source of total fibre production. Still, there exist some accounts of flax-harvest bees in the late nineteenth century. Back-to-back flax bees were held in Bosworth and Parker, near Drayton, in 1907, and both were an occasion to throw a party. After Harry Beal's flax-pulling bee on a Thursday night, "all resorted to the barn where they enjoyed themselves until the wee small hours." Then, on Friday, P. Priester had four acres pulled "in a short time," leaving plenty of time for socializing.[52]

Ottis Wright remembered flax pulling as a form of work he tolerated for two or three hours just to get the social and recreational reward of a free party afterward. "The last flax pulling bee I was at," Wright recounted,

> was on the second of Luther, over at the Murray's. My neighbour and I went and we started pulling at five or six in the evening until about ten o'clock. After pulling flax for two or three hours you were

ready to quit. Your back was aching and it was a hard job. At ten o'clock there was a big dance – dance until morning. We didn't quit at one oclock – we'd dance until daylight maybe. At some places you'd have lots of drinks and others wouldn't. Now, that's it – that's the way they were. The flax had to be pulled and then taken to the flax mill in Arthur. The farmer would get it all pulled for nothing by putting on a big night. And we were happy to go and pull for nothing for a big night. Sure wouldn't miss it for anything.[53]

One memoir of the flax industry in Maryborough Township claimed that bees were popular and were organized by the farmers. They gathered their neighbours in the early evening for the flax pulling, and then "[q]uite often the night wound up with dancing and Whoopee." Bonnie Elliott recalled that the bees were "huge" and attended by all of the neighbours, and the occasion was always celebrated with a party.[54] Terry Crowley's history of labour in rural Ontario also identified the importance of social events at flax bees; it seems memoirs of flax bees noted dances and harvest competitions as frequently as they noted the work itself.[55]

Although many people laboured in some capacity in the flax harvests, we know very little about them. Some flax-mill account books offer tantalizing glimpses of mill workers, but rarely describe their work. Agricultural censuses are essentially place biographies, telling us much about a field's produce and productive capacities but little about the energy and labour required to make them. The agricultural journals in the 1860s discussed flax production and profitability *ad nauseam* but were strangely silent on labour and especially flax pullers. The *Canada Farmer* published an article on flax in almost every fortnightly issue of its inaugural volume in 1864, but even the one article on "Harvesting Flax" offers only a few words on the technique of pulling and spreading and none on the availability and productivity of labour. Farmers believed that the press was ignoring important information. The promoters of the mid-century tended to dwell on the technical improvements that they expected would soon come and rid the flax industry of any last barriers. A popular topic was the elusive dream of a flax-pulling machine that would relieve the farmer of this difficult job. New mechanized flax pullers were frequently announced in the 1860s but were not perfected until after the First World War.[56]

The focus on field-labour productivity in the 1860s was appropriate, since farmers reported that pulling was the most costly aspect of growing flax, usually ranging from $3 to $7 per acre, and sometimes reaching as much as $20 or $30.[57] In 1865 a farmer named "E.M." wrote to the journal to refute its constant endorsement of flax, claiming that "[i]n my own case the crop did not pay for pulling." Another farmer retorted, questioning E.M.'s vigour. "The pulling which 'E.M.' seems to make such a bugbear of," he taunted, "is not a job for kid gloves, but I have heard many respectable farmers state that they would much rather pull it than cut it … as the weight of the roots pays for the extra labour." Another writer agreed with E.M. Calling flax the "gambler's crop," he claimed that twelve hands per acre were required to harvest in time and asked, "[W]hat small farmer's establishment is able to cultivate a large breadth of flax, since this is the case?" However, an examination of sources of information such as photographs, personal memoirs, the census, and the Perines' Waterloo County account books shows that even when farmers grew flax in small amounts, they relied on gangs of flax pullers from the nearest mill for help.[58] By the mid-1860s, the Perines drew on over 2,000 acres of flax in Waterloo County.[59] To put this in perspective, we should note that the average flax farm in this region grew only 2.5 acres in 1860.[60] Those amounts could be pulled by a couple of hands in less than five days, yet the account books reveal that many of the early flax farmers still relied on help from the mills to harvest their small crops.

The Perines hired a range of labourers to do this work. Some were well-known names in the accounts, and others, such as in the case of "a dutchman," were likely migrants. Workers could be employed for one day or for the full span of a harvest season. Other workers were members of farm families performing labour on land they had rented to the mill.[61] John Meyers described his childhood life on a Mennonite farm in Waterloo County in this period and recalled his father selling flax to the millers in an equally complex relationship. The millers would "get" his father to sow a field of flax, and they would bring "a gang" of flax pullers to pull and stook the flax. The Meyers would draw the flax to the barn, thresh it using the mill's equipment, and then spread it in the fields to ret. A male or female mill worker was sent to turn the rotting flax over (with a ten-foot pole); the gang would return to rake and bind the flax; and Mr Meyers would transport it to the mill with his horses. All this while, the mill owner checked on the flax to gauge its readiness. The miller then processed the flax straw and bought the finished fibre from Meyers after "deducting the price of the labor." Meyers kept some fibre for his wife and daughters to spin into thread.[62] In these intricate strands of the commodity web, the millers and their gangs visited the farms on numerous occasions. The farmers provided land, carried flax to and from the mill, and took flax seed from the millers in the spring. The Waterloo flax millers used a gang of workers to pull even the crops of these individual farmers in the early 1860s.

Flax pulling came into sharper focus in the late nineteenth- and early twentieth-century rural media. Newspapers reported on local flax-pulling gangs; farm journals described the work of pulling; Indian agents complained of First Nations harvesters leaving the reserve; and gangs of flax pullers became a focus of amateur and professional photographers. Women often worked in the flax harvest in Europe, and male observers claimed "no healthier women can be seen than those employed during the whole flax season in pulling, spreading and lifting it."[63] There is no evidence of gendered labour in North American flax fields: both men and women pulled flax. However, pulling flax was often poorly paid and always unpleasant, and since labour was in short supply, women and children were likely candidates.[64] In 1909 James Byrne stated that "the pulling is done by gangs," and the production of flax occurs "in conjunction with mills and flax workers in towns, who buy the product from neighbouring farmers by the ton, or rent land and cultivate it themselves." In most cases, according to Byrne, "a gang of 25 or 30 pullers" led the way and were followed by an older person who did the "shocking" – the process of tying the sheaves of fibre.[65] Flax pullers marketed their services to both farmers and millers, and were known for their skill in the field. Byrne claimed that farmers paid "from $5 to $7 an acre to a man or group of men and women, who make flax-pulling their summer trade." "Where millmen grow the 'stuff' on leased land," he reported, "they have gangs, consisting of men, women, and children, hired at ... [f]rom 25 cents to $1.50 a day ... according to 'pulling capacity.'" In most instances, the mill took the initiative and assembled a gang of willing workers, but in some cases, the gangs were regular itinerant teams, as we will see with the First Nations labourers. Byrne argued that millers could draw their labour from towns within eight or nine miles from the flax fields, even though they provided no meals but daily transportation on an austere "gang wagon."[66]

Rural photographer Reuben Sallows captured this perspective in a typical image of men and boys pulling flax in 1910 (Figure 2.3). His subjects work close together in the field, bending over, kneeling on the ground, and standing to rest their backs or move to the next stalk. They form a line that

Figure 2.3   Pulling flax, 1910

ensures they pull all the flax without treading on the standing crop, and their clothing offers protection from the sun and the thistles. The focus is on the work and the faceless gang – the photographer is not concerned with individuals or identities. Even the camera angle is practical, allowing Sallows to capture as many pullers as possible without getting too far from his subjects. The children in the picture are not set apart in any way, and they wear the same clothes and perform the same task in the same position.

An image of ten women and girls harvesting flax near Arthur, Ontario, tells us more about the flax harvest and about the role of children and women (Figure 2.4). Most of the subjects face the camera, except for one woman who continues to spread the flax in the field for drying and retting and a group of boys who watch the workers from the hedgerow. We do not know if the workers are spreading, turning, or taking up the flax. The focus is on the people, as if they were family or other identifiable persons. The photographer is not known, and it has been suggested that the picture was taken in 1915. No men are present, though this could be more a result of war than of a gendered division of labour. In most other documents, the pullers are described as men and women, adults and children. However, it is possible that the activity they perform in this picture was lighter work than the actual pulling, and there are no similar photographs available that show an entirely female gang of pullers.

Figure 2.4   Women and children gathering flax near Arthur, ON, c. 1915

Children are present in many photographs of flax harvesters, and their roles in the flax industry were important. Since the 1860s, the boosters had suggested employing children in flax pulling to solve the labour problem.[67] In Thamesford, newspapers reported "about 30 hands in the field spreading (mostly boys and girls) ... divided up into two gangs."[68] The children in each gang were under a boss's supervision and earned less than a dollar per day. The prominence of child labour was captured in great splendour one bright summer day by Tavistock-area photographer Lemp Studios (Figure 2.5). Twenty-one young children sit or kneel in the first two rows and at least a dozen more young people stand around them. Adults, wagons, horses, and other props frame the community's children who are clearly the focal point of the image.

Wages were not high, but flax workers potentially made more than general farm labourers. The Tweedsmuir history of Alma, another long-time flax mill location, stated that farmers sowed the flax, but it was "pulled by hand, by boys and girls, gathered into bundles and taken to the mill ... Sometimes as many as 40 to 50 boys and girls would work in the fields for 75 cents a day spreading the flax."[69] Children were usually paid at a lower rate than adult labourers. Mary Grant of Huron County, for example, recalled earning 25 cents per day pulling flax in her youth.[70] Byrne's estimation of $5 per acre and 3.5 as the average person-days required for

Figure 2.5 Tavistock Flax Mill Gang

an acre of flax suggests that adult flax pullers could expect at least $1.40 per day in 1909. According to Cecilia Danysk, "wages for farm work were notoriously low" and "extremely varied,"[71] but flax harvesters earned more than most agricultural labourers. The largest flax manufacturer in Ontario in this period was James Livingston, and business correspondence from his son-in-law, James McColl, indicated a higher wage in areas of heavy flax production.

McColl contracted with farmers in ten communities north of Yale, Michigan, to harvest about 5,500 tons of flax in 1907, the product of about 3,600 acres and the work of 420 labourers if he harvested for thirty days straight. McColl said little about the flax pullers themselves, but indicated at the start of the season that "[l]abor is scarce here, wages are high, and we may be obliged to cut about half of our crop." Back in Ontario, other fibre mills gave more notice when they switched from pulled to cut flax. John McGowan's flax business in Alma advertised flax seed for farmers in the spring and warned farmers that "[w]e now prefer flax which has been cut and bound in very small sheaves rather than pulled, and advise farmers to note the change."[72] McColl's problems were about maintaining a profit during the harvest; they were about competition as opposed to preference; and he had lost, he claimed, two or three men to his competition, who were offering $2.25 per day.[73] Later, he wrote that a competitor was paying one man $2.50 and two others $1.75 per day, noting with some satisfaction that their "flax is costing them something to stack."[74]

McColl had teams operating in each community, and although we do not know how many worked on each team, he did mention that they normally "baled" 22 tons per day and that exceptional leadership on one team helped it bale 28 tons per day. At this rate, it would take seven or eight teams to process the entire harvest in a month.[75] Whether a miller like McColl could find adequate labour for the flax harvest determined how he harvested and what he earned. Flax that was "pulled, properly bound, cured and handled and free of all foreign matter" was the most valuable type, worth $1 per ton more than cut flax.[76] McColl sold his flax for $17 per ton in 1907. Despite the labour shortages he was able to pull most of the crop, partly because the beet crop failed and the "beet cleaners" were able to help pull flax.[77] It is entirely possible that the wages mentioned by McColl were for skilled workers and that unskilled flax pullers were paid closer to the $1.40 per day mentioned by Byrne, but even this was well above the $5 per week that most farm workers earned in eastern Canada at the turn of the century.[78] These wages rarely included room and board, and even itinerant groups had to take that into their calculations. For local families who were transported to the field by wagon or truck, this was less of a barrier, but workers who travelled from harvest to harvest had a considerable amount of extra expenses.

First Nations

The most important group of workers were often First Nations labourers who had left the reserve for seasonal work on Ontario's farms and flax mills. First Nations workers have sometimes been considered "cheap labour," but there is no reason to assume that First Nations flax pullers were paid less than other farm labourers. In some places, First Nations harvest workers consistently earned more than non-Native workers.[79] In the early 1890s Indian agents for Six Nations, Oneida of the Thames, and Chippewa of the Thames began reporting that First Nations workers were leaving during the berry-picking season and returning after the flax harvest.[80] Working off the reserve was criticized by at least one Saugeen chief, and according to Peter Schmalz, "it was the Ojibwa who were against such moves because it discouraged farming among their own people."[81] One Six Nations agent reported that "[s]everal hundreds leave the reserve" for this activity.

The Thames agent claimed that "[c]onsiderable money is earned by pulling flax among the whites," and in 1909 the *Dutton Advance* reported that the Oneida and Chippewa continued to "earn a large amount of money by pulling flax, wood-cutting among the whites and berry picking."[82] First Nations flax pullers travelled widely, and Saugeen worked in flax fields from nearby Ripley to the more distant Floradale.[83] In Floradale they were greeted at the station by Edward Perine, a member of the family that had introduced factorship and the gang system to the local flax economy.[84]

Many millers relied on flexible and transient labour to do a job that most others preferred not to do. James Byrne pronounced flax pulling an unforgiving job because of "the heat of a July sun, a fierce, craving, tendency of the spinal column to crack," and the ubiquity of splinters. "The only consolation the puller has," Byrne assured readers, "is that his season is short."[85] The unpopularity of this work, therefore, could be due more to the nature of the job than to its rate of pay. Diane Newell demonstrates that British Columbia salmon-cannery owners relied heavily on First Nations labourers because they moved their families and belongings to the canneries for the season, and as one cannery owner complained, "[W]hite people would not do this." Non-Native families, in her opinion, would not operate in such distant, seasonal work, and thus cannery owners went to great lengths to attract First Nations people who would. Similarly, John Lutz notes that Indigenous labourers in British Columbia could only choose from a limited set of occupations, but still, they migrated for work often and on their own terms.[86] In Ontario, First Nations labourers might have found flax pulling a useful occupation for a variety of reasons, not the least of which was the lack of alternative occupations on reserves. In southwestern Ontario, Indian agents disliked off-reserve labour, as it supposedly distracted First Nations people from farming. The short season and the scheduling of flax pulling allowed them to be involved in other activities for the remainder of the year. The type of work chosen could mean that entire families could contribute to the harvest in various capacities, and the proximity of Ontario's flax regions to several First Nations reserves made the work more appealing. We read that, in some cases, workers were brought in from their reserves by flax millers, but even though Edward Perine and Jesse B. Snyder of Floradale met their workers at train stations, there is no evidence to suggest that their intra-regional transport was paid by interested millers. Figure 2.6 shows a group of First Nations flax labourers in transit. They gathered on the platform at the Drayton train station to

Figure 2.6 (*left*)   First Nations flax labourers on Drayton train station platform

Figure 2.7 (*right*)   Jesse B. Snyder and two men from the Floradale flax mill

harvest flax for the Floradale Flax Company; it is not known where they lived, where they had been, or what small-town flax mill was next on their itinerary. The photograph was likely commissioned by the millers themselves, who appeared in a separate image on the same day (Figure 2.7).[87] In Tavistock, the Livingstons began hiring First Nations workers as early as 1886. The *Stratford Beacon* simply noted that "[i]t was quite a sight when a large group gathered while awaiting their rides to the fields."[88]

The flax fibre belt moved frequently, and its price fluctuated wildly, so most itinerant work gangs accepted the location and wages offered to them. However, several First Nations groups found that their work sites carried cultural significance as well, and that their participation in the flax harvest allowed them to supplement and extend other seasonal summer migrations. Flax and other farm-labour opportunities extended the seasonal summer trips that some First Nations had been taking since before European settlement. This is most evident in the case of the Munsee and

Anishinaabe communities in the Thames and Ausable River valleys. Historians Johnston and Johnston argue that when the southern First Nations visited farms in the flax belt, they were more than simply responding to a help-wanted ad. Munsee families made regular visits to Perth County farms that echoed their seasonal hunting trips from decades before, and the locations they visited were deliberately selected for a mix of wildland resources, traditional campsites, and local farm-labour opportunities. For instance, one woman in the Carlingford area recalled arriving at her farm as a new bride in 1892 and seeing First Nations camps north across the 4th Concession in "Rohfreitsch's Bush," where the people wove baskets from black ash. Johnston and Johnston argue that "[t]here is little doubt that the Indians knew the site, and returned to it regularly, because black ash grew there naturally."[89] Another farmer from this area, Fred Clark, remembered Munsee families like Joseph Waddelove's arriving in his neighbourhood in 1910 to pull flax. According to Clark, the family camped in places Munsee had travelled to for hunting in the pioneer era, well before the flax industry.[90] Joseph and his wife, Laura, had four children in 1910, two of them boys, Francis, age nine, and Percival, age one. Their daughters, Bilva and Dora, were fourteen and thirteen, so it is likely that the three eldest attended the flax harvest with Joseph. Laura, who was pregnant with their fifth child, may have remained at home with the baby Percival.[91] Weaving baskets for sale was another frequently mentioned and photographed economic activity, and the Munsee's campsite selection near Carlingford in 1892 suggests that they returned to certain locations for wildland resources as well as labour markets.[92]

First Nations flax pullers typically camped near their work sites as they travelled to different locations. Figure 2.8 was taken by an unknown photographer near Arthur, Ontario, and shows how the harvesters lived when they were on location. The tents were large – at least twice as tall as the workhorses standing in front of them – and the three in the picture could accommodate at least the twenty-five to thirty people in the typical work gangs of this period. The tents appear to be either commercially available dwellings or structures made of ordinary canvas. Moving around for seasonal work was nothing new to most Anishinaabe bands, and the practice long continued. Well into the twentieth century the Lake Superior Anishinaabe "used tents during berry picking, [but] they preferred the wigwam in both sugarbush and rice camp."[93] The sight of these tents captured the attention of at least one photographer, but there was little about Ontario's

Figure 2.8   First Nations camp on field near Arthur, ON, c. 1915

First Nations flax pullers that seemed out of the ordinary to non-Native residents. Figure 2.8 and the reports of the Indian agent suggest one reason for this: these labourers had been coming to flax fields every year for five decades and they lived closely with non-Native rural people.[94] The farmhouse in the background and the outbuildings to the right of the tents mean that they camped between farms and probably interacted frequently with their neighbours. The "growers were in no way inconvenienced by having them on their farms," reported the *Farmer's Advocate*, because the workers prepared their own food.[95] In Berlin (Kitchener), labourers from the Six Nations Reserve at Caledonia set up their own tents and frame cookhouses on company-owned (or rented) farmland.

Paige Raibmon has shown how, in Puget Sound, First Nations hop-pickers became "Authentic Indians" and a source of entertainment for non-Native tourists, but in Ontario, flax pickers were initially observed with more caution.[96] When they first started pulling flax off-reserve in the early 1880s, their non-Native hosts eyed them with suspicion. One Carlingford resident, who was a child in the 1880s, recalled that the Munsee "usually stayed for several weeks and would camp in tents at the back end of a farm, on the edge of the farm woodlot. 'They were a hard-looking bunch but they wouldn't ever hurt anyone.'"[97] In Ailsa Craig, Ontario, residents remarked on the "other" in their midst, but the descriptions of

Munsee flax harvesters who arrived in 1880 were more temperance literature than accurate reporting.

> Messrs. Gunn Bros. Have imported a host of Indians from Muncey-town to pull flax for their Ailsa Craig Mills. It was found necessary to warn all the liquor stores and hotels on no account to give or sell the natives any intoxicating drinks. Mr. Alex Gunn positively asserts he will make an example of any dealer that violates the law in this respect. The first night or so after the arrival of the aborigines things were getting interesting at the "hub." One man came near being stabbed, and Billy Coulter felt uneasy about his scalp.[98]

Julia Roberts's study of Upper Canadian tavern-goers deflates the stereotype of the drunken Indian by showing that the presence and patronage of First Nations in drinking establishments were normal and relatively unremarkable.[99] The "drunken Indian" was not a harmless stereotype. In August 1905 a white man killed a teenage flax puller named Adam Seneca in Stratford by laying his unconscious body across the railroad tracks behind the flax yard. The young Anishinaabe man from Chippewa of the Thames was struck by the morning train and gruesomely killed. As well, the flax-harvest manager, a local Irish-Canadian man by the name of John Gamble, was found dead outside the workers' accommodations, and another white man was found washing himself of blood. The latter was accused of double murder, but ultimately his defence used an exceptionally weak argument: specifically, that the Seneca was a drunken Indian who killed the miller, knocked out the accused, and then stumbled across the tracks and caused his own death.[100] After several years of non-Natives' close contact with First Nations flax pullers, local reports became less sensational, although not always less degrading. "These Indians are usually law-abiding and fairly industrious," the *Dutton Advance* assured its readers, although "they do not make much progress."[101] In Woodstock, the *Sentinel-Review* announced in 1887 that "the flax pulling commenced last week and we have the usual influx of Indians. They are very quiet and inoffensive citizens if the pale faces will not give them any 'fire water.'"[102]

By the turn of the century, there was less uneasiness between First Nations labourers and their non-Native neighbours. Flax millers around Harriston rented land and hired First Nations family harvesters in the 1910s. According to one local resident, "it's about the only time I've seen

Figure 2.9   Peter Hill, St Marys, ON. The inscription reads: "Peter Hill. Muncey Indian, played in town band, came here to work on flax. He had a wife and 5 children – lived at corner of St. Maria and Church Street."

Indian people right here."[103] As workers, they remained segregated harvest gangs and continued to raise the curiosity of some rural onlookers, such as photographers Reuben Sallows and Tavistock's Lemp Studio.[104] Relatively little is known about the practice of First Nations migrant labourers and even less is known about their integration into the non-Native culture. Most of the photographic evidence suggests that First Nations families worked and operated as a group, but no doubt there was a certain amount of movement between groups, as in the case of Peter Hill. Hill was a Munsee flax harvester who moved into a house in St Marys, played in a local brass band, and adopted a decidedly Scottish appearance, at least for one photograph (Figure 2.9).

Other new immigrants and ethnic minorities were also quite common in the flax industry. Some immigrants had been given advice or direct assistance from flax millers, such as the McCredie Flax Company. Others were transient and followed the flax harvest, such as the family of Addie Aylestock, an African-Canadian from Glen Allan, in Peel Township, Wellington County, Ontario. Addie's father, James William, immigrated from the West Indies to Peel Township with her grandparents. He was a twenty-four-year-old labourer when he married his wife, Jemima Lawson, at her family home on the 5th Concession in March 1909. He worked in the sawmill and made $450 per year. Addie was born that September at home in Glen Allan, and by 1921 she had five siblings and was living in neighbouring Maryborough Township. Her father went on to work in the flax fields near Drayton, in Maryborough, and in other places throughout southwestern Ontario. Addie went on to become the first ordained black woman in Canada.[105]

The work of photographers like Sallows, Lemp, and others preserved a brief glimpse of First Nations flax pullers, although the documents are few and it is difficult to shed the lens of their settler creators. Figure 2.10, a photograph by Sallows, shows a group of First Nations men, women, and children pulling flax in 1910. There do not appear to be any non-Natives working with the harvesters; this group formed a segregated gang. The camera angle and activity of the subjects suggest that this photograph was more or less candid. The subjects were perhaps posing for the photographer, although several of the workers are continuing their work and at least two are not looking at the camera. Like Sallows's other image of flax pulling in 1910, the camera angle is practical and allows the photographer to capture most or all of the workers at once. The workers are closer together than those in Figure 2.3, maybe because the photographer needed to get the entire group in the frame or because the upright workers were actually working *behind* the pullers binding the flax in sheaves. If this was the case, it is interesting to note that the women and children performed the same work as the men. A gendered division of flax labour would mean that the women here should have been handling the pulled flax like the subjects of Figure 2.4 did.

The makeup of the typical First Nations flax gang had evidently changed very little since July 1888, when the *Woodstock Sentinel-Review* noted that

Figure 2.10   Pulling flax, 1910

"[a]bout twenty-five Indians, braves, squaws, and papooses, arrived in town this morning from Brantford. They are going flax-pulling."[106] Furthermore, the *Farmer's Advocate* used another Sallows photograph of the same First Nations gang when it published a picture of "A Group of Flax Pullers Ready for Work," in 1918.[107] It is possible that Sallows was commissioned to photograph various aspects of the flax industry, and he had no particular interest, personal or professional, in First Nations as a form of spectacle. Most reserves close to the flax-producing region of southwestern Ontario, particularly Six Nations, Anishinaabe, and Munsee, are indicated as sources of workers, which suggests that the flax harvest was a very common activity for First Nations families. In 1895, E.D. Cameron, the Indian agent for Six Nations, was "glad to add that this custom of seeking employment is not increasing, but the desire to remain on the reserve and cultivate land is gaining yearly."[108] Yet, First Nations participation in the flax harvest did not decline until the stagnation of Ontario flax production after the First World War. The flax fibre industry itself went into decline in the 1920s and many mills fell silent, but within a few years it revived and these labour arrangements resumed. One hundred Munsee First Nations worked in the Tavistock harvest of 1928, all employed by local flax miller A.A. McQueen.[109] In the 1930s, First Nations families continued to work for Howard Fraleigh and other flax millers in the Forest area. Young men like Ernest George and his brother Harrison were in their teens and early twenties when they travelled from the nearby Kettle Point Reserve to pull flax.[110] They often continued from there to harvest other crops for area farmers, although in 1942 a major land conflict diminished their ability to work in these markets.[111] The long tradition of First Nations families working in the flax harvest declined primarily because the industry itself vanished from almost every part of southwestern Ontario. The region's First Nation work gangs participated briefly in substitute industries such as sugar beet production in the Waterloo and Middlesex areas and celery picking in Lambton, but these activities also declined in the mid-twentieth century.[112] In many ways, the labour of First Nations and marginalized ethnic groups of the early and mid-twentieth century has been replaced by managed temporary migrant-worker programs. These started in 1966 for the purpose of bringing Jamaican, and then Mexican, temporary farm workers, and they have become both an indispensable and a contested part of Ontario's agriculture.[113]

Based on probable acreage, the person-days required to harvest flax meant that somewhere between 500 and 2,000 people were involved in pulling the crop in late July or August. One problem with gauging flax pulling in terms of "man-days/acre" was that it did not factor age and gender into the equation.[114] Economic considerations were not the only reasons women and children did or did not participate in the labour force. Diane Newell's research shows the significance of considering waged family units for seasonal employment, as they were important factors in First Nations working off-reserve in Ontario.[115]

The increased production of flax was contingent on the availability of flexible First Nations labour, precipitated by a lack of adequate agricultural and other opportunities on reserves. This helped create a labour pool ideal for flax pulling and other work off-reserve. First Nations families had become critical enough to the harvest to be able to negotiate their wages.[116] The Ontario flax story suggests that gangs of First Nations labourers were a critical element of the harvest, these workers considered flax pulling an important source of income, and their encounters with non-Native Ontarians were another form of everyday exchange in rural southwestern Ontario.

More generally, the participation of women and children in the flax harvest helped make the industry profitable in the late nineteenth century. The extensive photographic evidence reaffirms Craig and Weiss's argument

Figure 2.11 Tavistock flax mill

that "women and children shifted their time from unmeasured household tasks to market activities." It also suggests an early departure from both Marjorie Cohen's argument that women were retreating into the domestic sphere and Marvin McInnis's observation that "[w]omen didn't do field work."[117] Kenneth Keller writes that the sight of gangs of women and children pulling and stacking flax startled observers, such as Yale president Timothy Dwight when he travelled through Upstate New York, but in the late nineteenth century, women's and children's labour in the Great Lakes region was an ordinary part of the rural economy in a few southwestern Ontario towns.[118] This adds complexity to the myth of the gendered and self-sufficient production of homespun linen.

According to photographic evidence, the flax harvest labour was shared by men and women, and work in the small-town mills was performed predominantly by men. The censuses suggest that women over sixteen made up only 8 per cent of the workforce of flax-scutching mills, and girls represented 6 per cent.[119] Temporary workers were needed for processing the flax in the fields after the pulling was complete, and women were present for this field work as well. Figure 2.4 shows women gathering flax, and John Meyers specifies that the Perine mills sent men and women to the flax fields. As we saw in Chapter 1, women weavers played important roles

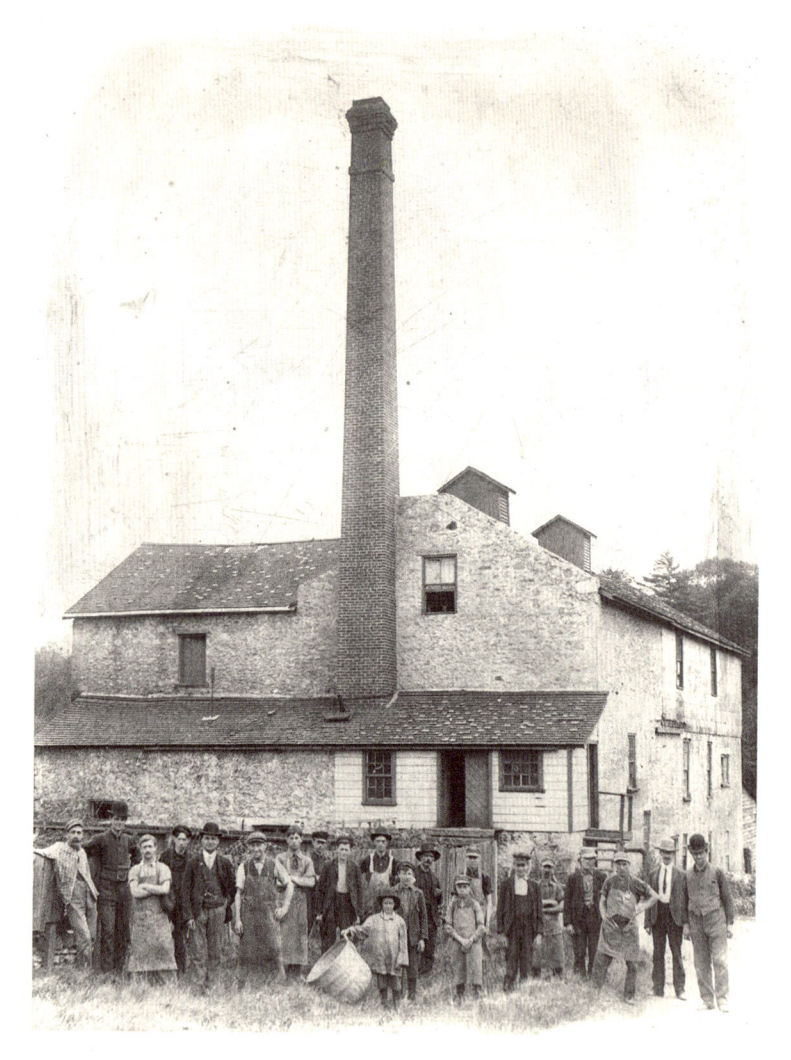

Figure 2.12　Aboyne flax scutching mill in Nichol Township, Wellington County, ON, c. 1890

in linen production throughout the region, but workers in scutching mills were mainly male. Figure 2.11 shows a group of flax workers in Tavistock. Lemp tended to pose his subjects carefully, and in this photograph the subjects are spaced evenly, with a young boy perched on a ladder and all but one worker holding a strick of scutched flax. Nevertheless the details of Lemp's photograph match rather well with photos of the flax mills in Wellington County (Figure 2.12) and St Thomas (Figure 2.13). Only one

Table 2.3 Flax milling in Ontario by workforce, 1871

1871 Industrial Census Observations for Ontario Flax Mills	1–5	6–10	11–15	16–20	21–25	26+	Total	Perine mill
Number of mills	2	4	4	6	6	4	26	
Workers	4	32	55	112	145	169	517	29
Workers/establishment	2	8	13.8	18.7	24.2	42.3	19.9	
Men and boys in mill	4	32	55	101	123	107	422	20
Women and girls in mill	0	0	0	11	22	62	95	9
Steam power	0	2	0	4	5	2	13	1
Power (mean hp)	8.5	13.8	16.8	14.0	15.7	23.3	15.8	30
Fixed capital	$2,000	$8,200	$19,000	$12,800	$32,500	$26,500	$101,000	$10,000
Mean F-Cap/mill	$1,000	$2,050	$4,750	$2,133	$5,417	$6,625	$3,885	
Raw material (tons of flax)	50	510	946	4,835	2,900	3,761	13,002	380
Mean flax/mill	25	128	236	806	483	940	500	
Mills listing product value	2	1	4	3	3	4	17	
Product value/mill	$780	$5,000	$6,330	$10,835	$8,500	$19,217	$9,809	$25,000
Mills differentiating tow	2	3	2	3	2	2	14	
Mean price/ton							$45.38	
Value of tow	$1,560	$1,634	$2,365	$4,846	$2,634	$5,500	$18,538	
Value of tow/mill	$780	$545	$1,183	$1,615	$1,317	$2,750	$1,324	

Note: Value is calculated from recorded quantities of tow and the average price of $45.38 per ton.

Source: Canada, *Census of Canada, 1870–71*, LAC, Industrial Schedules. Acknowledgment is gratefully given to Kris E. Inwood, University of Guelph, for access to these data from the 1871 census manuscripts.

woman is present in the St Thomas photo, and both mills apparently had peak workforces of twenty males, including a few boys. Similarly, the St Thomas mill employed thirty men and thirty boys at harvest time and twenty workers throughout the year.[120]

Just as the 1871 census manuscripts provide the last geographic snapshot of Canadian flax fibre cultivation, they also allow us to analyze the production and workforce of scutch mills. The twenty-five flax mills and one cordage mill using and producing flax are categorized according to the size of each mill's workforce (see Tables 2.3 and 2.4). The Perine cordage mill is juxtaposed with these and is not included in the total. The Perine mill was

Table 2.4 Cordage milling in Ontario by workforce, 1871

1871 Industrial Census Observations for Ontario

Cordage Mills	2–4	5–8	15–31	Total	Perine mill
Number of mills	5	4	3	12	
Workers	14	27	78	119	29
Workers/establishment	2.8	6.8	26.0	9.9	
Men and boys in mill	14	27	75	116	20
Women and girls in mill			3	3	9
Fixed capital	$1,410	$7,800	$21,000	$30,210	$10,000
Mean F-Cap/mill	$282	$1,950	$7,000	$2,518	
Value of product	$7,960	$20,280	$50,400	$78,640	$25,000
Value/mill	$1,592	$5,070	$16,800	$6,553	

Source: Canada, *Census of Canada, 1870–71*, LAC, Industrial Schedules.

among the larger mills in terms of workers, and its fixed capital was over twice the average for flax mills and almost four times that of the average cordage mill. The average number of all workers in scutch mills was twenty, as the mill photos suggest. These mills had nearly five more employees than the wool mills, twice that of other cordage mills, and over eight times the number employed in carding and fulling mills – arguably the industry closest to flax manufacturing in nature. Flax mills had more fixed capital on average than cordage mills and more valuable products. Women and girls represented 19 per cent of the labour force in scutch mills and only 3 per cent in cordage mills. Still, this was significant considering the virtual absence of women in the three scutch-mill photos shown above. The Perine mills were anomalous in that women represented close to a third of the workforce.

The escaped slaves who worked in the St Thomas mill included Nathan Ryan and his family. Born in Maryland in 1840, Ryan lived on northern St George Street, up behind the flax fields, with his wife, Frances (Stewart), and their three children until his death in 1903 at the age of sixty-three.[121] The Ryans were five of the fifty-eight St Thomas residents who identified as African in 1881, when the census began to record "Origins of the People."[122] At age fourteen, Ryan's eldest son, William, was a labourer in the flax mill, according to the 1891 census.[123] Frances Ryan and her daughter reported no profession, but they performed domestic service in the

Figure 2.13　Workers at Keith's St Thomas flax mill, c. 1890, with Ryan Sr on left and Ryan Jr sixth from right

city. Nathan, his son, and the other men and boys who worked in the St Thomas flax mill are visible in a photograph of William Keith's second flax mill, taken sometime between its construction in 1887 and its destruction by fire in 1894 (Figure 2.13).[124] Like the marginal people in many industrial economies, many of these mill workers, Nathan included, were edge-dwellers, living and working in the interstitial spaces of southwestern Ontario's mill complexes.[125] Flax work was not an attractive occupation, and as we have seen, the worst of it was performed by people with few options, such as First Nations labourers like Adam Seneca and African-Canadians like Nathan Ryan and William Aylestock. Nevertheless, Ryan became a farmer in his own right, owning a modest parcel of land on St George Street and self-identifying as "farmer" in the 1891 census. He outlived the flax mill by a decade (and Keith by three years), and his wife lived to be eighty-six. The Ryan family continued to live on their property until Talbot Realty developed a new suburb there in the late 1950s. One of the new streets surrounding the Ryan home was named "Nathan."[126]

Flax fibre mills were certainly uncomfortable work sites, with constant dust from the straw shives and constant noise from the machinery, but they could also be deadly places. The mills and straw storage sheds were prone to fires, and mechanical brakes and scutchers were physical threats. The manager of James Livingston's Tavistock flax mill, J. Hoffmann, lost a piece of his finger in the mill equipment in March 1890, and the Essex flax mill owner A.H. Raymond lost his finger in a scutcher in May 1905. Raymond died shortly thereafter, though desperation due to his heavy debt and recently vanished co-owner may have led him to commit suicide, an industrial hazard of a different kind. Frederick Rominger, a forty-two-year-old Waterloo man with no wife or family, suffered a fractured skull in an accident in January 1911; the second-generation German-Canadian died two hours later. Rominger had been a farm labourer since at least 1901, when, according to the census, his annual income was $100. He was unable to read or write, so taking up work in the flax mill would have been one way to augment his salary, but clearly it also put him at significant risk. James Livingston's employees faced this risk in mills across the region, and whenever a "serious accident" occurred, as it did in Yale, Michigan, in 1907, it failed to spark any further comments in the correspondence between Livingston and his manager, McColl. Both millers and the media accepted this reality throughout flax's history in Ontario. Workers continued to experience serious injuries until the fibre industry came to an end, and as late as 1948, after an Elmira flax mill worker lost a leg, workplace hazards remained a problem.[127]

The accounts of Chris Yantzi, a flax mill worker in Tavistock, reveal the pattern of work in flax fibre mills toward the end of the fibre industry (Figure 2.14).[128] The accounts were kept very consistently for the better part of the 1940s and into the 1950s as well. They indicate the days and hours worked and usually whether the labour was in the Tavistock or Seaforth flax mills or "in the field." Yantzi's work was surprisingly steady in the early 1940s, but later in the decade he supplemented his flax work with employment at other places; presumably this occurred even more as his hours at the mill decreased. December was a particularly slow time in the flax mill, even in good years. Yantzi earned about 45 cents per hour in the summer of 1944, and since most of his workdays were between ten and twelve hours long, he made from $4.50 to $5.50 per day. This was the highest rate of pay in any flax record, indicating that Yantzi was more than a general field labourer and may have been directing gangs of flax workers.

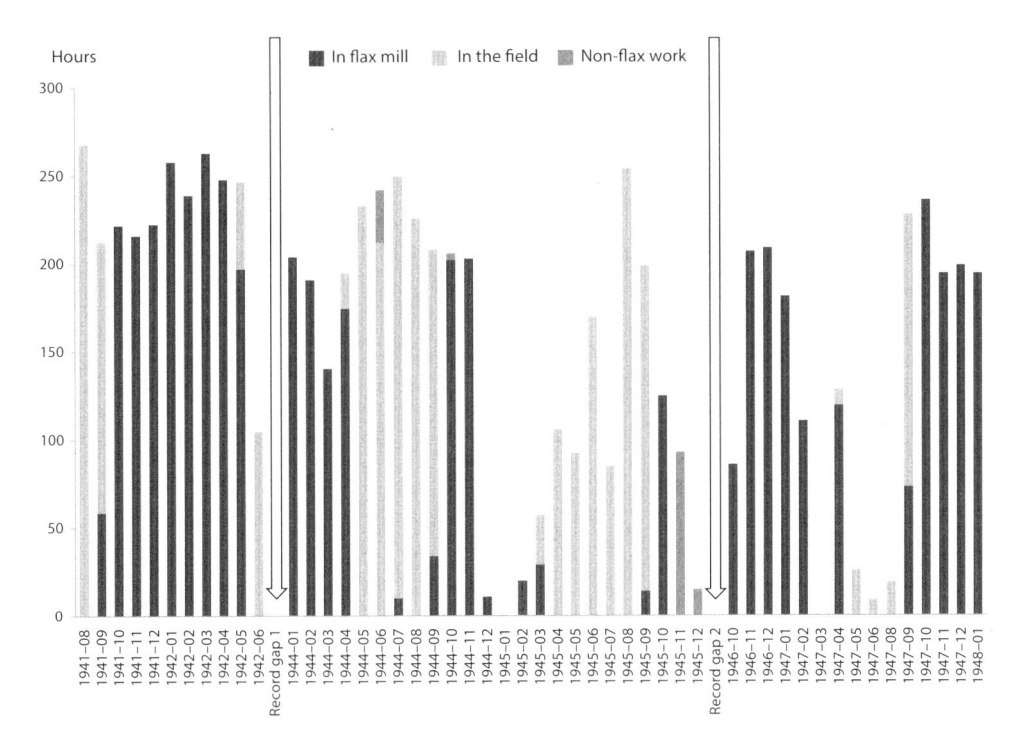

Figure 2.14 Chris Yantzi's flax mill work, 1941–1948

Sources: Chris Yantzi account books, A, B, C, and D; personal collection of Mrs Yantzi, Tavistock, ON.

He frequently noted whose field was being harvested on a particular day; the fact that most farms were only recorded once points to there being a large group of labourers. The labourers were not identified in his accounts, so we do not know if they were local friends and family or itinerant gangs from further afield.

The major problem with fibre production in the late nineteenth century was that pulling flax crops by hand was a particularly arduous task. Still, the value of flax caused farmers and millers to create new labour relationships. They capitalized on "gang" labour for the flax harvest and employed diverse teams of workers, which included landowning flax farmers, permanent mill workers, transient labour, and entire groups of First Nations families. Ontario farmers had always been reluctant to grow flax in the absence of the capital, labour, and knowledge inputs of local millers. When a concentrated flax industry did emerge in Ontario and continued for over a century, it was because small-town millers organized most of the field work and because a growing market for seed offered more value and flexibility for a field of the blue flowering plant. In 1869 St Thomas farmers saw an advertisement posted by flax millers such as James Livingston and William Keith titled "Flax! Flax!" Ads like this assured the farmers not only of the miller's "long experience in Flax operations" but also of his "wide acceptance with the farming community." And although they surely took all the hype with a grain of salt, most of them travelled past the mill with every visit to the city, learning more about the millers with every trip, about flax with every season. As the St Thomas-based *Canadian Home Journal* argued, the flax mill would add "an additional source of revenue, upon which they can draw when, as in the present year, other crops fail them."[129] These marginal and interstitial mill complexes were the nodes connecting multiple flax-commodity webs and coupling the region's farmers to expanding global markets. Mill complexes were critical because they were the places where risk-averse farmers gained credit, purchased goods, and experimented with new forms of investment for their land, labour, and other inputs. These places were unequal and unstable, but they were also transformative for both their materials and their makers.

It is worth considering a final group of actors in the fibre web of the Great Lakes region before we turn to flax production in the West – the promoters. Promoters were highly visible people who contributed the most to the policy and public opinions of the flax web. As we will see, public opinion and fibre enthusiasts' extensive promotion contributed little to the web itself. There was a dramatic disconnect between the promoters and the farmer-millers in some periods. Fibre promoters knew little about flax factorship, the region's agro-ecosystems, or the superiority of the seed market for flax farmers. Conversely, farmer-millers and labourers were only lightly involved in shaping flax policy and promotions. But because the writings of promoters have lasted much longer than a crop cycle and even longer than the dynamic geography of urban peripheries, it has been difficult to appreciate the everyday exchanges of marginal men and women like Nathan Ryan.

CHAPTER THREE

Flax Fabrications

Selling the Promoter's Plant

In December 1912 inspectors raided the offices of the Sterling Debenture Corporation in Madison Square, Manhattan, and arrested four men on a mail fraud indictment. Two more people were arrested in New York on the same charges, as well as another man in Chicago and one in North Brookfield, Massachusetts, later that night. This was no sweepstakes scam. Post office inspectors had been building a case against Sterling Debenture for about six years, their curiosity piqued by the $1 to $2 million per year earned suddenly by a small group of financiers. The investigators found that the corporation had sold over $33 million in stocks, most of them in fraudulent companies, and the most egregious example was a flax-processing scheme.[1] Benjamin Cushings Mudge, a ninth suspect who briefly evaded arrest that weekend, had organized Oxford Linen Mills several years earlier and sold its shares through Sterling Debenture. Oxford Linen was founded and promoted on the false claim that Mudge could use special chemical processes to make fine linen cloth out of rough North American flax fibre and thereby revitalize a northern textile industry based on local raw materials.

The Sterling Debenture fraud was the most public and most egregious example of a long history of flax and hemp fibre promotions and mishaps in Canada and the United States. There was often a fine line between promotion and fraud. During the First World War but shortly after these arrests, the Canadian government instigated flax fibre initiatives that evolved into a full fibre division in the interwar period. Perhaps unbeknownst to the government, federal governments had tried this before. While they were not misleading, Canadian fibre promoters misunderstood the commodity web and failed to expand, much less save, the fibre industries in Canada. Most northern flax fibre was only produced in the factorship system, and the seed was becoming the more important half of the flax commodity web. Tracing some of the many ways promoters got flax wrong and some of the ways swindlers spun the crop, however, reveals some persistent misunderstandings about the plant and how people thought it should be produced. Many such commodities have been focal points for scholars of capitalism, globalization, and the other social-ecological factors that defined the "great transformation" – even the Anthropocene.

Studies of Canadian commodities have been foundational, their focus ranging from mercantilist thought to the staples thesis and, more recently, to dependency theory. One prevailing notion has been that Canada was primarily an exporter of unfinished commodities, or staples. To promoters, flax seemed to be a perfect export commodity, and exports were indeed important in some parts of flax history.[2] However, the full examination of a commodity's web reveals the changing roles of markets and teleconnected systems, local actors and environments, and commodity policies and promotions. Flax provides a case study in land use and crop promotion, revealing how "commodity blindness" has masked the industry's most successful contexts and contributed centuries of scams and failed interventions in agriculture.

The Promoter's Plant

As humanity's oldest manufactured crop, flax understandably had a deep hold on the popular imagination. North American flax industries and improvements have been promoted at various points since the early colonial period. Much as in other industries, centres of fibre production and processing operated according to local forces of supply and demand, while others ran, with great trouble, on personal or state promotion.[3] A pattern of state intervention in flax and hemp production stems back to the beginnings of North American colonization. The earliest episodes included unsuccessful attempts by the Virginia Company of London and the Massachusetts Bay Company; state-subsidized prices for hemp in New France in the 1720s; municipal support for a poor-relief linen factory in Boston; a flax colony in 1770 in what became Prince Edward Island; and premiums offered for hemp production in Nova Scotia in 1788.[4] In the early 1800s, hemp production was promoted, unsuccessfully, in Nova Scotia, New Brunswick, and Upper Canada. Lord Selkirk planned to export flax and hemp from the Red River Colony in the 1810s, and Governor George Simpson started a short-lived flax and hemp export company on the Assiniboine River in the 1820s.[5] Canadian and US promoters sought to create a northern equivalent of the fibre colonialism that characterized so much of the cotton empire in the Global South.

Later schemes were designed to promote domestic manufacturing as opposed to an export-based industry. These included the Canada Company's flax promotion efforts in the 1840s, the US Congress's "investigations to test the practicability of cultivating and preparing flax or hemp as a substitute for cotton" in 1863, and the Canadian Department of Agriculture's distribution of flax seed, literature, and related funding to agricultural societies in the 1860s. There was, as we have seen in southwestern Upper Canada, a great deal of interest in flax production in this decade, but whereas the literature covered the Great Lakes region, the actual production was always more localized. The *British American Magazine* published a twelve-page editorial in August 1863 urging Canadians to consider the crop now that King Cotton had been dethroned by the US conflict, and the *Gazette des Campagnes* carried a similar promotional campaign in that same year in Quebec. The Canadian seed distribution was a failure, however, and very few farmers took advantage of these offers from the colony.[6] Whether undaunted or just forgetful, Canadian officials continued to explore the possibilities of flax promotion in the 1880s, and in 1915 they formed an entire division for this purpose. Their promotional efforts trickled into the provincial departments of agriculture in Ontario, Quebec, and the Maritimes, and the division was maintained throughout the interwar period.[7]

South of the border, state-supported fibre industries still existed in Oregon in the late nineteenth and early twentieth centuries, but production eventually ceased when processors lost access to penitentiary labour and funding from the state and local boards of trade.[8] The US Department of Agriculture formed a fibre division in the early 1900s, promoting the commercial production of flax and other non-cotton-fibre plants and even growing fields of flax and hemp on Arlington Farms.[9] The Canadian federal division lasted for the better part of the twentieth century, but flax producers did not respond.

In the mid-nineteenth century, North American flax promoters focused heavily on the need to escape dependence on imported textiles and monoculture land use, which they understood was common. Scientists and agricultural specialists believed that flax promotion would result in more scientific farming and greater botanical inventories; many thought flax should become a crop extensively grown by farmers in every region.[10] They argued that flax was well suited to the Canadian climate, that it was a very marketable export commodity as well as a substitute for some of Canada's substantial cotton imports. Finding substitute fibres was part of a broader political discourse among cotton manufacturers in New England and northern Europe who called for economic independence from cotton suppliers in India and America. Additionally, the world's industrial powers promoted colonial cotton production in places like Togo to increase fibre supplies and stabilize prices.[11] This kind of devastating colonial expansion proved more successful than redirecting free labour-commodity production in the North, but many British and North American promoters believed technological improvements in flax processing would permit northern American and Canadian farmers to supply a strong domestic linen industry. Free labour markets did develop in the Canadian fibre industry, but, as we will see, industrial promoters such as James McCracken and Howard Fraleigh renewed the call for federal intervention in the flax fibre web. In their racially tinged views, the factorship system had become too heavily dependent on First Nations labour, and they argued for research subsidies to help mechanize the fibre harvest.

The suggestion that new technologies would allow primary processors to manufacture flax as cheaply as cotton electrified Victorian industrialists and promoters, who felt that industrialized regions relied too heavily on trade with cotton-producing states. Up to this point, the labour and other costs of North American flax production were too high relative to northern and eastern European cultivation, and the cost of processing the raw material was too high relative to cotton. The best flax for manufacturing long fibres (line) had to be pulled from the ground, a process that was not mechanized until the mid-twentieth century. Promoters like Allan Cameron noted that North American farmers sometimes cut the crop, which reduced the costs of harvesting but produced a shorter fibre.[12] If technologists could improve mechanical and chemical fibre processing then the length of the fibre would become immaterial. Farmers in Canada and the northern United States could then grow much larger quantities in a more flexible crop calendar, and they could reap the crop as they did wheat or any other grain. In effect, mechanical and chemical innovation promised to change the biological characteristics of the flax plant so that it could be harvested like grain and manufactured like cotton. This material innovation would benefit northern producers and textile manufacturers alike and provide alternative inputs to the cotton imported from the tropical world.

Many Victorian inventors attempted to "cottonize" flax, but the most famous attempt was undertaken by the Danish amateur botanist and flax charlatan Peter Clausen. After immigrating to Brazil in the 1820s, Clausen had brief careers in the military, commerce, and agriculture, but he was primarily interested in natural history. He met and worked with visiting naturalists like Peter Wilhelm Lund and Francis de Castelnau, and he contributed plants and fossils to European herbaria between 1834 and 1843.[13] Struggling with poverty and mental illness, Clausen declared himself a "practical chemist," took on the name "Chevalier Claussen," and moved to England shortly before the Great Exhibition. Claussen claimed to have developed a chemical method of turning short fibre flax into a material that resembled cotton. His so-called flax-cotton could be spun on ordinary cotton spinning equipment, but it would produce a superior linen thread. In 1851 he displayed some flax at the Exhibition and published a pamphlet on flax-cotton that was featured in the *London Journal of Arts and Science*. He conducted public demonstrations of the process at a "flax works" in Stepney Green, and according to the *London Spectator*, he "set all the world in raptures about the new invention."[14] Allan Cameron subsequently promoted Claussen's method "to the agriculturalists and capitalists of the United States," hoping that Americans would make use of the technology and the commodity. His description matched the sense of rapture observed by the *Spectator*. Visitors to Claussen's demonstration witnessed a four-stage process that prepared flax for carding and rough textile and cordage production. First, the flax went through a mechanical breaking and cleaning process. Then, it was steeped in a new chemical combination for at least three days, alternating between hot and cold water, and then immersed in caustic and acidic solutions. The third stage, called "cottonizing," consisted of pressurized soaking in a cold vat of soda and water followed by submersion in a vat of sulphuric acid. And this is where the raptures began. At that point, according to one visitor to Stepney Green, "the mechanical action of chemical forces is so beautifully illustrated." And "the effect was almost instantaneous. The character of the Flax fibre became at once changed … to a light expansive mass of cotton-like texture … The result was generally and loudly cheered by all present." After the plaudits had finished, the fibrous clumps were dunked in a vat that prepared flax for spinning and allowed it to be processed "in the same manner as ordinary cotton." In a fourth and final exercise, the flax-cotton was bleached, and "[t]he rapid change in the color, as in the previous instance, of the texture of the substance, was warmly applauded."[15] According to the inventor, the process took only three to six weeks in real time compared to the six-month process of preparing flax for the hackles without the help of chemicals. Claussen subsequently travelled to the United States to promote his method. John Ryan, a chemistry professor at the Royal Polytechnic Institution, facilitated the demonstrations and wrote a book titled *The Claussen Flax Process Illustrated*, which was advertised along with Claussen's own pamphlet throughout the 1850s.[16]

Agricultural chemistry was a new tool for an old experiment in the material transmutation of fibre. Cottonization reflected not only the excitement and imagination of promoters, but the speed at which agricultural and practical chemistry was applied to a range of commodity products. The mid-nineteenth century was a critical period of growth for this science, which led many to think of agricultural and nutritional systems in terms of atoms, "the capital of nature."[17] The Claussen method was one of several attempts to chemically transform flax into a more valuable commodity, and the Harvard economist Walter S. Barker later wrote that "the retting process has probably attracted as many and as brilliant minds as any of the textile processes."[18] Earlier inventions for preparing flax for use on cotton spinning frames appeared in the late eighteenth century. In the early nineteenth century the "Lee process" used some chemical substances, and in 1846 an American by the name of Schenck patented a retting process involving pressurized vats similar to Claussen's.[19] The chemistry professor, John Ryan, claimed that these other experiments had indeed made flax "cottony," but only Claussen's method "obtained a commercial scale of magnitude."[20]

At the commercial scale, there were simply no examples of millers who had set up a successful operation based on Claussen's method. However, some processors who operated on the factorship model did experiment with products that resembled flax cotton. The Perines in Doon began selling some cottonized flax to Canadian textile millers such as the Barber Brothers. They did not offer this product in the first few years of their business, so it would seem that when the American Civil War reduced regular cotton supplies, they produced flax-cotton for local mills as a replacement.

However, the Perines sold no flax-cotton to US or British mills, and the value of flax-cotton sold locally was quite small compared to the value of cotton imports. The Perines' sales of flax-cotton to the Barber Brothers was a short-lived experiment, not a substitute for the many flax products discussed in Chapter 1. The Claussen method required extensive infrastructure, including large soaking tanks, chemical additives, and fuel for heating the solutions, and there is no evidence that the Perines or other mid-century mills invested in this equipment. There were, nevertheless, other ways to reduce the processing time of flax fibre, and although Canadian millers usually dew-retted their flax, the vats set up at two mills along the Credit River (Mitchell and Gooderham & Worts) and the large stacks of green fibre at Kettle Creek (Perine & Young) suggest that these companies, too, were experimenting with new industrial techniques. As we will see through the private and state flax promotional campaigns of the 1850s and 1860s, Canada was an early player in the nineteenth-century process and practice of industrial chemistry.

Individual Flax Fibre Promoters

Flax fibre promoters were very active in Canada and across the Canadian border. Like the Perine brothers, and like other promoters who appear in later chapters, they brought knowledge and experience of the plant to Canada from Europe and the United States. However, the promoters' main audiences were high farming agricultural societies and state officials – not the processors and the farmers who actually grew the crop. In an address printed in the *Canadian Agriculturalist* in 1855, John Wade presented a brief history of the first generation of Canadian agricultural societies, acknowledging that most of its members were elite and gentlemen farmers. To Wade, the greatest progress in agriculture – advancing food systems from "simply a subsistence" level to one of "unlimited [power], both morally and physically" – had occurred in the previous quarter century. He argued that it was no coincidence that the study of geology, chemistry, and physiology, the sciences most beneficial to agriculture, was about the same age as the province's agricultural societies. It was the societies' responsibility to present scientific knowledge to ordinary farmers in ways that would change their practices and obtain "the desired results of improvement."[21] This belief was common throughout British North America. Rod Bantjes speaks of these improvers as agents of modernity, and Daniel Samson's work on agricultural societies in Nova Scotia concludes that "[s]ociety members ... knew their primary mission was to change the behaviour of those around them."[22]

Canadian agricultural societies were curious about flax and the chemistry of cottonization, and the Canadian Board of Agriculture commissioned in 1852 a report by Alexander Kirkwood on flax methods in northern Europe. The following year the minister of agriculture, Malcolm Cameron, commissioned William McDougall of the *Canadian Agriculturalist* to visit flax-growing locations in New York and determine whether, in the "spirit of improvement," the board should assist with the importation of flax pulling, scutching, and other processing machines for use by society members. McDougall poured a little cold water on the board's initiatives, concluding in his report that farmers could decide for themselves whether flax was suitable for Canadian soil. "The Canadian farmer may freely choose from the world's seed store," McDougall argued, and "[f]or what purpose then should Government interfere?" As for the mechanical flax puller that would mitigate the problems of difficult and expensive field labour, McDougall could only say that he had not seen any in the United States and that although they were purported to exist in Great Britain, no one had actually seen one and his source's "data were somewhat vague and unsatisfactory."[23]

If McDougall was pessimistic about the outcome of promoting a flax industry with state funding, the crop's most histrionic optimists in Canada were Alexander Kirkwood, John A. Donaldson, and Henry Youle Hind. Hind's flax-promoting career was highly public but brief as he traded in his editor's position at the *British American Magazine* for international work in geology and surveying.[24] The other two men were both Ulster Protestants who brought trade knowledge and observations from the flax industries in Ireland and New York State. Both were public servants who devoted part of their time to flax promotion. Kirkwood was a native of County Antrim who farmed briefly in New York State before immigrating to Canada and taking a clerical position at the Crown Lands Office. He authored an article on flax that was published in the *Canadian Agriculturalist*, the 1852 report on flax to the Canadian Board of Agriculture, and a pamphlet titled *Flax and Hemp*.[25] He later became a leader in Ontario natural resource management and helped create Algonquin Park, the province's most famous park and forest reserve. Kirkwood promoted growing flax to supply Irish spinning mills, which would provide a market for all the flax that Canadians could hope to produce. He believed that eastern Ontario's Ottawa Valley had a particularly ideal geography for flax growth.[26]

A relative latecomer in flax promotion, Donaldson was an emigration agent and the author of *Practical Hints on the Cultivation and Treatment of the Flax Plant* (1865) and numerous flax-promoting articles in the *Canada Farmer*, *Ontario Farmer*, and *The Irish Canadian* over the next twenty years. He argued that flax in Canada grew as well as any he had seen cultivated in Ireland and that flax, with the growers' proper knowledge, could be more remunerative than wheat.[27]

Kirkwood's lengthy report to the Canadian Department of Agriculture aggrandized the new role of science in flax processing. One of Kirkwood's Irish sources had observed Claussen's method for cottonizing flax in London the year before. This man believed the investment in machinery and "scientific men for its superintendence" demonstrated its likely success in the world of fibre manufacturing.[28] Others doubted the usefulness of chemically reducing flax straw, because, according to Barker, it produced a conglomerate fibre that was "suitable for upholstery or mattress filling only."[29] Indeed, although a few vats at the Credit River mills indicated the need for a more intensive retting process, there is no indication that millers used chemicals with any success. Like the Perines, however, other millers knew there was value in the shorter fibres if they could only find a market for them. In 1864 Donaldson reported that a cottonizing factory had recently started up in St Catharines. He called it "an important step in flax manufacturing," as "it opens a market for coarse qualities of flax, and converts the tow or refuse ... into an article" of use.[30] Still, opening markets was a difficult art, even during the fibre famine, and the St Catharines mill soon failed. The Perines' business accounts revealed something that many promoters and some start-ups did not foresee: new markets for tow, cordage, twine, and batting took time to develop, and success in the flax fibre business required networks that extended from local fields to the distant urban customer.

Nevertheless, efforts to promote chemical flax processing swept across Britain and North America during the wartime cotton famine. The crisis for cotton manufacturers and factory workers was caused by the decline in Southern cotton exports, first by Confederate embargo and then by Union blockade. Manchester and Lancashire were hardest hit, but as Sven Beckert demonstrates, the famine and the calls for alternative sources of fibre were global in extent.[31] Some newspapers and journals, like the *London Daily News*, harkened back to Chevalier Claussen's efforts to cottonize flax. As Manchester mill closures caused widespread unemployment in 1862, the editors claimed that if the cotton manufacturers had adopted alternative sources when they had the chance, then Britain "would now be at our ease about our great manufacture." However, the short-sighted "mill owners ... would not attend to any resources but the American plantations," despite many warnings that the "ordinary supply must fail, sooner or later." Across the Atlantic, newspaper and journal editors were convinced that flax "is providing employment and subsistence for a large number of American operatives, who no doubt owe their raw material to [Claussen]."[32] In reality, North American flax milling was never in a position to replace cotton, but this did not prevent a similar wave of flax promotion from appearing in North American newspapers and farm and mechanics journals. Less than a decade after McDougall's negative assessment of a Canadian flax industry in the *Canadian Agriculturalist*, the journal's successor, the *Canada Farmer*, printed at least fifty-two articles and letters related to flax in its inaugural volume in 1864. It devoted more space to flax and the writings of Donaldson and other correspondents than it did to most other crops or animals. The following year it pursued its interest in flax and, specifically, the ways it could be substituted for cotton. Chemical methods gradually fell out of favour with writers, but flax-cotton was replaced in promotional literature with a fibre called "flax-wool that can be mixed up to 30% with wool spinning." Essentially another version of cottonization, this new process would supposedly enable flax to "take the place of cotton in all mixed fabrics," the *Canada Farmer* declared triumphantly. First, the raw straw was processed in a breaker and became high-quality, short-staple tow. The tow was put through a larger machine, producing a uniform staple about 2.5 inches in length that was packed into bales and shipped to wool-spinners.[33]

Flax fibre promotion continued to evolve throughout the 1860s. In the first half of the decade, promoters in rural periodicals agitated tirelessly for flax cultivation, and thanks to the wartime flax boom, they believed they had helped initiate a fail-safe new industry. The promoters briefly highlighted a few of the prominent millers, such as Perine brothers in Waterloo, Colonel Mitchell in Halton, and Gooderham & Worts in Streetsville, and they printed testimonies from farmers across Upper Canada. But most of their writing was for, and about, farmers. If the agricultural press could be believed, there were frequent meetings of farmers and other businesspeople interested in starting flax scutching mills in 1865 as a result of their promotional activities. The journals also encouraged joint-stock companies of farmers to "derive larger profits than" they would get "by merely selling their straw at the flax mill."[34] The Perines' experience suggests that this would not have worked. In the flax commodity web, millers became farmers, farmers did not become millers. Local agricultural societies placed ads such as the one titled "Flax Mill Wanted" by the president of the Ops Township Agricultural Society, and other articles, such as "Stratford Flax Mill," announced new mills that had been set up in part as a result of interest created by the *Canada Farmer* and other promoters.[35] Promoters did not understand the flax factorship system and the way it integrated Canadian land and labour into urban and US markets. But promoting flax had other benefits for the press. Journals like the *Canada Farmer* began to attract outlandish new advertisements for flax-specific fertilizers, machinery, and seed.[36] Thinking and writing about flax were personal interests of most of its early promoters, but given the considerable effort spent by Kirkwood and Donaldson, they were also ways to gain professional credibility and career advancement. For Kirkwood, promoting paid dividends, and frequent publishing in his areas of interest – agriculture, settlement, and forestry – boosted his reputation in the department and presumably among other Canadians.[37]

The prattle of flax specialists does not produce a reliable picture of what a flax farmer performed, much less understood. As W.H. Graham reminded us, "it is easy ... to judge the farmers of the time as lazy or incompetent or ignorant or indifferent, but their view from the stable was very different from ours."[38] Historian R.L. Jones wrote that the farm periodicals had only about 8,000 subscribers, so it would be "safe to state that not one farmer in ten ordinarily read an agricultural periodical" at mid-century.[39] The farmers in Waterloo made business decisions on the same bases as Perines did, working closely with millers and adapting quickly to a system of production that was new in Canada. Arguably, the individual fibre promoters faced a variety of funding and dissemination limits; the promoters who spun flax to Canadians from the state agencies were a different matter.

Spinning Flax to Canadians: State Promotion

Flax promotion spilled over from the private sector to capture the interest of colonial governments. Agricultural societies were closely intertwined with the state, and the government of Canada had promoted flax and hemp production in the past. Alexander Kirkwood's report was intended for government readers, and the US Congress was in the process of reviewing flax promotional work in the early 1860s. In 1861 the Canadian Board of

Agriculture and Statistics, a division of the Provincial Agricultural Association, assembled a small set of agricultural questionnaires it had been circulating among the agricultural societies as an information supplement to the Census of Agriculture. In it, the minister of the Canadian Bureau of Agriculture asked select respondents to provide information on all agricultural activity in their region, but the questions pertaining to flax and hemp were particularly detailed. Kirkwood wanted to know about the extent of flax cultivation and about increases or decreases in output. In particular the questionnaires asked whether flax was processed mechanically or by hand and how much flax "ready for the spinningwheel" an acre would produce.[40]

All of the respondents were elite members of their communities and agricultural societies, most owning two hundred acres of land or more. The two dozen respondents from Upper Canada reported no hemp and very little flax. The respondent from Elgin County confessed, "I cannot give either the weight nor the quantity of seed per acre as I have never seen an acre sown in one place in Canada." An agricultural society president claimed there was "no machinery to prepare it and no market to encourage it." The only three responses from Lower Canada indicated there was no machinery for processing flax and no export market to receive it, but that there was a flax industry "préparés à la main, s'il était possible d'avoir un moulin ou machine propre pour préparer le lin cela serait avantageux."[41]

The bureau was interested in flax before the distribution of the questionnaires and even before Kirkwood's reconnaissance of the flax industry in northern Europe, but it did not begin to invest in promotion in earnest until the 1860s. Early in 1862 the bureau commissioned one of its emigration agents, John A. Donaldson, to purchase six flax-scutching machines in Ireland to be divided between the Upper and Lower Canadian Bureaus of Agriculture for demonstration purposes at Canadian exhibitions.[42] Two of these machines were shipped to Benjamin Walker in St Thomas for use in the Elgin Flax Association's demonstrations. The following year Donaldson was back in Canada, and although he was still an agent, flax promotion was a personally funded side project of his. In that year he published his first treatise on flax cultivation, delivered lectures on the topic, and accompanied the government's new flax scutchers to fairs for the demonstrations.

Around this time, there developed an interest in importing flax seed for distribution to the agricultural societies at a subsidized rate. Donaldson joined in agitating for this activity by informing the bureau that the recently formed Belfast India Flax Company had sent £2,000 worth of seed to India with instructions for its cultivation. In addition, the bureau wrote to Walker's Elgin Flax Association, asking that they share information about the practice of offering subsidized seed in Europe.[43] Bureau members discussed the best variety of seed for use in Canadian fibre production. Agricultural societies were asked for the quantities they expected to dispose of, and arrangements were made with Leeming Brothers to purchase the total amount from Russia. At first they imagined importing between 1,000 and 2,000 bushels of seed, but after petitioning agricultural societies, they decided to limit their order to 1,000. Georges Leclerc replied on behalf of the Quebec agricultural societies and explained why they spoke for such a disappointingly low amount of seed. He argued that local societies had tried this before and failed. The problem, in his opinion, was the risk of receiving poor seed and the reality that even if it was good seed, the local farmers still had no machines to process the fibre or markets to receive it.[44] Leclerc may have been referring to hemp and any number of promotional campaigns for that fibre in eighteenth- and early-nineteenth-century Quebec. One campaign that the bureau apparently knew nothing about had been conducted through agricultural societies along virtually the same lines in 1790. The mercantilist governor of Quebec, Baron Dorchester, ordered 2,000 bushels of Russian seed, hired a Russian expert, and guaranteed prices to producers. Only fifteen farmers responded, and only twenty-nine bushels were ever sown.[45]

Undaunted by warnings like Leclerc's, the Bureau of Agriculture moved ahead with its plan to distribute flax seed across the colony. Its records show that many of the agricultural societies did show at least a passing interest in receiving flax seed for distribution, thinking that this promotion of flax might work better than the failed fibre promotions in the past. There was initial support from the private sector as well; the Perines and likely Gooderham & Worts ordered the largest amounts in Upper Canada (Figure 3.1). But even with the Riga seed project in motion, there was some disagreement over the form of production that Donaldson encouraged and, by extension, the bureau, especially with French Canadians. In the summer of 1862, E. Simays of Montreal had written to A.A. Dorion, the provincial secretary, with the first major critique of the English Canadian factory system of producing linen. He suggested "de consolider la culture du lin par la fabrication interne (indigène) manuelle de ses produits en enseignant aux liniculteurs à les convertir eux mêmes en toiles fines (à domicile)." This

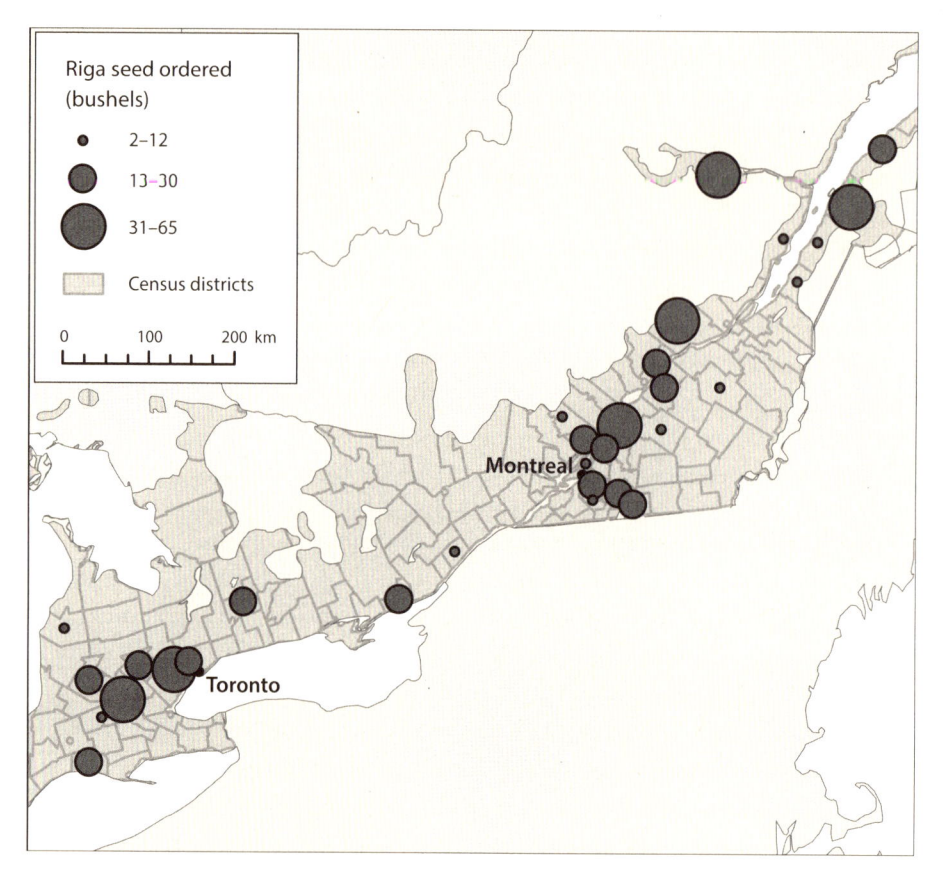

Figure 3.1 Riga flax ordered by agricultural societies, 1866
Source: LAC, H.J. Joly, 5 May 1866, RG 17, Vol. 10.

would be best accomplished, he argued, using Belgian techniques and training and would ultimately insulate Quebec farmers from industrial production, "protéger les tisserands agriculteurs contre la concurrence désastreuse des monopolisantes machines anglaises."[46] Shortly after Joseph Charles Taché was appointed deputy minister, he began to criticize the government's interest in importing scutching machines. One withering critique claimed that these machines had the effect "d'ouvrir la porte au système manufacturier des boutiques qui entraine à sa suite les agglomerations de population, la misère, le prolétariat et surtout la démoralisation."[47] Political activists like William Lyon Mackenzie had criticized the European style of mechanical flax scutching over three decades earlier, but by the middle of the 1860s flax boom, Taché was one of the few remaining opponents of the factory scutching system.[48] In 1865 he wrote that "in Lower Canada especially amongst the French population the operations of preparing the fibre & of putting it into yarn & linen are exclusively done by the hand at home, a mode of manufacture which in my humble opinion [is] much the better system."[49]

Nevertheless, Taché gained control over his department well after the Riga seed project was in motion, and it was his duty to see it to completion. Unfortunately, almost everything was against him. When the seed arrived in Montreal, it was discovered that many of the barrels had been destroyed in transit and the seed was wet and dirty. The shipment also took longer than expected and arrived at some destinations too late for distribution. The price of seed in Russia had been high owing to poor harvests, and even with the Riga subsidy, local farmers found they could get better-quality seed locally for similar prices. The province's largest flax millers, the Perine Brothers, wrote from their St Thomas mill to say they were returning their order of almost nineteen bushels for two reasons. First, "it arrived 'too late' not for sowing but for leasing land for sowing it, and second the seed is so dirty that although we were to screen it we could not induce any farmer to allow us to sow it on his land for fear of introducing foul weeds."[50] This response from the province's largest flax miller reinforces the argument that Upper Canadian farmers and farm societies campaigned even more aggressively against weeds than their British counterparts.[51]

What were the results of all the promotional efforts of experts, agents, editors, and officials in the 1860s? By the end of 1866, Donaldson possessed a surplus of over 1,000 copies of his flax booklet and resorted to petitioning the colonial government for an annual stipend to remunerate his educational efforts in flax promotion. Benjamin Walker also petitioned the bureau for renewed support in 1865. The government was left with hundreds of bushels of dirty flax seed that it could not sell even at a loss.[52] Donaldson continued to assure the bureau that the seed would greatly benefit the country. Ontarians were left with dozens of abandoned flax mills, because, as the *Montreal Trade Review* explained,

> [t]he years 1866 and 1867 were undoubtedly bad ones for those engaged in this occupation, whether they simply performed the part of the scutching, or manufactured the raw material into linen, ropes, &c. Not a few losses occurred, and several large manufactories which had been erected and fitted up with machinery at a very heavy

expense ... not only stopped work but in some cases the machinery was sold off at a sacrifice, and the enterprise abandoned altogether.[53]

The reasons for this temporary failure were varied and certainly not all the fault of over-enthusiastic promoters. The end of Reciprocity in 1866 did have a negative effect on startups which were only designed to export semi-processed flax to mills in the United States. However, many of these had recovered by 1871, and others, such as the Perine linen mill in Doon, were in the process of consolidating their interests in secondary manufacturing. Several other flax and linen businesses, like Gooderham & Worts' mills in Streetsville, were destroyed in fires and not rebuilt, although with strong suspicions of mid-winter arson, one wonders if local residents had grown weary of flax and Gooderham & Worts. Similar tactics were used openly in Ontario in the 1880s.[54]

The Canadian Bureau of Agriculture had not always been unanimously in favour of promoting flax cultivation in Canada. In 1862 a clerk wrote to J. Manning ("Great Flax Agent") in England to warn him that "the intending emigrant is not advised to risk his capital in flax" until more mills and machinery had been established for its production.[55] There was also a general disagreement about how flax should be promoted. Taché preferred the system of household production in place in Quebec, believing that exposure to increased education and seed distribution could only benefit the producers. Donaldson, on the other hand, promoted the industry to millers and scientific farmers, believing that the best way to assist the public was by providing machinery and other technology, by educating farmers across the colony, and by helping communities establish their own scutching co-operatives. Ultimately, mid-century millers chose a different path, which entailed providing seed to farmers or leasing their land, and by marketing their products to a variety of household, local, and foreign markets.

Some flax millers were also unimpressed with the state's efforts to promote the production of their main raw material. Donaldson had been canvassing local councils in the early 1860s for their support for a new position created to officially promote flax production. But support was not to come from Waterloo. Charles Hendry, an established miller in Conestogo[56] and future partner with the Perines, moved in a session of the Waterloo County Council "that no action be taken with respect to the petition sent for signature by John A. Donaldson, Esq., regarding his appointment *as* special agent for encouraging the cultivation of Flax, as this Council is of opinion that private enterprise is much more preferable than Governmental support."[57] The motion was carried; the state had rejected flax promotion in the middle of the flax boom and at the very heart of the fibre belt.

Nevertheless, state and provincial governments promoted the flax crop in this period. Donaldson claimed that New York State had recently spent $20,000 to promote flax industries.[58] In the 1850s, flax promotional efforts in Nova Scotia had highlighted the crop's potential for "clothing farm families but also [for] … 'industrial' uses" and for easing the labour problem.[59] In 1866, the same year that Leeming Brothers went to Russia to buy flax seed, Prince Edward Island invested a slightly more modest £150 to import flax seed for its agricultural societies. Then, in 1869, land commissioners in Prince Edward Island established values for the purchase of lands owned by absentee landlords, and they took it upon themselves to recommend areas where tenants should be further assisted by the promotion of flax. The commissioners suggested that the government help "by importing and setting up, in some central situation, flax breaking and scutching machinery," so that "dressed flax might shortly become an important article of export."[60]

Both state and private actors promoted flax in many parts of midnineteenth-century North America, and their efforts reached a fever pitch during the American Civil War. As we saw in Chapters 1 and 2, for over a century flax production was important in certain areas in the Great Lakes region. It was well established on both sides of the border before war caught the attention of flax promoters, and it would continue this way after wartime rhetoric had subsided. In the post-bellum period, flax enthusiasts and state officials gained no more clarity on how the region's small flax fibre industry worked. A US Department of Agriculture report claimed, dismissively, in 1879 that US dressed flax "is a ridiculously small amount at best."[61] Although "an enormous impulse was given to the flax industry during the war of the Rebellion," one US author admitted that after the war flax cultivation was insignificant. Save for a little "North River Flax" in Rensselaer, New York, and a tiny bit grown in New Jersey, "the bulk of American flax is fit only for paper-stock or upholsterer's tow, and only a small amount is good enough for even the coarsest kind of bagging." An examination of the Perine accounts shows that this was exactly the kind of market flax millers were targeting. It is true that flax fibre production declined slowly in parts of the northern United States, but in some ways the crop had simply moved around as it had done in Upper Canada.[62] The mid-nineteenth-century fibre promoters were mistaken about how a commodity like flax could work in Canada and the northern United States. By the 1870s most voices that had "set all the world in raptures" – like Chevalier Claussen's – fell silent for a season. Claussen himself purportedly died in a British insane asylum in 1872, although *Scientific American* declared that he died a decade earlier.[63] The main distinction between these flax promotions and those that came later was the nineteenth-century promoter's focus on practical chemistry. Although chemical processing theories never completely disappeared, they were generally displaced by the twentieth-century's fascination with mechanized harvesting. Although flax pulling machines were mentioned as early as McDougall's 1853 report, even Donaldson's chauvinist writings recognized the limits of this technology. By the turn of the century, mechanized harvesting had become the next generation of fibre promoters' greatest challenge.

Machine Learning: Promotion and Fraud in the Twentieth Century

As the flax fibre industry in the Great Lakes region stabilized in the late nineteenth century – producing steady amounts of fibre but moving slowly westward along with the small-town fibre mills – the promotion of flax temporarily subsided in both the state and the private sector. Yet, as subsequent chapters demonstrate, the seed components of the commodity web drove a new and more extensive kind of flax cultivation, mostly in the Midwestern and Prairie farm regions. This fanned the flame of fibre promotion once again, and toward the end of the nineteenth century a whole new wave of promoters appeared in Canada and the United States. Some promoters revived interest in chemical fibre processing, but the new flax enthusiasts generally advocated mechanical innovations. They argued that flax fibre harvesting could be mechanized, which would increase labour productivity for farmers and decrease the cost of raw materials for millers. Chevalier Claussen's chemical crop innovation – displayed at the Great Exhibition in London – was forgotten in just a couple of decades, but another agricultural inventor, Cyrus McCormick, achieved fame there for his grain reaper that echoed through at least a century of western settlement. One by one, each of the major North American crops was improved though the application of new mechanical technologies that increased their productivity, McCormick's reaper being one of the most famous.[64]

In the early twentieth century, flax fibre was still being harvested by hand. The high labour costs, and even some racist resistance to First Nations labourers who did much of the work, prompted some Great Lakes–area promoters to imagine a new mechanized industry that would finally make flax fibre a lucrative crop for all of the region's farmers.

The twentieth-century promoters were often local politicians or scoundrels, sometimes both. As the fibre industry expanded across southwestern Ontario, many small towns were tempted to subsidize the construction of a local flax mill. In the case of Essex, a town of 1,400 people in Essex County, this proved a costly mistake. Flax milling had moved into peninsular southwestern Ontario (Essex, Lambton, and Chatham-Kent) relatively late. In 1899 there remained only one mill, in Wallaceburg, but by 1904, two more had appeared in Dresden and Tilbury.[65] A manager from the Perth Flax and Cordage Company, A.H. Raymond, moved much further west to open a mill outside of the flax belt in the town of Essex. In 1903 he received generous tax breaks from Essex, which, like many small towns, hoped a flax mill would help it become an industrial centre. Raymond contracted 300 acres from farmers around Essex and a similar amount in nearby Belle River, where he operated a second mill. He was joined by S.J. Cohen from Philadelphia in January 1904, and together they attempted to convince the town of Essex to grant them more concessions and bonds so that they might open a linseed oil mill that would employ thirty workers and process 1,000 bushels of seed per day. The town heard Cohen's pitch and granted the request, finding him "a capable, clear headed and straightforward business man [who] would make a very desirable citizen." Essex was apparently a poor judge of character, because seven months later Cohen and Raymond were gone and their accounts in arrears. The exact reasons are unknown, but a winter fire had destroyed a warehouse full of flax and Raymond had been injured, fatally it seems, in the mill that May. By July, the town faced creditors for machinery from Toronto, flax seed from Crediton, Stratford, and Toronto, and rented land and flax crops from farmers around Essex and Belle River.[66] Raymond had attempted to replicate James Livingston's multi-output facility, but with quick money and a farm landscape outside the flax belt.

The failed venture cost the town of Essex dearly. However, what followed with the standing flax crop reveals that some elements of the rural mill complex were still in place. On 27 June the town's barrister, E.A. Wismer, advertised a two-week opening for tenders to buy the flax growing on

300 acres around the town. The crop was in "excellent condition and ready to pull about July 13th." He also offered the use of the flax mill at Essex on reasonable terms. In very short order, the town's most successful flour miller, Charles E. Naylor, was able to step in and purchase the crop and pull it with the help of Munsee First Nations flax pullers. The harvest gang moved just as quickly, and by 19 August they had finished pulling, threshing, and spreading the crop.[67] Clearly, small-town millers had somewhat interchangeable skill sets, and although Naylor did not front capital and assume the risk of starting a mill in Essex, he was able to implement some of the networks required to bring flax from field to market.

The most dramatic example of early twentieth-century flax promotion frauds took place in Massachusetts, in the laboratory of lay chemist and Oxford Linen Mills founder Benjamin Mudge. Mudge was the flax promoter behind the 1912 Sterling Debenture Corporation stock fraud in Manhattan and North Brookfield, Massachusetts.[68] He was also the ninth suspect to be arrested, briefly evading police that weekend in North Brookfield. The roots of the postal fraud story stem back half a century, when Mudge was born to a prominent family in Lynn, Massachusetts, and came of age during the civil war. He graduated from Massachusetts Institute of Technology in 1867, where the wartime flax fervour penetrated even to the institute's laboratory instruction. One of Mudge's lab experiments had to do with flax processing, and the lesson perplexed him for years. Over the next four decades he was engrossed with the idea of reducing the processing time of flax.[69] In a 1903 patent application the chemist explained how he could remove both the visible and the microscopic shivs from flax straw through a process of mechanical carding followed by a series of caustic acid and alkaline baths. This was followed by an intensive form of bleaching. Together the processes took what to most textile manufacturers was a fibrous waste product and replaced it with a valuable fibre.[70] In short, Mudge was promoting a US linen industry built on cottonization, a fusty variation of processes that had been used by inventors like Chevalier Claussen since 1851.

There was nothing particularly innovative about this "secret process." Reporting on one of Mudge's contemporaries, an article in the *New York Times* noted that some of the city's mills owned "scrap books filled with accounts of ventures of this sort which have not been successful."[71] Yet less than a decade after filing his patent, the fifty-nine-year-old was president of Oxford Linen Mills, a company in North Brookfield, Massachusetts, worth $2 million on paper. Unfortunately, it was worth much less in reality, and just as Cohen had mysteriously disappeared from Essex, Mudge found himself fleeing North Brookfield in the night to avoid arrest for defrauding stockholders.

The secret to Mudge's and the Sterling Debenture Corporation's brief success was also grounds for their arrest. Using large-format brochures and booklets, the company advertised aggressively across the nation, arguing that Mudge's system reduced the processing time of linen thread from three months to twelve hours. They also featured photographs of flax fields from Minnesota to Puget Sound, juxtaposing them with images of the immense infrastructure required for retting, drying, hackling, and bleaching in various parts of Belgium and Ireland. One pamphlet claimed that "millions of dollars [were] burned annually because there was no way known by which that straw could be converted, economically, into profitable fibre!"[72] The linen company used Department of Agriculture bulletins and other official reports to argue that the waste straw in the West was of the same quality as the fibre produced in Europe. Mudge's system specialized in recovering this waste, and with language like "complete utilization," "waste turned to profit," and securing "the full value of the flax," potential investors were led to believe that the flax of the US Northwest could be the raw material for a US linen industry.[73]

Elwyn A. Barron, another one of the accused in the Sterling Debenture Corporation scandal, was the man with Sterling Debenture's silver tongue. The author of most of the company's literature, including Oxford Linen Mills' advertising, Barron was an author and playwright. In addition to writing several books and plays, Barron was formerly the drama critic for the *Chicago Inter-Ocean* and the Paris correspondent for the *Chicago Times-Herald*.[74] The linen company's propaganda was no less a work of art. Structured around the passage of fibres from field to fabric and set in the racially charged nationalist language of the day, the company presented a persuasive argument for linen's superiority to cotton and the feasibility of producing it in North America. On the cover of "From Flax to the Fabric," a young man in a suit poses with a strick of pulled flax plants in one hand and a fine linen towel in the other. The subtitle reads, "A Process of Hours Instead of Months" (Figure 3.2).

Not only was linen just as feasible as cotton, according to Barron, but "among cultured men and women" cotton was an inferior substitute for linen in "white goods." He argued, "It was not the superiority of cotton

From Flax to the Fabric

A Process of HOURS Instead of MONTHS

Figure 3.2   Oxford Linen Mills: a process of hours instead of months

that made cotton king, but it was the discovery of a process by which the raw material was better handled. Had the processes of the Oxford Linen Mills been discovered at the same period that the cotton gin was brought out ... [linen] would never have given way to an inferior product of the soil."[75]

Barron appealed to white, middle-class North Americans' preference for wheat over other grains, arguing that "linen is to cotton what wheat is to corn, what gold is to silver."[76] An advanced nation did not have to settle for inferior commodities in food, according to Barron, and they should not have to settle for inferior fibres such as cotton and jute. Benjamin Mudge's persistence where others had failed was going to make him the Eli Whitney of the linen world – where "many men have tried," Barron triumphed, "at last one grandly succeeded."[77] In just a few years, Mudge would be serving a four-year sentence in an Atlanta penitentiary.[78]

The sudden reversal of fortune for the Oxford Linen Mills, and for Mudge, was mainly because the mill expansions promised in the promotional literature were not carried out and because the financiers paid dividends from their savings to keep up the illusion of profitability.[79] However, another reason the Oxford Linen Mills became a pariah across the country was because it failed to advance US flax fibre production. The *Boston Evening-Transcript* reported that the fraud lay in the claim "that the Oxford mills owned exclusive patent and secret processes that would manufacture linen from US-grown flax at one-third the cost of foreign linen."[80] The *New York Times* titled its description of an Oxford Linen Company Minneapolis subsidiary "Mattresses amid Flax Fields," claiming that the company falsely promised to make mattresses from American flax tow in Winona, before sending the longer fibres on to the Oxford Mills. The inspectors confirmed that these plans did not materialize.[81]

Always the wordsmith, Barron created carefully phrased pamphlets that never actually promised that the mills would use American linen. The courts and the press, however, felt this was implied through the photos and the location of new plants. Other millers confirmed the company's intent in personal correspondence. Writing to flax specialist Henry L. Bolley in North Dakota, William J. Robinson of the United States Linen Flax Corporation in New York City claimed that he had "been all through their Plant." The Oxford Linen Mills, he stated in 1910, "are not using one pound of American Straw." Robinson went on to suggest that "the Mudge Process is a permanganate of potash bleach and absolutely ruins the fibre." The only cloth the mills were producing was some common huck towels, and those were mostly from a cotton warp.[82] Bolley also received a letter that year from the Sterling Debenture Corporation, which was seeking access to government mailing lists and describing the work of the Oxford Linen Mills. The North Dakotan plant pathologist knew the company was only trying to target potential investors, and he was outraged when the letter confirmed that the linen mills had only used imported European flax tow. Bolley treated these letters as evidence in a *Review of Reviews* editorial, and he undoubtedly contributed to the investigation. Before the arrests were made, Bolley argued that the company's greatest transgression was aggressive capitalization based on what appeared to be a fabricated plan for stimulating local supply. Special processes or not, he argued, honest millers should set up near a potential source of raw material and simply start milling.[83]

Unlike Mudge and Raymond, most of Ontario's millers simply chose a location within the flax belt and repurposed a lumber mill with relatively low start-up costs. Many flax fibre mills did receive municipal support, and indeed town leadership and mill ownership were often closely intertwined. The Wallaceburg mill was owned by James W. Steinhoff, a lumberman and shipbuilder who became the town's first mayor in 1896. Steinhoff's junior partner, Hinnegan, was likely the Thomas Frank Hinnegan who became mayor himself in 1905.[84] Like the Perines in the nineteenth century, the most successful mills operated within the slowly migrating flax belt, and they took advantage of local labour, knowledge, and the intersections with overlapping webs. Like the flax districts further east, these mills followed on the heels of a local timber industry. The 1904 Consular Report from US Commercial Agent Holmes in Wallaceburg noted that a very high quality flax was the most valuable export from this district to the United States. Logs and cordwood were second. At that point, Hinnegan managed all three mills, processing flax from 300 acres around each one. The company sold most of the long fibre to US consumers and most of the tow in Canadian markets.[85] It drew raw materials from across the region, including Essex, where Hinnegan and his new partner Daniel Burns travelled in November 1904 to sue for purchased flax seed that had been destroyed in Raymond's mill fire.[86] By 1910, Hinnegan had relinquished ownership of the flax mills. The Tilbury mill was owned outright by Burns, Dresden's by Wilson and Burns, and Wallaceburg's by the Wallaceburg Flax Mill Company.[87]

In 1906 one of Canada's most ardent young fibre promoters wrote about the more permanent of these small-town Ontario mills, essentially blaming them for what he considered the "surprising continuance of old-fashioned methods in Ontario." To Austin L. McCredie, a twenty-nine-year-old student at the Ontario Agricultural College, fibre millers had been dew retting their flax for too long, and he figured they ought to switch to a form of water retting. He felt that dew retting was a slower and less scientific process that produced lower-quality fibre and often risked damaging the crop in bad weather. McCredie was very familiar with the traditional methods, as he had been raised on a farm in South Dorchester, Elgin County, near the upper reaches of Kettle Creek. His own family likely grew flax for the St Thomas mill, and they would have certainly witnessed the mill's use of

Figure 3.3   Wilted and non-wilted flax plots in bloom at the Central Experimental Farm, 1924

family and First Nations labour. McCredie argued that their method of dew retting was too labour-intensive, and that "this labor is becoming increasingly hard to obtain."[88] Through a little accounting that included the cost of labour and the occasional damaged crop, McCredie argued that it was cheaper to produce water-retted flax over the long run. There is no evidence that millers changed their practice in the years following the writing of his thesis, and eventually McCredie tried a different tack in lobbying for mechanized harvest methods. After a brief career editing the *Canadian Countryman*, he was encouraged by wartime prices for fibre, and he took on a flax factorship of his own in St Marys.

Similar to the temporary boost it received during the American Civil War, the flax fibre web benefited from war again in 1914. Canadian fibre promotion gained tremendous momentum during the First World War, when the Canadian and British government combined to promote Canadian flax production for the Irish seed crop. The Central Experimental Farm built flax storage barns and processing equipment in 1914 to demonstrate fibre production and breed pedigreed flax varieties (Figure 3.3). By 1915 this wartime demand for linen helped increase Canadian flax production from 4,000 to 35,000 acres.[89] The flax heartland was still in southwestern Ontario where, according to Austin McCredie, millers could find "the remnants of skilled labor" that were required to produce the linen of any quality. McCredie argued that this labour remained the bottleneck for flax millers, and he advocated that a viable mechanical flax puller be developed.[90] Late in 1914, the federal Department of Agriculture commissioned a young flax promoter, James A. McCracken, to report on flax fibre in Canada. His conclusion stressed the importance of developing a mechanical means of harvesting flax in order to reduce the labour costs of production.[91] When the Experimental Farms director, J.H. Grisdale, offered his annual report in 1917, he drew extensively on McCracken's research and encouraged the Department of Agriculture to create a new division to focus on fibre and encourage flax production.[92]

When the federal Department of Agriculture established the Economic Fibre Division later in 1917, R.J. Hutchinson became the director and promptly organized a series of experiments and federal-provincial partnerships. One of the initiatives was a 1918 flax crop on the Willowdale Experimental Farm, north of Toronto. There was also a summer "Flax-Pulling Festival" that they promoted as a war effort and a fundraiser for the Red Cross. The festival was very well attended, largely because of the Toronto-built Curtiss "Jenny" planes that buzzed overhead. By 1918 the Allied forces had also encouraged increased linen production for use on airplane wings; this was exactly the kind of technological promotion that Canadians could appreciate. Austin McCredie, the former student from Ontario Agricultural College, presented flax to the farm press in terms of equivalencies – one acre of flax equaled one set of airplane wings for the war effort. The *Farmer's Advocate* argued that flax production was increasing in 1918 so "that the allied supremacy in the air may not wane," and even up in the "Bruce Peninsula remote from any mill [they] observed a small field as a 'war crop.'"[93] In Windsor, Ford Canada and the Border Manufacturers Farmers' Association grew flax for the war effort under the "Greater Food Production" campaign in 1918. One of the directors of the association, George E. Rason, committed at least a dozen acres of the 200-acre project to flax, with the understanding that the fibre would be used to produce "linen for aeroplane wings." However, at harvest time they were short on labour, and Rason had to issue a public "appeal for a large party to go out to the farm on the old Kiltie camping grounds tonight to pull flax." He already had several gangs of men harvesting each night that week, but now he needed

Figure 3.4　Pulling flax in Willowdale, ON, 1918

families to volunteer "for a patriotic cause" or the fibre would be useless for the war effort.[94] Children were even more important to the flax harvest in wartime. In 1918, for example, a group of twenty-five boys and girls were employed by the Department of Agriculture to pull flax in Willowdale. They were paid an impressive rate of $15 per acre.[95] A photograph of five boys pulling flax at the Willowdale farm offers the best available close-up image of children employed in this work (Figure 3.4).[96] A Curtiss Jenny plane passed low over their heads, completing the spectacle. As it happens, the photo was a fake. The photographer combined two images from the festival in an effort to demonstrate how the fibre commodity web stretched from field to fabric, from Ontario's soil to Europe's skies.[97]

What the photographer and the Economic Fibre Division missed was that farmers in all areas – outside of Huron County – were reluctant to grow flax, and when other farmers grew the crop, it was entirely for seed. The victory plots and family festivals made for excellent publicity, but these groups were far less important than the organized harvest gangs, particularly the First Nations labourers. By focusing on the cost of labour, promoters typically came around to commenting on the kinds of labour that Ontario flax millers relied on in this period. "It takes skilled labor to produce flax for wings," McCredie declared, and if Canada wanted to become a "nation of flyers," it would have to produce its own airplanes and, by extension, its own linen. McCredie believed that this could only occur through "the transfer from hand to machine work," citing a flax-pulling machine that some Ontarians tested in 1918.[98] McCracken presented the same case, calling field labour the paramount issue and hoping for good results from a flax-pulling machine designed by Charles Vessot in Ottawa. In McCracken's case, dwelling on the labour problem resulted in a racist appraisal of the First Nations workers. The problem was more than a shortage of labour, it was the "insufficient supply of the right sort of labour" and specifically the "scarcity of white pullers" that limited his vision of the industry. By relying for decades on "the American Red Indian from the various reserves," Ontario flax millers supposedly "postpone[d] the inevitable era of machine harvesting." According to McCracken, the problem with "the Indian" was "not the quality of the work he performs, but manifold troubles, such as preliminary coaxing and dickering and disappointment, his lack of responsibility, the coddling he demands for himself and family, his train of paraphernalia, and similar nuisances." The fibre promoter was likely tempering his true prejudice for the report, since in farm periodicals like the *American Thresherman*, he was even worse. "Unfortunately for flax men," McCracken told that journal, the Native farm worker had "come to realize his importance in the flax harvest. He demands and gets advanced money, transportation to and from the flax towns, liberal credit at the store guaranteed by the flax factor, and, in some cases, free housing, fuel, furniture, and household utensils."[99]

What McCracken described as the increasingly prevalent "problem" of accommodating itinerant field workers was actually the stable and satisfactory labour arrangement that various First Nations communities had developed with Ontario's millers over three decades. In no other description or case were First Nations workers described as irresponsible or troublesome, and in some cases, millers indicated that they preferred dependable and flexible labour from the reserves over non-Native labourers.[100] Significantly, the one stereotype that did appear in some news outlets, alcohol abuse, was not mentioned in either of McCracken's accounts. McCracken represented the latest in a wave of promoters who offered theories for expanding the fibre industry but who lacked any apparent experience in the most successful aspects of the commodity web. McCracken was from outside the region in West Gwillimbury Township, Simcoe County, where his father, Alexander Hugh, had lived since at least 1881. He conducted extensive research for his flax report, including correspondence with millers across Ontario and plant pathologists like Henry Bolley.[101] He was also an executive member of the recently formed Canadian Flax Growers Association, and Grisdale apparently had plans to employ McCracken in the Economic Flax Fibre Division and eventually make him its chief. However, the war opened other paths for McCracken, and shortly after writing the report, he enlisted in the 157th Simcoe Battalion and was eventually wounded at the Battle of Vimy Ridge in 1917. After returning to Canada, McCracken spent a career in business in northern Ontario.[102]

Promoters like McCredie and McCracken frequently brought labour and mechanical solutions to the foreground, but as director of the Experimental Farms, Grisdale produced a more judicious summary of the flax fibre situation in 1917. It did not dwell on mechanical or chemical innovations, and it barely mentioned First Nations other than to say that they were often important to the workforce. Grisdale ignored the farm press and other flax promoters, and he reported directly on small-town mills based on visits that he conducted in southwestern Ontario and Michigan. The central message that emerged from these reports was not the importance of labour, but the importance of knowledge. According to Grisdale, his interviews with millers revealed that their most urgent wish was that the government provide technicians who could help farmers improve their flax cultivation and grading. One miller, Howard Fraleigh, said that the most useful assistance would be "a man, who could immediately go out and help" farmers understand the different grades of fibre. In his final recommendations, Grisdale asked the department to focus on hiring expertise for plant research, personnel for the fibre division's mill and outreach, and technicians who could help farmers with grading and processing in the near term.[103] The two issues Grisdale did *not* mention in his report were the well-worn topics favoured by the fibre promoters: chemical processing and mechanical harvesting.

Similar to the post-bellum decline in the number of flax experts, the loudest fibre promoters decreased in number with the close of the war. Still, the war's effect on fibre prices was significant, and Grisdale believed this would continue for some time after, since such a large proportion of the Belgian flax workforce had been "killed or scattered."[104] In Canada's Parliament, the southwestern Ontario MP Samuel Glass attempted to present a motion in the House of Commons to introduce flax protection and standards for the recently expanded industry. The motion failed and production dropped to its pre-war levels, but the federal Department of Agriculture continued its flax fibre research and promotion through the Economic Flax Fibre Division. The division's primary work was to market Ontario's crop for farmers and help set up cooperative flax-growing societies in Quebec. Howard Fraleigh, an Ontario MPP from Forest with an avid interest in fibre farming, became the region's best-known flax and hemp grower. He developed the first mechanized hemp operation in North America, which was also the last, after Prime Minister Mackenzie King added cannabis to the restricted substances on the Opium and Narcotic Control Act, in 1938.[105] The Economic Fibre Division continued its work through the interwar years, and in 1939 it reported that Canada was "in a position to compete with most European" flax-producing countries.[106] When the dust of that war had finally settled, however, Canadian flax fibre production was gasping for life and struggling for customers. The industry's customers never returned. What it received instead was a fresh injection of cash from Parliament for the development of a new flax fibre research and development facility in Manitoba at the Portage la Prairie Station. Completed in 1945, the mill operated for a few years, but it was soon converted to a vegetable-crop research station.[107]

One problem the fibre division never solved was how to mechanize flax pulling. As we saw in Chapter 2, First Nations labourers continued to work in the flax harvest until at least the late 1930s, although Indian agents continued to oppose the practice. One particularly overbearing agent was another McCracken, a man born and raised in the flax belt. Morley William McCracken was ten years old when he appeared in the 1921 census on a farm in Caradoc, Middlesex County. He lived with his parents and two older sisters on a farm that had been in the family since at least 1878. The farm was thirteen kilometres south of the Gunn and Murray flax mill in Strathroy, and it was only one concession road north of the Chippewa of the Thames First Nation Indian Reserve No. 42.[108] It is quite likely that the McCracken family worked with First Nations in the flax fields. In 1939 Morley was an Indian agent on Christian Island in Georgian Bay, but he soon moved back to Middlesex to be the agent at the Munccy Reserve (now Munsee-Delaware First Nation). In 1942 he agreed to trade posts with the agent in Sarnia, whose office included Kettle Point and Stony Point Reserves. These communities had worked extensively in the flax harvests for Fraleigh and other millers. On his arrival, McCracken was tasked with facilitating the expropriation of Stony Point for the construction of Fort Ipperwash and with relocating its residents to neighbouring Kettle Point. The move fragmented the community socially and economically, and McCracken deliberately restricted their ability to work on surrounding farms.[109] The land dispute was not resolved until after the Ipperwash crisis and the killing of Dudley George in 1995. Ernest and Harrison George, the two young brothers from Kettle Point who worked for Fraleigh, were part of this community and part of Dudley's extended family. Racism was apparently not a significant problem for the First Nations workers in the early days of the fibre web, and there are no accounts of confrontations between these workers and the mill owners. The most discriminatory people in this part of the web were the occasional newspaper editor and flax fibre promoter and, more significantly, the Indian agents themselves.

Many of the fibre promoters were either outright fraudsters like Benjamin Mudge or just naive enthusiasts like the business and government agents who attempted to grow an acre of flax for every plane needed in the war effort. Men like A.H. Raymond and Austin McCredie represented the fine line between promotion and fraud in the flax business. Raymond had a knowledge of flax and likely did not set out to defraud the town of Essex; he just quickly found himself overcommitted and unable to attend to every strand of the flax and linseed oil web. McCredie, too, knew a great deal about flax, likely from living in the flax belt and studying the plant at the Ontario Agricultural College. He was, however, first and foremost a writer, and his commitment to research and development was not matched by an ability to organize a mill over the long term. Why did flax draw so many promoters, and why have so many failed across its three hundred years of history in North America? It seems that for most of this period promoters made similar mistakes. They assumed that Canadian flax would grow like Ireland's, which was fibre based, and would supply British textile markets. Instead, Canadian farmers treated flax more like India's flax, which was seed based, and they adapted to North American markets for linseed and rough textiles on the fly. Despite policies that favoured Britain, Canada's export market and entrepreneurial base were increasingly found in the United States. What emerged instead was the continent's first oilseed commodity web and a regional network of oilseed expertise, which are explored in the following chapters.

The recurring story of flax fibre promotion has masked the plant's most important history in North America. Promoters returned to fibre as they might to a bad habit, promising that it could replace cotton and appealing to racist, nationalist, and economic language as evidence. Fibre's periodic appearance in over two centuries of promotional literature and state initiatives made it a spurious subplot. Officials and other promoters dreamed of a national staple, but in the concentrated regions where flax was grown for fibre, this was because millers and about 25 per cent of local farm operators used First Nations and other family labour to produce it.

It may seem surprising that the century-long history of small-town flax mills, with conspicuous buildings and extensive properties, groups of itinerant workers, and complex land-renting agreements, has not attracted the attention of professional historians, but there are two good reasons for the lacunae and persistence of alternate images. First, this industrial crop was an unexceptional part of life for rural people. Their records show that they did not treat flax as extraordinary or worthy of promotional fervour, but as one small piece in a highly diverse system of products and their relationships. Second, wherever ink was spilled for flax promotion, the crop was usually disregarded by farmers. Promoters adumbrated a version of the flax story that aligned with their own Eurocentric interpretations of Canadian business and agro-ecosystems, often missing the market, format and geography of the fibre industry and usually ignoring the seed industry completely. The practices of renting land to millers for flax, organizing family and First Nations groups for harvesting, and planting flax on the new breaking in the Prairies seem extraordinary today, but in most flax farmers' diaries and memoirs these everyday exchanges were not worthy of mention. Historians who explored these rural worlds through the expert's lens missed these relationships because North American flax promoters thought about an entirely different commodity and almost only about fibre. It would be wrong, however, to imagine that fibre's routine failures in the East and worthlessness in the West made it somehow part of a different story. Fibre and seed are always produced from the same plant, and the plant itself comprises the foundation and frame for the flax story in North America.

CHAPTER FOUR

Covering the Earth

*North American Flax
and Paint to 1878*

The Industrial Revolution was and continues to be a global and ongoing historical phenomenon. One of the most important outcomes for colonial and environmental history was the introduction of many new or significantly expanded commodities, often produced with raw materials from the tropical world. The best-known crops include non-comestible plants such as henequen, stimulants such as tobacco and coffee, and industrial goods such as rubber, palm oil, and gutta-percha.[1] However, not all of the new commodities emerged from the tropics. Linseed oil and its consumer goods were not exactly new, but, like other commodities drawn into the mainstream by changes in consumer taste in the eighteenth and nineteenth centuries, flax was an industrial crop.[2] Linseed oil was primarily used in luxury goods like paint and varnish, and the consumption of these luxuries was an example of what Arjun Appadurai called "tournaments of value," or symbolic as opposed to material consumption.[3] As the flax-paint web expanded in the late nineteenth century, millions of acres of prairie came under cultivation in part to satisfy these markets. And urban manufacturers benefited the most. The 1860 US Census of Manufacturing shows that at $13.4 million ($6 million from linseed oil; $3.5 million from paint, $3.2 million from lead, and $700,000 from oil floor cloth) the flax-paint web was the largest triglyceride industry on the continent. From a domestic manufacturing standpoint, the size was already quite significant. It was larger than the hardware goods industry, and over twice the combined value of one of US history's favourite examples of domestic manufacturing, the gunpowder and firearms businesses. Moreover, this was at the outset of the nation's industrial ascendancy and before flax's irruption into the Northern Great Plains.[4]

Just like fibre, the oil side of flax milling initially took place in mill complexes around the eastern Great Lakes and the seaboard. This chapter examines those places, tracing them up to the point of most rapid transition, which occurred in the late 1870s. At that point the mill complex began to lose its centrality in the flax industry, partly because of developments in industrial growth and organization, but also because of a change in the commodity itself. The appetite for fibre continued, but it would soon be overshadowed by a new mass-produced output from the web – ready-mixed paint. This chapter demonstrates that the demand-side drivers of the paint web were not residential painters, but rather industrial paint shops, particularly manufacturers of carriages, sleighs, ships, and eventually the gamut of steel vehicles. Chapter 2 provides a glimpse of how farmers responded to this change in the flax belt around Berlin (Kitchener) in 1861; the transition from fibre to seed markets was already under way and producers were adapting. These developments hastened the social and ecological disturbance in producer–consumer relationships, the growing distance between flax farmers and end users, and the reduced role of the rural mill complex as an important space for farmers, millers, and other makers.

Also similar to the flax fibre, flax seed disappeared into a world of goods. Before the late nineteenth century, "paint" was not an off-the-shelf consumer good but rather a service hired from the master painter or an ingredient to be combined with other hardware products. When people began to use these colours and coverings extensively, they had no reason to connect paint to flax seed. Even after paint and varnish had become important objects of urban consumption, linseed oil remained "to the general public ... very much of an abstraction."[5] This opacity was common in many industrial commodity webs, and for flax it had at least two effects. First, paint, lead, and linseed oil became deeply embroiled in campaigns for the standardization of industrial ingredients. Processing became much more technical, mills hired chemists for quality control and marketing purposes, and, like many industries in the Gilded Age, linseed oil proprietors hired managers to replace them on the shop floor.[6] At the same time, manufacturers began branding intermediate goods in the paint industry (especially white lead and linseed oil) as products with a past. Second, as explored in

Chapter 6, consumers did not connect painting with farming, and they were not aware that the raw materials for their consumption came from, and helped resettle, the most fragile dryland ecosystems on the continent.

Early Uses of Linseed Oil and Paint

Linseed oil was used directly by consumers as a protective covering, leather softener, and medicine,[7] or it was processed in other shops that made paint, ink, lubricants, liniments, enameled shoes, oil cloth, and soap.[8] The oil has been used in inks applied to everything from papyrus to Johannes Gutenberg's printing press.[9] It became a drying vehicle in paint and enamel as early as the sixth century, CE, and was used regularly by twelfth-century clerical painters.[10] Jan Van Eyck was wrongly credited with inventing the technique of oil painting, but he was likely the first to use linseed oil and mineral pigments as a varnish for protecting the oil paintings of the Renaissance.[11] Linseed oil was common in lamp oil in Turkey and seventeenth-century Germany, and presumably the mixing of vegetable oils, fatty oils, and petroleum was part of the lamp oil industry before petroleum oil appeared in the 1860s.[12] Linseed oil cake was a by-product of the seed-pressing process. Its production required so much machinery and power that industrial milling was necessary and the manufacturing of consumer goods, especially paint and varnish, from the oil required highly specialized skills.

In the early linseed oil industry, producers, or crushers, processed the farmer's flax seed and sold it to master painters and glaziers, where it was further refined and ground with pigments and driers to produce paint. The earliest North American linseed oil mills were probably small operations built by painters. For example, Edward Pell's "Oyle Mill" was built in 1726 and valued at more than his paint shop or house.[13] However, specialized linseed oil crushers began to supply the urban paint trade, and in the mid-eighteenth century a water-powered mill appeared in Bethlehem, Pennsylvania. It was a replica of the German and Dutch technology for wind-powered linseed oil mills in northern Europe. The mill's account books show that the Moravian Brethren consumed between 200 and 300 gallons of linseed oil in 1745. After seven years, output increased to between 750 and 1,550 gallons. This could be thought of as one gallon for each of the town's inhabitants, but in reality most of the oil was sold in nearby cities. The oil was used for making paint and preserving wood, as well as in lamp oil, printing inks, and medicines. The mill transported most of the oil to consumers in surrounding towns and as far away as Philadelphia and New York. Milling technology varied even in the first two decades at Bethlehem, but all such mills used some form of mechanical leverage, including wedge and screw presses.[14]

The Brethren were a unique community, but their oil mill resembled many others in the colonies. Three more linseed oil mills were established near Lancaster, Pennsylvania, and one in Easton, Massachusetts in the late eighteenth century; the larger mills could process up to 2,000 gallons per year.[15] Presumably, these mills were like the Moravians' in that they produced a range of goods in mill complexes. Local linseed oil markets supported little more than a few months of activity in these small scattered mills, and their main customers were other craftspeople who converted the oil into varnish or paint. These master painters used a mix of local and imported ingredients. They imported pigments until at least the early nineteenth century, but with the development of colonial mills and flax production, they had access to domestic linseed oil. In 1810 most of the oil in Pennsylvania's German immigrant communities was exported to Philadelphia paint manufacturers.[16] White lead manufacturing began in the United States following the American Revolution, although it remained the product of one or two small mills until after the War of 1812. Then annual production rose, from 370 tons in 1810, to 9,000 tons at mid-century.[17] According to the economic historian Williams Haynes, the most important chemical products were paints, medicines, and gunpowder, and "almost every chemical maker of the first generation became also a producer of dry colors and pigments."[18] Domestic production of dry colour pigments began in 1807 with two pinks, a green, and a blue, and within two decades, mills were grinding dozens of colours from US minerals in Boston, Philadelphia, Brooklyn, Albany, and Rensselaer County, New York.[19]

Mixing and applying paint was possible with imported colours, but for various reasons, including its high cost, painting was a sybaritic luxury before the 1790s. Abbott Lowell Cummings and Richard M. Candee assert "from both physical and documentary evidence that a majority of rural houses remained unpainted on the outside for much of the eighteenth century."[20] By the 1790s, according to paint manufacturer Daniel Tiemann's history of the US paint industry, "the use of paint had become common" in towns. If "the ordinary householder ... was too poor to indulge in the luxury of an outside coat, the interior woodwork at least was painted, and the churches and public buildings all showed the work of the painter." Apparently white was not commonly chosen as exterior paint by colonial Americans, but the growing importance of white lead might be reflected in Tiemann's claim that "[t]he white house with green blinds was ... the single type of ultra-esthetic [*sic*] decoration."[21]

Canadians at least read of this ideal type: "how different the lively green and white of a well painted house," a Canadian farm journal printed, "from the dark, gloomy appearance of unpainted weatherboarding."[22] And paint was used to fashion a sense of respectability for white, middle-class Canadians. One sign of First Nations preacher Peter Jones's gentility was his "neat frame house, painted white," and the appearance and decorations of his home were what convinced one interviewer that this "rough and brawny son of the forest" had become civilized.[23] Yet like most Upper Canadian houses, Anishinaabe homes were unpainted, at least if the neat frame house clad with unpainted shingles in Peter Schmalz's *The Ojibwa of Southern Ontario* is representative.[24] Cummings and Candee confirm the popularity of white homes among the country elite in Federal New England, and they argue that these few facades were completed by having cheaper red paint on back walls and that they only interspersed a countryside of "many small unpainted houses."[25] Similar roughly decorated houses are visible in photographs of rural Ontario as late as 1903, such as the images of W.W. Carter, who photographed people in front of their homes in the St Marys area.[26] In Figure 4.1, two men sit in front of a house with a whitewashed facade and unpainted sides and porches. The two-storey home behind them was relatively large, but the decorations, trim, and grounds were anything but lavish. These rural people decorated some things and not others; the window frames and trim appear to be painted, but the screen door and pump casing probably contained more paint and ornamentation than the rest of the exterior combined.

But how common was the use of paint in the early nineteenth century? Paint and varnish certainly began covering some wood and walls in North American cities, but would the average rural farm family ever have painted their house or carriage? References to country homes in New England specifically noted the complete lack of colour throughout the eighteenth century and as late as 1833.[27] "The earliest settlers," according to one history, "had neither the time, the energy, nor the money to indulge in the use of paint."[28] One optimistic painting guide tried to convince farmers to learn the skill because, it admitted, "[t]o send for a professed painter,

Figure 4.1 A house in the St Marys area, c. 1903

and have his three-coat work measured by the yard at the usual rate is a pretty expensive business."[29] Linseed and cottonseed were among the first non-fibre crops grown entirely for off-farm industrial processing. In fact it was virtually impossible for families to produce a gallon of paint from flax seed or a bottle of salad dressing from cottonseed. These goods were parts of complicated systems of production, and thus paint had to be purchased from imported or local manufacturers. Douglas McCalla has shown that consumers in Upper Canadian country stores only began to buy the oil, lead, and other hardware required for painting in the 1850s, and even then at very high costs.[30] Although paints mixed in oil were present soon after they came on the market, these were some of the most expensive chemical purchases and were likely only used on important projects.

Nineteenth-century painters and glaziers combined pigments, white lead, resins, and other driers with flax seed (linseed) oil to make paint and putty, mainly for a wealthy and relatively small segment of society. Sometimes they advertised as portrait artists, but were likely active in other visual arts.[31] In the federal era, houses made up a small part of American painters' incomes. Of course, as glaziers they mixed putty when installing and repairing windows, but painters like Daniel Rea and Thomas Johnston of Boston usually earned much less than a third of their income working on houses. The rest came from painting vehicles, signs, and small objects and from preparing paint and paint supplies for others.[32] Hezekiah Reynolds's guide, *Directions for House and Ship Painting*, emphasized marine painting in 1812, and Haligonian John Merrick painted both for the Royal Navy and private customers in the early nineteenth century.[33]

In 1850 New York State was home to 111 varnish makers and 7,435 painters and glaziers.[34] These trades often went hand in hand, as linseed oil, driers, and pigments were the standard materials of painters and glaziers and paint manufacturers alike. Glass framing putty was also made with linseed oil combined with ground chalk (whiting).[35] The master painter prepared paint by boiling linseed oil and grinding white lead or other whiting such as zinc into the oil until it was dissolved. Then, the white liquid paste was diluted with more linseed oil, and if colour was desired, the white base was tinted by mixing in more ground pigments. Since the work of painting was time consuming and often depended on weather conditions, the paint was mixed with driers such as Japan or litharge to shorten the time between coats and thinned with turpentine to facilitate brushing.[36] Painters and glaziers protected their trade secrets carefully and rarely even advertised. The first instructional guide written by a professional was Hezekiah Reynolds's *Directions for House and Ship Painting* (1812), a beginner's guide to making and using oil paints. Reynolds's intended audience was not the general public, but rather other tradespeople who used paint.[37]

The complex task of paint preparation became easier with several improvements throughout the nineteenth century. The first ready-made products were paint pastes made from linseed oil and pigments as early as 1830. Because records of the products "lead" and "white lead" seldom differentiated between the dry powder and the lead set in linseed oil, it is difficult to know when the first combined paint base was available. In 1846 Canadians could "buy twenty-five pounds of white lead from an apothecary, for two dollars and twenty-five cents; it comes ready ground and mixed with oil; all you have to do is to rub it up on a stone or in a paint mill with an additional quantity of linseed oil until it is thinned to the proper consistency."[38] In case that sounds easy, the stone referred to here was a specific painter's slab and grinding stone, and the process was physically demanding.[39] Yet rural periodicals figured that these technological advances would help preserve and bring colour to many country homes. In the 1830s and 1840s, journals stressed "the economy of covering wood work with paint. The most economical people in the world do it universally."[40] But how would this appear to consumers who had probably never attempted to suspend metal pigments in oil or add just the right amount of driers? The journal introducing paint as "ready ground and mixed with oil" explained how the druggist's concoction should be diluted with linseed oil on a grinding stone, combined with turpentine, and mixed with lead colours and lampblack "until the proper shade is obtained." Surfaces required cleaning and repair, and brushes would be ruined without care. Painting in this writer's opinion was "so simple that any boy may be learned to do it in a couple hours," yet the journal's lengthy instructions imply that most people had very little experience with the process.[41]

By Confederation, the image of a respectable country house included the use of paint and varnish. According to *Canada Farmer*, homeowners should

> [b]rush, stop, rub down, prime and paint the whole of the exterior wood-work of the house, with four coats of good white lead and linseed-oil paint, finished to harmonize in colour with the materials of which the walls are built. The interior wood-work will always be finished to suit the tastes of the proprietors, but we would recommend graining oak and varnishing as the best and most lasting finish. Glaze the windows with 21 oz. sheet glass, well bedded in linseed oil putty.[42]

Another rural Canadian claimed to enjoy regular painting, and one year he wantonly "commenced painting the house." For painting floors, he claimed, "I now find that no paint but the best white lead, tinted to stone or other colour will answer, and this must be used with good boiled linseed oil, with plenty of litharge or dryers in it combined with some turpentine." He recommended three coats and warned that wear spots would have to be repainted each summer. For people who "liked the appearance of the grain of the wood," he recommended "raw linseed oil, put on quite hot." He argued that "it far exceeds paint, as nothing will wear it off."[43] Raw linseed oil was more durable for floor and exterior paints, but since it took seventeen days to dry under ideal conditions (and longer on certain woods), it required the addition of strong driers.[44] He admitted that boiling the oil was a difficult and dangerous process that should only be attempted outdoors – and away from children.

Most people would not have had the time, resources, or patience to perform these stages on much more than a few pieces of trim, furniture, or floors. Walter Needham's *Book of Country Things* recorded stories of everyday life as told by his grandfather. The older man recounted all sorts of handmade objects, but he claimed he "wouldn't recommend the job of grinding paint by hand to anybody; I found it took quite a long time to get anything done."[45] The availability of lead ground in oil facilitated paint mixing somewhat and likely encouraged some experimentation on farms. However, the technological advances were too small and the cost of paint and cost of time too great in the early nineteenth century for any of these guides of 1812, 1833, and 1846 to have much influence on general consumption.

Farmers often experimented with water-based variations on oil paint. Thrift guides encouraged the use of water-based lime paint or "whitewash" that "answers as well as oil-paint for wood, brick, or stone, is much cheaper, and will retain its lustre for years."[46] Whitewash had been the standard interior and plaster paint in colonial America.[47] It was sufficient for temporary colour and cleanness, but oil paint was the strongest covering and the only one that could suspend metallic pigments for vibrant colours. Oil paint also made a superior floor covering in high-traffic areas, and it was appreciated by anyone who disliked scrubbing bare porous wood by hand.[48] For a much rougher covering on large surfaces, some farmers tried diluting paint with sand or cement. In 1850, under "General Science and Miscellany," farmers were taught how to make paint with lead, oil, sand, and driers.[49] There are few references to rural paint making in this period, and farmers would likely have been frustrated with the quality of this gritty and still expensive paint. Some of the Canadian research on nineteenth-century paint restoration shows that paint was used selectively and whitewash more generally in homes like Joseph Schneider Haus's in Berlin (Kitchener) in the 1850s. The house demonstrates the importance of using paint as an accent and whitewash as a covering for larger surfaces.[50]

The paint economy began to change in the mid-nineteenth century, but up to this point, paint was used selectively, especially in predominantly rural Canada. Rural periodicals, which tended to encourage farmers to invest in farm and home upgrades even when they seemed clearly impractical, had little else to say about paint until the end of the century.[51] Even rural beautification, which became a major theme in the farm press, was rarely mentioned.[52] Julie Harris and Jennifer Mueller show how the

Central Experimental Farm and its physical layout were designed to display the progressive farm science and attractive face of rural Canada, although the large and brightly painted buildings belonging to "Canada's Biggest Farmer" were out of reach to most people.[53] A contributor to the *Farmer's Advocate*, in an article in 1890, treated mixed paint like a novelty and used laymen's terms for very basic painting techniques.[54] By 1908 the journal described painting as "neither complicated nor troublesome," but farmers might have found the long description of mixing paint from lead paste, oil, driers, and turpentine complicated and the arithmetic of costs a little troubling.[55] In "A Plea for Ready-Mixed Paints," one farmer suggested a way to get "something better than the whitewash of our fathers." His solution was mixed paint, which reflected a belief in the superiority of "paint made and mixed under a laboratory formula." He explained some of the different kinds of paint available, promising that there were some cheap paints for barns, but all of these required some thinning with linseed oil and presumably with turpentine.[56] But even at the end of the century, most paint was found in urban areas, and rural people coloured their worlds with other coverings and smaller surfaces.

Ready-Mixed Paint

In the 1850s a product appeared on the shelves of general stores that was previously unknown in the world of colour – ready-mixed paint.[57] Liquid paints became increasingly popular in the late nineteenth century, especially as tin cans were more readily available for storage.[58] One early user of this new technology was Mennonite farmer Christian B. Snyder of Waterloo County. In 1856 he noted in his diary that he had "haul[ed] brick from Waterloo [village] to repair our school house" on one day and travelled back "to Waterloo for nails, paint, and oil for school house" on the next.[59] This was one of his few hardware purchases of the year, and it suggests that paint and oil were still bought separately and that the paint was so important it usually only appeared on a community project or on high-traffic or small surfaces, such as the floors or wood trim on this brick schoolhouse. Still, in the United States the consumption of paint and linseed oil increased rapidly after mid-century. In the 1850s the value of oil output tripled, and over the period from 1850 to 1900 the value of oil output increased at a rate of 5.4 per cent per year, outpacing the rate of growth of gross domestic product (GDP) by more than 1 per cent (Table 4.1).

Table 4.1 US oilseed product values and GDP (current dollars)

	Linseed oil ('000)	Paint, varnish ('000)	Linoleum, oilcloth ('000)	Cottonseed oil ('000)	Oeleo-margarine ('000)	All oil products (incl. margarine) ('000)	GDP ('000,000)
1849	1,949						2,400
1859	5,982	3,500	700			13,400	4,380
1869	8,882	21,924					7,850
1879	15,394	29,112	5,815	7,691	6,893	64,904	9,360
1889	23,534	54,234	5,481	19,336	2,989	105,574	13,900
1899	27,184	69,562	11,402	58,727	12,500	179,375	19,500
1904	27,577	90,840	14,792	96,408	5,574	235,190	25,700
1909	36,739	124,889	23,339	147,868	8,148	340,983	32,200
1914	44,883	145,624	25,568	212,127	15,080	443,281	36,500
1919	120,638	340,347	68,110	581,245	79,816	1,190,155	78,300
1921	71,032	274,310	62,314	217,225	39,177	664,058	73,600

Sources: United States, Department of the Interior, *Seventh Census, 1850: Embracing a Statistical View of Each of the States and Territories, Arranged by Counties, Towns, etc. ...* (Washington, DC: Robert Armstrong 1853), lxxii; *Eighth Census, 1860*, Vol. 3, Manufactures (Washington, DC: GPO 1865); for outputs in 1850, 1860, and 1870, see United States, Department of the Interior, *Ninth Census, 1870*, Vol. 3 (Washington, DC: GPO 1872), Table 10 – Oil Vegetable, 618–19; *Tenth Census, 1880*, Vol. 2, Manufactures (Washington, DC: GPO 1883); *Eleventh Census, 1890*, Vol. 6, Manufactures, 3 Parts (Washington, DC: GPO 1895); *Twelfth Census, 1900*, Vol. 7, Manufactures: United States by Industries (Washington: United States Census Office 1902), Part 1, 323; *Thirteenth Census, 1910*, Vol. 8, Manufactures: General Report and Analysis (Washington, DC: GPO 1913), Table 7, 68; United States, Bureau of the Census, *Fourteenth Census, 1920*, Vol. 10, Manufactures, 1919, Reports for selected industries (Washington, DC: GPO 1921), 748–52.

In Canada, developments in paint followed a parallel track but lagged behind the United States somewhat. In 1868 Lyman, Elliot & Co. won prizes for the "[b]est paints, ground in oil, and putty," and the following year it advertised "Paint Ground in Oil" among other colours and driers.[60] Even so, in 1870 the *Canadian Pharmaceutical Journal* lamented "the extent of the chemical manufactures of Canada," claiming that "[a] more meagre display it would indeed be difficult to imagine. Of articles strictly chemical there were none, but under the head of 'preparations' and 'extras' there were a few entries, embracing samples of white lead, ground in oil; putty; lubricating oil; inks; and essential oils."[61]

It is still unclear when paint became widely available for application by ordinary consumers. Tiemann's claim to the invention of the first prepared paint in 1852 was actually a reference to a coloured paste that certainly required dilution in oil. The first colour sample sheets for ready-mixed paint were offered by the Averill Chemical Paint Company in 1866 and by F.W. Devoe and Company a couple of years later. The Averill company promised over a hundred shades and claimed that "these paints are prepared for immediate application, requiring no mixing." In 1876 it distributed sample cards that included two pictures showing a century of paint making: one was of a painter hunched over a dirty paint mill and the other, beside the first, showed a contemporary painter standing with brush and can next to a barrel labelled "ready for immediate use."[62] Averill struggled with making a consistent product, and other sources point to Sherwin-Williams as the first company to offer a truly standardized mixed paint in the 1880s.[63] Yet as late as 1908 "prepared paint" required mixing with oil, and various dilutions were expected for each coat.[64] The real advantage of mixed paint might have been standardized tints matched to reference cards that the master painter could present to clients. Shopping for paint could then be done at greater distances, and rural customers could request paint sample sheets by mail to sample the colours of Johnson's Pure Paints.[65]

Paint underwent two major changes in the late nineteenth century. The first was a gradual division in the intermediate preparation of paint between prepared pigments and ready-mixed paint. The second change was a broad divergence between the value of linseed oil and finished goods like paint and varnish. This spread increased in the early twentieth century, indicating the effect of the first change – that is, fewer people purchased linseed oil just to mix with prepared colours. In other words, paint manufacturing grew in complexity and in terms of value added. Still, preparing and applying even ready-mixed paint was a complicated task, and demand for the master painter grew rapidly. In Canada the value of the painters' and glaziers' outputs evolved almost in tandem with that of the paint industry.

Painters and glaziers contributed the final input in the flax-paint web. When North Americans wanted to have something painted in the mid-nineteenth century, they hired a painter, chose from a small selection of colours and varnish, and might have expected to see a finished product in a few days at the very soonest. The growth in mixed paint was relatively slow as a proportion of all paints. Old methods were quite functional, and the painters' and glaziers' trades were not suddenly threatened by the rise in consumer goods. Consumer-ready paints probably did not penetrate markets to any extent until the 1860s, although various forms of the technology had been available previously.[66] Only one Canadian painter specified the use or production of "mixed paint" in the industrial schedules of the 1871 census, although it was presumably used by more than one painter. Others used mixed paint without taking much note of the new material, as a house painter in Caledonia did who used "paint furnished by employer."[67] As druggists improved prepared colours, painters could spend less time grinding and more time painting. When ready-mixed paint production increased, the products still required certain modifications and skilled application, and the master painter painted even more in the early twentieth century.[68] Pittsburgh Plate Glass Company's large colour catalogue, with a complete line of consumer-ready mixed paint and varnishes, contained an equally elaborate selection (in dull trade language) of dozens of pastes, compounds, fillers, and "finely ground colors in oil." Most of these products required "manipulation by the Master Painter," and that usually meant combining their wares with even more linseed oil of various types and quantities.[69]

Painting was an urbanized craft in mid-nineteenth-century Upper Canada,[70] probably because growth in the trade followed growth in industry and construction, especially in cities. Through the census manuscripts, we can explore the geographic pattern more closely. In 1871 the industrial schedules of the Canadian census recorded 152 establishments that reported paint and painting as a product in the four confederated provinces. Nine establishments worked in specialty painting such as stained glass, tin, or sign painting. Sixteen were manufacturers of paint and varnish. The largest of these were almost entirely urban and included six in Quebec, four on the island of Montreal and two in the Quebec City area; two in Ontario (in Toronto and Brockville); and two in the Maritimes (in Saint John and Colchester). There was one large white lead mill in rural Nova Scotia (near Economy), and some smaller colour grinders operated near local mineral supplies (e.g., Sarnia and Woolwich Townships); these paints made their way into local and distant stores, painters' businesses, and carriage and furniture shops. Although Quebecers were evidently not the largest paint consumers, most of the value of paint was generated in Quebec and Montreal was paint's metropolis. Of the remaining firms involved in painting, 29 probably worked mainly on carriages and less than 100 firms worked on homes and other projects.[71] Painting and protecting furniture was a difficult process, and covering outdoor implements such as carriages and sleighs was an immensely complex procedure. In one painter's exhaustive description of nineteenth-century carriage finishing, "[T]here are comparatively few, even of good workmen, who can varnish a car properly."[72]

The first year the census differentiated between the paint and oil trades in 1871, separating them into four categories: painters, painters and glaziers, paint and varnish manufacturers, and linseed oil manufacturers. The divisions are imperfect, as it is likely that many painters and glaziers were involved in manufacturing product and some paint manufacturers painted. Linseed oil manufacturing sometimes occurred in paint and varnish works, and vice versa. The census categorizations for paint producers and users in 1871 do not permit a clear division to be made between paint and varnish manufacturers and painters and glaziers. The data for the Maritimes were very anomalous. In Nova Scotia, four painting establishments operated in or near Halifax; one firm in Truro specialized in painting carriages, sleighs, and signs; and two mills in Colchester had complementary products, raw materials from the northern shore of the Minas Basin, and a ready market in at least one large paint shop in Truro.

If we exclude the largest businesses, which evidently were manufacturers, we find that local consumption of painting services was higher in Ontario than in the rest of Canada combined. In fact, paint use might have been insignificant in all but the largest Canadian cities and a few towns in Southwestern Ontario. Fully half of the money spent on professional painting, varnishing, and glass work was connected to Montreal, Toronto, Ottawa, London, Saint John, Hamilton, Truro, St Catharines, and Halifax. The importance of urban and small-town Ontario went beyond the major centres. Of the top forty-seven census districts, which together represented 90 per cent of Canadian painter and paint shop outputs, only seven were outside Ontario.[73] Paint in the small districts was markedly different from the commodity in Ontario, and the patterns reflect regional demand. Painters in Quebec rarely lived in towns like Sherbrooke or Trois-Rivières, but were often found at rural addresses and sometimes in the network of villages servicing consumers in both rural areas and the larger centres.

Just as in the flax fibre side of the commodity web, Canadians used linseed oil and paint in a wide range of establishments that were not clearly identified in general census categorizations. The painters who specified work on carriages were mostly from rural areas in 1871, although a small share of the paint business in London, Toronto, and Montreal was held by similar shops. Of all the objects that rural and small-town people paid to protect and decorate, vehicles were far more important than houses. The carriage-making industry in Ontario was mostly a rural and small-town industry, with over 90 per cent appearing outside of Toronto, Hamilton,

Kingston, Ottawa, and London.[74] Prince Edward Island did not have linseed oil mills or paint manufacturers, but it did produce a large number of wooden ships in this period that presumably would have required even more paint than the carriages would. In 1871 PEI customs officials recorded £1,146 of linseed oil and £1,625 "packages" of paint, which likely meant lead ground in oil. Most of the island's paint and oil came from Britain; US and Canadian shipments were much smaller. It is not possible to determine where the paint was used, but a good bet would be mostly on ships, carriages, and a few Charlottetown buildings. Painting in this period was as much about manufacturing as it was about decorating.

By 1891 the rural-urban divide had become even more pronounced. Fully 88 per cent of the value of Canada's paint products appeared in ten cities and six (usually densely populated) districts. Only in Ottawa, Montreal, Winnipeg, and Toronto did the value of professionally applied paint and glasswork amount to $4 or more per capita, and these amounts include painters who worked for different industries in these places. One third of the Dominion's population lived in districts with no record of painters' and glaziers' activity at all. When painters and glaziers did appear in rural areas, their outputs were not large. Thus, 84 per cent of Canadians lived in districts where painters and glaziers produced less than 80 cents of goods and services (including glasswork) for each person locally.[75] The wholesale value of oil and paint materials imported for Prince Edward Islanders twenty years earlier amounted to only 12 cents per person; residential consumers likely received a much smaller share than that.

Rather than assume that 80 cents per year per person was too low to paint anything with, we should imagine how much one could paint with, say, $3.50 per family. As long as painting did not occur every year, that amount could indicate that some trim, or a room, or a floor was painted in rural homes. Of course, dollar value might have been entirely concentrated in a few lavishly painted homes in small towns, and the 1871 data seem to point to this, at least in Ontario. In Quebec where more painters lived in the countryside, 80 cents might have been more broadly distributed. If rural families hired painters for anything, it was more likely to paint objects and carriages than large walls. Painters and glaziers worked on more than houses and other buildings; they painted signs on wood and store windows as well as ships, carriages, furniture, and other objects.

It is highly unlikely that many people covered and decorated their wooden objects, large and small, with paint before the late nineteenth century.

Doing so would have required the expansion of refined linseed oil processing, white lead manufacturing, mixed paint factories, and the employment of people skilled enough to use and apply these goods. The objects and surfaces that were painted and varnished likely came from a manufacturer or from custom work by painters and glaziers. Paint remained an urban luxury, and while most painters and glaziers lived in rural counties in 1891, the output in these areas was small. Half of the wealth in the trade was generated in three Canadian cities by a small group of painting establishments.[76] Before moving to the commodity webs that created bright kitchens in 1860s Waterloo or 1920s Pittsburgh, we need to understand why this was so inconceivable in the centuries before. Just like the Perines' role in the fibre story, the linseed oil millers and their products are critical to understanding the entire flax-paint web.

Linseed Oil and Its "products too various to describe"

The linseed oil–processing industry grew in scale in the late nineteenth century and made obsolete the small mills that operated to supplement small-town mill complexes. Industrial reorganization served to eliminate small operators, but the linseed oil trusts could not completely eliminate competition. The new large and efficient businesses were either established in the East, where as we see in Chapter 6, they had access to urban markets and overseas supplies of flax seed, or in the West, where they followed the raw material and situated themselves in the frontier myth and the image of flax as a frontier crop. The most successful mills in both Canada and the United States remained in or close to lake ports where they could sell to the region's cities and still buy seed from Duluth, Fort William, or overseas markets that arrived through the Lachine or Erie Canal. Canada's transition away from the mill complex was much slower, and Baden's James and John Livingston bridged East and West by having direct relationships with Great Lakes and Manitoba farmers. They were able to remain competitive by using a combination of local, Prairie, and imported raw materials, and by operating both fibre and oil industries. Like other Upper Canadian mill complexes before them, the Livingstons dealt in lumber, twine, carpet yarn, thread, oil cloth, linen, linoleum, heavy machinery, stoves, animal feed, pre-popped popcorn, puffed cereal, soya bean oil, and soy flour.

The material processes required to produce a quart of linseed oil reveal why early millers required such a broad knowledge of the flax plant and its potential markets. Linseed oil is produced by pressing flax seeds to extract the oil from the seed's embryo and endosperm.[77] Early operations involved crushing and grinding the seed, placing the ground seed in bags, and pressing them with weights, wedges, or screw presses at room temperature. This produced a high-quality oil with no mucilage, but it left much of the oil in the leftover seed cake. A linseed crusher produced two main products, oil and cake (or meal) from a bushel of flax seed weighing 56 pounds. Eighteenth-century wedge presses could drain from 10 to 13 pounds of oil out of a bushel of seed, and the rest was linseed cake.[78] In the early period of hydraulic presses, a bushel yielded 18 pounds, or about 2.3 gallons of linseed oil, and more efficient methods of crushing in the mid-twentieth century increased the yield of linseed oil slightly.[79] To increase the percentage of oil extracted, crushers heated the seed before pressing so that the cell structures broke down and released more of the seed's oil.[80] Heated pressing extracted more oil but also released gummy mucilaginous residues known as "foots," which had to be removed by settling and filtering.[81] Mucilage is produced in the seed's epidermal cells.[82] The highest-quality oil was oil that had been properly filtered and had settled over a long period of storage before being sold. Once the crushed oil settled, it could be sold without any further treatments as raw linseed oil.

The quality of the linseed oil was tied to the characteristics of the seed. Flax seed of different varieties and different maturities gave various qualities to the linseed oil, and all linseed oil was affected by the presence of any foreign seeds or substances mixed in with the flax seed. Elliot & Company in Toronto reminded farmers and flax scutchers that the company valued "clean, screened Flaxseed, without adulteration or admixture."[83] The point of origin also affected seed quality, and flax seed was more variable by source than other oilseeds.[84] Certain regions produced crops with higher oil and iodine content, and storage facilities ranged from Canadian elevators, to Argentine dockside storage sheds, to below-ground storage pits in Calcutta.[85] In 1900 Canada added flax to its seed-grading Act, making only dry, sweet flax seed with less than 10 per cent moisture a top-graded seed. Second-grade seed could contain up to 20 per cent moisture, and anything damper would be classified "rejected."[86]

Linseed oil in its raw form had limited use as a covering. Although it hardens as it absorbs oxygen, raw linseed oil dries slowly. Thus it was usually modified for use in manufactured coverings, and it became the base for a number of boiled and refined (or special) oils. Producing the most common refined oil – basic boiled linseed oil – was relatively straightforward. It involved heating the oil to at least 107°C and adding small amounts of chemical driers, usually a compound of lead and manganese, which were catalysts for the oxygen-to-oil process. The refining required a tightly controlled environment, but with the right treatment, linseed oil's unique drying properties were increased in order to achieve a faster drying speed.[87]

Even a rather basic product like linseed oil had many varieties. Most *boiled oils* were prepared either in steam-jacketed tanks or in traditional kettles. *Treated oils* could be heavy or pale, which altered their viscosity and prepared them for suspending heavy or light-colour pigments. *Varnish oil* contained resins that had been added at extremely high temperatures and under close observation so that the oil did not break down as ordinary boiled oils would under such heat. The problem with boiling was that careless refining would cause colouring and opacity, a serious problem in a transparent product like varnish. Conversely, a *refined oil* was one that had been bleached to a light-yellow colour so that it would not tint white paints. Like raw oil, it dried more slowly than boiled oil, and it was often mixed with driers. Finally, the leather and linoleum industries relied heavily on *aged linseed oil*, which, after partial oxidization, dried a little more quickly than raw oil.[88] The technology required to produce special oils was established relatively early in the commodity's history. Elliot & Company, which produced paint, chemicals, and "products too various to describe" in 1871, was offering raw, refined, and pale boiled linseed oils by 1880. This drug company had the advantage of sharing the strong credit, technologies, and refining capacities of large processors in Montreal.[89]

In the mid-nineteenth century, when the Perines began investing in flax processing, North American linseed oil production was not large. The US census of 1850 reported that 168 mills were crushing linseed oil, and the average mill's output was not much over 10,000 gallons.[90] Thirty-two of these linseed oil manufacturers and 92 other oil makers operated in New York State; altogether they were fewer than the number of match makers. Pennsylvania, which had been a paint and linseed oil centre in the eighteenth century, listed fewer oil makers and did not specify any in the business of crushing linseed oil. In mid-nineteenth-century Canada, a few small operations, such as Elias Eby's in Bridgeport, Upper Canada, operated intermittently and usually only as small components of larger grist and sawmilling complexes.[91] Other early mills lent their names to the natural environment. "Oil Creek," Port Colborne, was not idly named, and it contributed marsh waters to a linseed oil mill before the Welland Canal displaced both marsh and mill.[92] There were only two other rural "oil mills" in Upper Canada at mid-century and a fourth in Toronto. In Lower Canada seven mills produced oil, all but one in Montreal. In 1861 there were only two mills in Upper Canada and six in Lower Canada. The numbers of mills continued to drop until only Quebec reported any oil production in 1881.[93] The census showed only a partial picture, missing some of the smaller mills that had crushed seed in certain years and then turned their attention to other outputs for prolonged periods. Moving beyond the censuses, we find an interconnected transnational world of production and knowledge. Eventually, all of the small mills gave way to a few large Canadian crushers.

Oil had been a small part of the Bridgeport mills since the 1820s when Jacob S. Shoemaker left his family's mills in Montgomery County, Pennsylvania, and immigrated to Canada. By 1835 his Bridgeport mills produced flour, oil, lumber, and fulled and carded wool.[94] At mid-century, Shoemaker owned the largest milling operation in Waterloo County, but financial problems compelled him to sell the complex and some land to Elias Eby in 1850 for about £8,000. Eby and Barnabas Devitt were listed as millers, lumber merchants, carders and fullers and oil manufacturers.[95] In 1851 the Census of Canada described Eby and Devitt as proprietors of an "Oil mill for Making Linseed Oil by Water Power … and Capable of making 40 Gallons of Oil per day." At this rate, Eby and Devitt could easily have produced oil equal to that of the average mill in the United States, but difficulties in procuring raw material meant that they only operated "about 30 to 40 days in the year."[96] Production expanded with the increase in local flax cultivation, and by 1864 the millers had a dedicated linseed oil foreman and produced about 6,000 gallons of oil. They dealt in flax seed and rape seed, and they manufactured raw linseed oil, boiled linseed oil, linseed cake, and rape seed oil.[97]

Just as the growth of a US flax fibre industry had ripple effects in Canada, the expansion of oil and drug production had parallels north of the border. Originally from the United States, the members of Lyman & Company established a major linseed oil operation in Montreal in the early 1850s.[98] It was one of only a few Canadian companies producing oil, paint, and pigments, and certainly the largest and most experienced. Throughout the 1860s they dominated the oil and chemical competitions at the provincial exhibition. In 1863 they won first prize for their oil cake and the majority of the "Chemical Manufactures and Preparations" prizes, including essential oils, linseed, rape and other expressed oils, and varnishes.[99]

In 1864, roughly ten years after the first major expansion of linseed oil and paint production in Canada, we get an early glimpse of several intersections of the commodity web under one roof, or at least in one city. In a highly integrated system, Lyman, Clare & Company loaded flax seed directly from canal boats and then cleaned, crushed, pressed, and refined it before grinding it into paints and other products and selling it in their retail and wholesale stores. The commodity moved through several storeys of several buildings, in a long process of bifurcation and combination with other ingredients, before it came to customers in tins of colour ground in oil, tins of linseed oil, and probably in oilcloth and other products. Manufacturing at this scale required elaborate machinery for loading, conveying, cleaning, elevating, crushing, heating, and pressing seed. The production of paint and especially linseed oil demanded much of the company's machinery, power, and storage space both in their large downtown retail store and its factory on the Lachine Canal. Seed was cleaned in at least three different places, and oil was stored in outdoor towers before being pumped to a separate building for refining. Storerooms, pipes, settling tanks, and cisterns were needed for the products, and refining the oil required a "series of boilers in a fire-proof building, where it is boiled and purified by temperatures ranging from 150 to 600 degrees. In this latter process the greatest caution is necessary."[100] Production was large by any standard, with four presses and between 50 and 9,000 gallons of oil per year in the 1860s, at least twice the probable output of the 1850s.[101]

Most of the company's raw material came from domestic supplies.[102] In 1863 Lyman, Clare & Company crushed some 38,000 bushels of Canadian flax seed, and they bought all the surplus seed from the Perine mills in Waterloo County.[103] Meanwhile, Lyman & Clare was also importing and producing paint, and in 1872 they dealt in imported "painters' colors, oils and dye stuffs."[104] At that point the Montreal company was the largest paint and oil manufacturer in Canada, and together with its drug company, it was one of the most heavily capitalized businesses in the city.[105] They crushed 40,000 bushels of seed in the Lachine Canal basin, or almost half of Quebec's crop. Their products included linseed oil, cake, and paint, as well as plaster, powder, and dye wood.[106] Family relations were critical to this business, and they also took on several other partners. In 1869 Lyman,

Elliot & Company of Toronto were wholesalers of "Paints Ground in Oil, White Lead, various qualities; White Zinc, Colors, and Patent Dryers."[107] When the company's senior partner, Benjamin Lyman, died in 1878, they controlled the largest wholesale drug company in Toronto and the largest linseed oil mill in Canada.[108] Their competitors, in the third quarter of the nineteenth century, were small and in decline, including Eby's mill complex in Bridgeport. During the proliferation of flax farming in the early 1860s, other oil mills – including Gooderham & Worts mills in Toronto, Woodstock, and Preston, another mill in London, and more in Quebec – grew up and then disappeared so quickly they were not recorded by any census.[109]

The year Benjamin Lyman died (1878) marked a new era in Canadian linseed oil manufacturing. James and John Livingston, two brothers from the flax fibre belt of southwestern Ontario, expanded into linseed oil production in the small town of Baden, and the wealth from this branch of the chemical sector began to concentrate in Ontario. In 1870, the Livingstons operated both flax mills and linseed oil works, and by 1878 James was able to build a lavishly decorated mansion in Baden.[110] In the years that followed, a flurry of expansion and growth (for which there are unfortunately few records) led to Livingston Linseed Oil (later called Dominion Linseed Oil) being the country's largest producer by 1891.[111] Between 1869 and John's death in 1895, the brothers' properties in Wilmot Township increased from $950 to $60,000 (Figure 4.2).[112] At the settlement of his estate the Livingstons' joint property was assessed at over half a million dollars – almost twice as much as the Lyman oil, paint, and drug business had been valued at in 1878 – and the business included fibre operations around Yale, Michigan, and in ten southwestern Ontario towns, and linseed oil mills and depots in Baden, Owen Sound, Toronto, Montreal, and St Boniface, Manitoba.[113]

In the flax story, the Livingstons are a rare example of a family business that was able to crush linseed oil profitably and on a large scale and maintain a long-standing flax fibre industry. They are also a case study of Canadian business expansion and vertical integration in the late nineteenth century. The business relationship of John Livingston and his younger brother James began in the mid-nineteenth century. In 1854 the brothers immigrated from Scotland with their widowed mother to Waterloo County, Upper Canada. James was sixteen, and he began working in the Perine brothers' flax scutching mills a couple of years after immigrating. By

Figure 4.2 Dominion Linseed Oil on left and Livingston foundry on right on the Baden mill pond, Baden, ON, 1920s

1861 he was a mill foreman, and he married Louisa Liersch that summer.[114] As the Perines began expanding their network of satellite mills, James and John built their first establishment near the family home in Wellesley.

After the Perines turned primarily to manufacturing cordage, the Livingstons bought the network of mills that had formerly belonged to the Perines and their partners. The growing company built a linseed oil mill in Baden shortly after Elias Eby sold his Bridgeport mill complex; its buyers apparently did not continue linseed oil production and may have sold parts of their equipment to the Livingstons.[115] Also, by 1891, when Livingston Linseed Oil became the country's leading supplier, there was only one other linseed oil mill in Canada (Winnipeg Linseed Oil), and the former oil giant Lyman & Company had focused its attention on drugs.

The system of flax factorship that the Perines had developed with local farms continued in the Livingston era, and James had learned it first hand by supervising field work under the employ of the Perines. One biographical sketch reported that James and John cultivated 3,000 acres of flax, or enough to produce about 825,000 pounds (413 tons) of scutched fibre.[116] In 1871 the Livingston mills in three neighbouring towns, Listowel, Baden, and Wellesley, required about 2,000 of the 13,000 tons of raw flax used in

Ontario scutching mills (Figure 2.2). The Livingstons were growing flax mostly for fibre during their rapid expansion in the 1870s, and since we know that millers continued to rent land, the Livingstons probably needed at least a couple of thousand acres of their own flax in cultivation.

The Livingston mills had competition in this flax belt, such as Charles Hendry in Maryborough, Robert Strang in Wellesley, and A.D. Shantz in Wilmot, but these mills were small. Hendry was probably a competitor in 1871, but within a few years he was a partner with the Livingstons. In 1872 the brothers had fibre mills in seven towns in Wellington, Perth, and Waterloo Counties, and they had likely completed their linseed oil mill in Baden.[117] In 1878 they added two fibre mills in Huron and Middlesex Counties.[118] That gave them about a third of the province's flax mills. For the remainder of the century it appears that their priority was to expand in areas like Michigan and in the linseed oil business.[119] After John's death the company's extensive Ontario fibre operations were sold in 1897, but an extension of the family fibre business continued in the region around Yale, Michigan.[120]

As James and John Livingston expanded their operations to capitalize on both the fibre and the oil webs, their business moved increasingly toward linseed oil and their catchment of seed suppliers became the open global market. Like the Perines before them, the Livingston brothers were initially dedicated to working in small-town mill complexes. These spaces allowed them to work with producers and take advantage of the rich intersections between the nodes of various commodity webs. However, the Livingstons soon recognized that success in the flax commodity web would require both rural and urban networks. Benjamin Lyman was now dead, and with Canada's largest linseed oil miller gone, the Montreal and Toronto markets were ready for a newcomer. James and John began expanding aggressively into the urban centres and drawing on the expanding production base of Manitoba and the Midwest for their seed supply. With their growing global network of suppliers and urban consumers, the Livingstons' network of small-town mill complexes in southwestern Ontario quickly began to lose its advantage. Caught between the rural and urban expressions of the flax web, the Livingstons had to act locally and think globally.

CHAPTER FIVE

Saving the Surface

*Flax in the Urban
Industrial Complex*

In December of 1877 three flax millers from Waterloo County, Ontario, boarded the train to Chicago, the Twin Cities, and eventually to Emerson, Manitoba, the end of the new line. From there they staged on to Winnipeg – ogling the progress of the Pembina Branch – in order to visit the recently established Mennonite settlement at the Manitoba East Reserve. One of the travellers was James Livingston, who as we have seen, was now an important figure in the linseed oil industry. He and his companions – Charles Hendry, a former competitor and partner with Perine Brothers, and a Mr Church, reported as associated with Gooderham & Worts – together represented the major players in the Ontario fibre industry.[1] They were there to assess the willingness of local Mennonites to grow flax, but why would they branch out this far for a reconnaissance tour? Although they came from Canada's other Mennonite enclave in Waterloo County, they were not Mennonite or even German. Like many of their contemporaries, they considered Manitoba and the West as potential sources for economic development in the fledgling Dominion. They were there to search for new sources of flax, and they believed the recently arrived Mennonite settlement would grow it as a cash crop and ship it to Waterloo.

In 1878 James Livingston returned from Manitoba to a slightly different world. His own business would begin to expand rapidly and focus increasingly on the seed. His raw materials would come from southwestern Ontario, from Manitoba, and increasingly from the US Midwest. The development of ready-mixed paint and the expansion of paint consumption would keep his mills running at full capacity for many years. And he would respond by building new mills in Toronto and by attempting to break into the Montreal market. With Canada's largest linseed oil miller, Benjamin Lyman, now gone, the Livingstons began to expand aggressively into the vacuum left behind. Manufacturing in both the flax fibre and oil webs would have been encouraged by John A. Macdonald's National Policy of protective tariffs and his government's continued interest in expanding into Manitoba and the Northwest.

An even larger sea change was occurring south of the border, where the centre of flax seed production and processing had moved to the Midwest and grew larger with each season. Processors integrated and promoted industry standards, but competition continued in different ways on both sides of the border. One response was to market this new global commodity as a "pure" and old-fashioned product and to hire chemists for internal research and development. By 1930 the linseed oil business, the location of flax farming, and the prairie landscape would be entirely different.

The Livingstons' linseed oil was sold mostly in Canada, although certain products were sold in northern Europe, Africa, and the West Indies.[2] In the early twentieth century, James sold linseed oil directly to merchants and farmers by the pint, quart, and gallon, but his main customers must have been painters and paint and varnish manufacturers.[3] This raises the question of why the country's largest linseed oil mill remained in a very small town, in a county where the average person had little use for paint. Waterloo South only reported two painters in 1891, and the neighbouring census districts Waterloo North and Oxford North reported no painters at all. Thus Livingston's linseed oil was largely shipped out of the county and across the country, and at periods of peak production he likely had more linseed oil in his storage tower than painters in the surrounding towns and countryside applied in years.[4] Labour supply was seldom a critical factor for a linseed oil mill's location, but a local market for the products of linseed oil was and Baden residents did not buy much paint. If proximity to markets was so important, it is curious that Livingston kept his mill in Baden. It would have been difficult for Livingston to get into the linseed oil business, especially from remote Baden, when Lyman & Clare had such an integrated system on the cheapest possible shipping line and next to the largest market in Canada. What was it about the integrated system that failed to keep out competition?

The answer is found in part in the nature of flax and the flax seed business, in the Livingston brothers' experience in the fibre industry, and in their own integration and entrepreneurialism. Industrial growth and an increase in demand more generally were crucial to their success, but these do not readily explain the geography. The Livingstons remained competitive by running a closely monitored system that used a local supply of raw materials and by having an intimate knowledge of the flax plant. Flax was a problematic crop, but in Waterloo County, millers found that it lived up to its name as a most "useful thread." In a time when many capitalists focused on processing a diverse range of goods in an urban context, the Livingstons focused on place and on what they could get from producing this one organism. James took advantage of knowledge learned from the Perines, and he echoed their system of a central operation with satellite scutching mills in small towns co-owned with trusted local businessmen. Just like the Perine brothers, employing family members was both a strength and a vulnerability. They provided the trust necessary for management in a diaspora of mills, and the geography of production helped keep brothers and cousins at a comfortable distance from one another.

Building a Canadian Flax Empire

James Livingston's fortune was made not entirely from flax; his worldly connections certainly helped him expand his business to an international scale.[5] He and John operated a small foundry and repair shop in Baden for their mill and local customers, and according to local legend, they even built five automobiles after hearing Henry Ford speak to potential investors in Berlin (Kitchener).[6] He entered politics in the late 1870s as reeve of Wilmot Township and member of the Waterloo County Council, and in 1879 he was elected member of the Ontario Legislature for Waterloo South (Figure 5.1).[7] His tenure in the provincial legislature was brief. Then, from 1882 to 1900 he served as the area's Liberal member of Parliament.[8] James Livingston's interest in finance was perhaps more lucrative than his interest in politics. He was president of four small banks in the United States, including one in Yale. In Yale he supplemented his flax and financial interests with a board membership in the Yale Lumber and Coal Company and the Yale Woolen Mills, and he was director of the Waterloo Mutual Fire Insurance in Waterloo.[9]

Figure 5.1 James Livingston, parliamentary photo (William James Topley Studio, 1883)

The Livingstons' business benefited from their many nodal connections across flax's commodity web. They were also shaped and influenced by other actors. They worked in the Perine fibre mills, learned about linseed oil from local processors, and tracked and emulated the spread of paint-production facilities in Montreal, Buffalo, and Toronto. James was a prominent figure in Waterloo County, but his business was also part of a deeply interconnected world. It included rural and urban people, Mennonites and non-Mennonites, Canadians and Americans, suppliers from Argentina to the Great Plains, and customers from the Maritimes to Manitoba. He has been called the "Flax King of Canada" in local histories, but to contemporaries his mills were simply another feature of small-town landscapes in southwestern Ontario. Even the process of millers renting land or hiring large groups of family and First Nations labourers seemed ordinary.

Ontario offered consistent amounts of flax fibre for the Livingston and other mills in the decades between Confederation and the First World War, but one element of the linseed oil industry in short supply was flax seed. Although some flax farmers specialized in seed production as early as 1860, the province's farmers produced meagre amounts and tended to harvest their flax before the seed formation deteriorated the fibre. The Bridgeport linseed oil mill often sat idle because of this shortage and because the Lyman mills in Montreal and Toronto were able to buy a large part of the Canadian crop.[10] Despite the prolific growth of flax farming in southwestern Ontario, there were fewer local linseed oil mills in the late 1860s and 1870s. When James Livingston added linseed oil to his mill complex in Baden, he was a latecomer compared to oil millers in Quebec and the northern US cities.

The Canadian census first reported flax seed in 1871, and at that point the counties near Lyman's Montreal mill were the main producers (see Figure 1.3). The mills in these counties were not large fibre producers. Flax could be grown mainly for fibre or mainly for seed. Whereas flax fibre production tended to appear around pre-existing mills, the decline or relocation of a linseed oil company was sometimes hastened by local flax seed shortages. The centre of Canadian production had shifted from Quebec to Ontario by 1891, when Livingston's linseed oil mill was the nation's largest. It quickly moved again. By the turn of the century it was centred in Manitoba, by 1911 Saskatchewan, and by 1916 the western half of Saskatchewan. The cause of this migration became a critical question for Livingston and other linseed oil millers, and it is one explored in the remaining chapters of this book.

In the 1890s the Livingstons crushed up to 500,000 bushels of flax seed each year and supplemented Canadian and US supplies by importing around 35,000 bushels per year from Argentina and some from Calcutta.[11] The Livingstons bought flax seed from Duluth merchants in most years, as Canadian supplies were too low, but they continued to draw from an extensive network of Ontario mills where scutchers and threshers had removed the seed from the fibre.[12] The amount they bought in 1897 was close to the provincial production levels of 1901 and suggests that the Livingstons were crushing most of Ontario's flax seed (Figure 5.2). However, the Ontario crop represented less than 10 per cent of the company's annual seed requirements.[13]

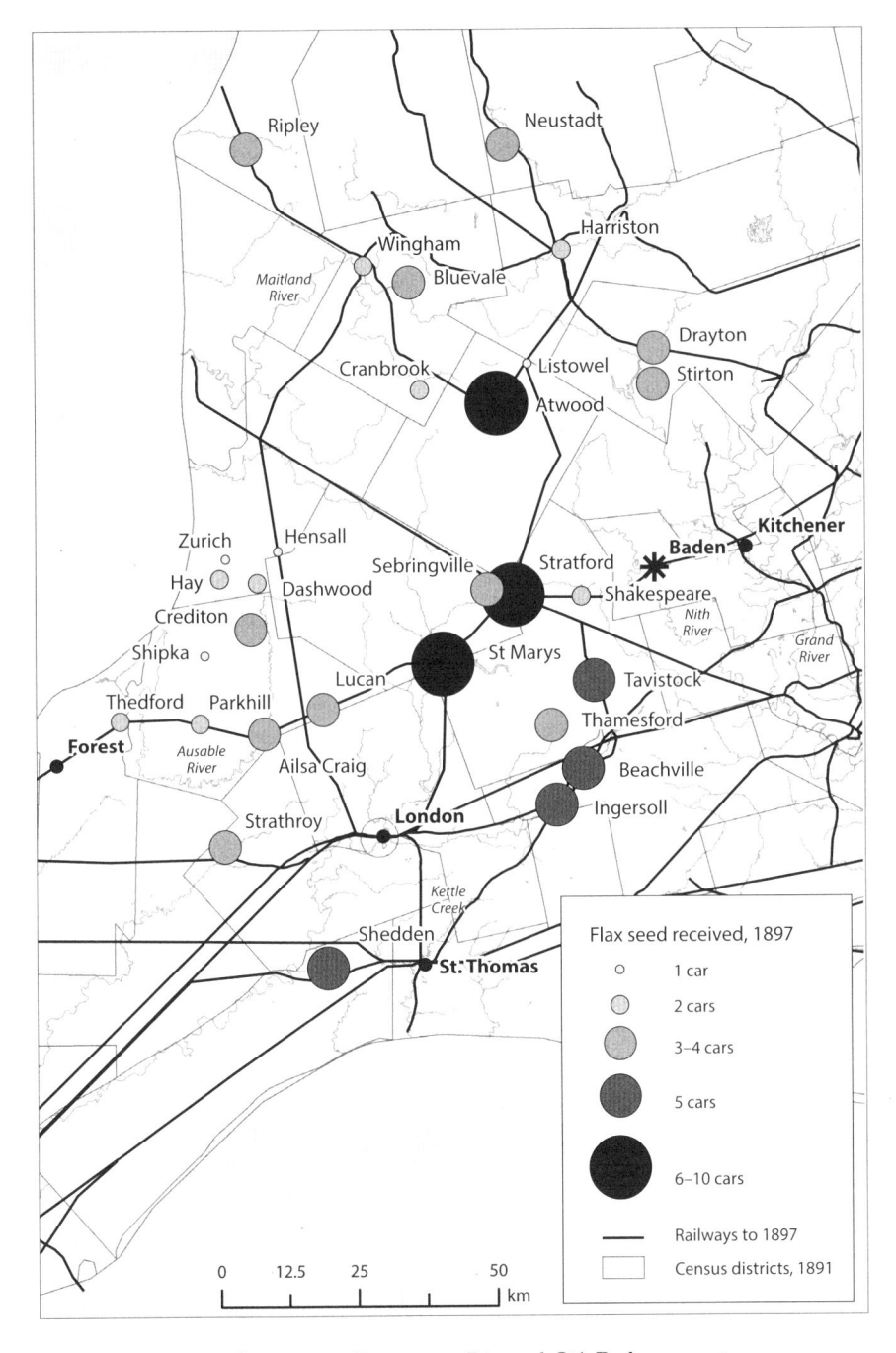

Figure 5.2 Flaxseed receipts at Livingston Linseed Oil, Baden, ON, 1897
Source: TWA, James Livingston, "Letter Book," 1897.

Figure 5.3   Carl E. Johnson, *Linseed Mill*, 1934

At the turn of the century, Livingston Linseed Oil continued to expand its oil production. "The mill at Baden employs about forty hands," a student from the Ontario Agricultural College reported, "and runs night and day. There is a seed storage capacity there of 200,000 bushels."[14] At its peak the mill probably employed a dozen office staff and operated a dozen linseed presses continuously, with up to three workers per press. The Canadian linseed oil mill labour force was so small that few records survive to describe the lives of these workers. We know that most Canadian presses were hydraulically operated, so the oppressive mill noise from the days of wedge presses and seed pulverizers was no longer an issue. The few images that exist of linseed oil mill workers indicate that oil work was characterized by high temperatures and slippery surfaces. In addition to the heat generated by the presses and their engines, crushed flax seed was usually heated before pressing and the oil was always boiled at least once in the refining process. A rare glimpse of the workers is provided in *Linseed Mill*,

Figure 5.4   Alberta Linseed Oil Mills fire (October 1914), Medicine Hat, AB

an etching by Carl E. Johnson, which shows two labourers, shoeless and bare-chested, operating the highly technical controls of linseed-crushing machinery, likely in Minneapolis (Figure 5.3).[15] Workers presumably went barefoot because of the heat and the oily floors. Being covered in raw linseed oil would have had no adverse health effects for workers, but the main workplace hazard was the constant threat of fires and explosions.

Linseed oil and flax fibre mills were among the worst industrial fire hazards in North American mill towns. Many mills went unremembered and unnoticed by local media except on the occasion of seasonal shutdowns and fires. The main threat to linseed oil operations was spontaneous combustion caused by oxidized oil on rags or oil-soaked clothes that had been exposed to the air.[16] In 1896 the W.P. Orr Linseed Oil mill in Piqua, Ohio, partially exploded when one of its boilers caught fire. The blast caused the largest fire in the town's history, destroying one end of the mill, a nearby train bridge, and a flour mill. The oil mill was never rebuilt.[17] In Medicine Hat, the Alberta Linseed Oil mill burned shortly after it opened in 1913, drawing many neighbours and at least one photographer to witness the conflagration (Figure 5.4). The Livingston linseed oil mill in Baden also burned at least once. As we will see later, fire and fire insurance were a major concern for linseed oil millers and one of their justifications for forming trade associations.[18] Livingston was never part of a trust, so he protected his properties in other ways. For instance, it is probably not coincidental that he joined the board of the Waterloo Mutual Insurance Company. We know that the Baden property was well insured, and under a later owner, the larger complex of buildings was even inventoried and described in detail by the Canadian Underwriters Association's "Special Risk Department."[19]

In other sections of its mill complex, the company boiled and refined oil, washed, painted, and labelled barrels and drums, tested and produced by-products, including meal and soap, and stored and distributed the products. A foundry and repair shop allowed the company to maintain many of its own parts.[20] Linseed meal was important enough to the company's sales that they performed feeding tests on livestock and poultry in three barns on the mill property. Keeping livestock was also a way to market linseed meal to local farmers, and at least one other Canadian linseed oil mill kept milch cows, hogs, and poultry in a sort of urban "demonstration farm."[21]

The Canadian Urban Industrial Complex

From this rural headquarters with its pungent scent of chickens and lye, James Livingston attempted to break into the hardware industry in Montreal. The disappearance of Lyman's paint and linseed oil operations had left a number of companies scrambling to define a niche in that city in the late 1890s. That was when Walter Cottingham, a local paint manufacturer, merged with Cleveland's Sherwin Williams and became an executive in and eventually president of that company.[22] Three paint wholesalers also formed the Canada Paint Company in this decade to manufacture white lead and colours "for domestic, carriage, locomotive, and paper staining purposes."[23] Like many paint companies in this period, they were "oil refiners and importers"; still, there had been no crushers in Montreal since Lyman dropped their linseed oil operations.

By 1890, eight of the city's ten paint firms were clustered in the north "Mile End" of the city, partly displacing the earlier concentrations around Lyman Clare and Company and others in Griffintown on the Lachine Canal. Brandram, another firm on the canal, merged with Henderson and Potts in 1906 and constructed the Brandram-Henderson lead and paint works in Mile End the following year.[24] However, the Griffintown and western suburbs continued to attract oil and paint companies, likely because the western Great Lake trade offered an increasingly important source of raw material. Sherwin Williams built a new factory on the canal in Cote Saint-Paul. Livingston continued to crush linseed oil on the canal, and companies like Canada Paint and R.C. Jamieson & Company (specializing in paints and varnishes) remained there as well.[25]

As we will see, the linseed oil industry formed one of the major US trusts; however, in Canada, integration vertically usually translated into intense competition horizontally. In 1915 Montreal was home to a dozen paint manufacturers, and by the late 1920s, 40 per cent of Canada's paint output was produced in Montreal, making it, in Robert Lewis's words, "Canada's paint centre, in part due to the significant vertical integration within the industry."[26] Suburban plants like the US branch plant of Martin-Senour in Outremont and its British neighbour Brandram-Henderson in Mile End experienced the most rapid growth, the latter opening plants in Halifax, Saint John, Toronto, Hamilton, Winnipeg, Medicine Hat, and Vancouver.[27]

In 1901 Livingston built Canada Linseed Oil Mills in Montreal, operated by his brother-in-law E. Liersch, which could crush 100 barrels per day.[28] By 1907 he was also operating a branch depot of Dominion Linseed Oil in Montreal, under his son H.P. Livingston.[29] However, the paint manufacturers were also entering the crushing business, and the following winter, the young manager wrote that a suspected competitor "Canada Paint are now offering [linseed meal] so they must be crushing."[30] Even without this interference linseed oil was available from other sources. A few days prior to learning about Canada Paint, H.P. Livingston wrote that he had lost a major customer's order because his price of 53 cents per gallon of oil was too high. "I see no other policy but to drop back to 52 ½ [cents] to get business … our competitors simply take orders away from us."[31]

One rapidly growing source of linseed oil was Winnipeg, and even before Canada Paint installed their own crushing mills in Montreal, they had close connections with a linseed oil manufacturing business in the prairie metropolis. No one knew better than James Livingston that the future of Canadian flax seed was in the West, but his involvement there had been as a seed buyer. Other paint and oil manufacturers saw the West as source for oil: Winnipeg Linseed Oil appeared in the 1880s; Manitoba Linseed Oil was a newcomer in the late nineteenth century; and Alberta Linseed Oil began in 1913. Winnipeg Linseed Oil produced $38,000 of linseed oil in 1885, but they complained that Canadian freight policies made it unprofitable to ship their product to cities further west.[32] The mill tripled the value of its production by 1891, but still it produced less than half of Livingston Linseed Oil's output. Nevertheless, rumours of a second mill in Manitoba abounded, and the West gained importance both as a producer and processor of flax seed.[33]

Winnipeg Linseed Oil was originally owned by Body & Noakes of Winnipeg, but by 1910 it was the property of Canada Paint and had moved its headquarters to Montreal.[34] That paint company's backward integration into linseed crushing was typical of Livingston's new competitors. The paint industries in Winnipeg shared managers with the newer Manitoba Linseed Oil Mills. The oil mill's vice-president, Melbourne F. Christie, was also secretary-treasurer (and a family member) of G.F. Stephens & Company, paint manufacturers. R.W. Paterson, managing director of the Winnipeg Paint and Glass Company from 1903 to 1914, was also president of Manitoba Linseed Oil Mills in 1911. His wife's uncle, F.W. Drewry, spent five years in malting and brewing in St Paul and was also a manager of the Manitoba Linseed Oil Mills by 1911.[35] These partnerships reveal the importance of family capital and of the industrial knowledge learned in mid- and northwest US cities.

Few Prairie towns were as remote from urban paint consumers as Medicine Hat, Alberta, yet if boosters, homestead entries, and the two new provinces' Departments of Agriculture were to be believed, this town was already becoming the metropole for flax seed producers when Alberta Linseed Oil Company formed there in 1913. Unfortunately, with the real-estate crash of that year, the plunge in Canadian flax seed production during the war, and the localized droughts of the 1920s, this company did not capitalize on raw materials as it had hoped. The sign posted to the sides of the elevator in Figure 5.5 probably says something similar to the company's poster in Figure 5.6. It seemed at first that it would pay to promote and process flax seed in southeast Alberta, but by the 1930s, the company had shifted to an industry with guaranteed local demand – irrigation.[36]

Canadian paint manufacturers and lead manufacturers were usually one and the same. Both were linseed oil consumers. Lead makers typically sold lead ground in oil, which according to the *Commercial* (a Winnipeg publication) was a paint product with "at least 8 percent linseed oil in each tin of white lead."[37] In the 1890s and the early twentieth century, the paint and linseed oil industries operated in a context of fierce competition. This was a period of association and trust building, but the paint industry, like most associations, was difficult to organize.[38] Although paint manufacturers had

Figure 5.5   Alberta Linseed Oil Company, Medicine Hat, AB

associations, the latter could not control competition and fell apart. According to the *Commercial*, when the "White Lead Association gave up the ghost," it only took about three weeks before the price of lead in every Montreal plant dropped substantially. The Paint Grinders' Association fell almost immediately thereafter, and "business in the paint and oil trade is in a demoralized state. There is no reliable range of values to go by … [and] buyers themselves hardly know what to do."[39] Horizontal integration had been tried and abandoned in Canada. In the United States, manufacturers successfully used trade associations and trusts to share trade knowledge and market information and to fix prices.[40]

The North American oil manufacturers in this story fall roughly into two categories: the older eastern businesses and the start-ups in the West. Western companies such as Archer-Daniels Linseed Oil, Midland Linseed Oil, Minnesota Linseed Oil, and Red Wing Linseed Oil appeared especially in Minneapolis to take advantage of seed supplies and reduced freight costs (Figure 5.7). The more established eastern millers included the companies making up the American Linseed Trust and their major competitor, Spencer Kellogg and Sons, in Buffalo. Lyman & Company

CANADA

NEEDS MORE FLAX

Southern and Central Alberta and Saskatchewan can grow the finest quality Flax. When sown under right conditions Flax is the best cash crop you can grow.

We offer a ready market right at your door for large quantities.

SEED A CROP THIS YEAR THAT WILL BRING YOU BIG RETURNS

Canada has in the last two years imported over 2,500,000 bushels of flax at a cost of $4,000,000.00. This is your opportunity to grow a crop that is not dependent on the export market.

For best results, Flax should be sown on good clean summerfallow or new breaking at the same time as other grains.

Recleaned flax suitable for seed can be secured at very reasonable prices at our elevator.

WRITE TODAY FOR QUOTATIONS AND INFORMATION TO

The Alberta Linseed Oil Co. Limited
Manufacturers of Raw, Boiled and Refined Oils, Oil Cake and Oil Cake Meal

PHONE 2206   MEDICINE HAT, ALBERTA,

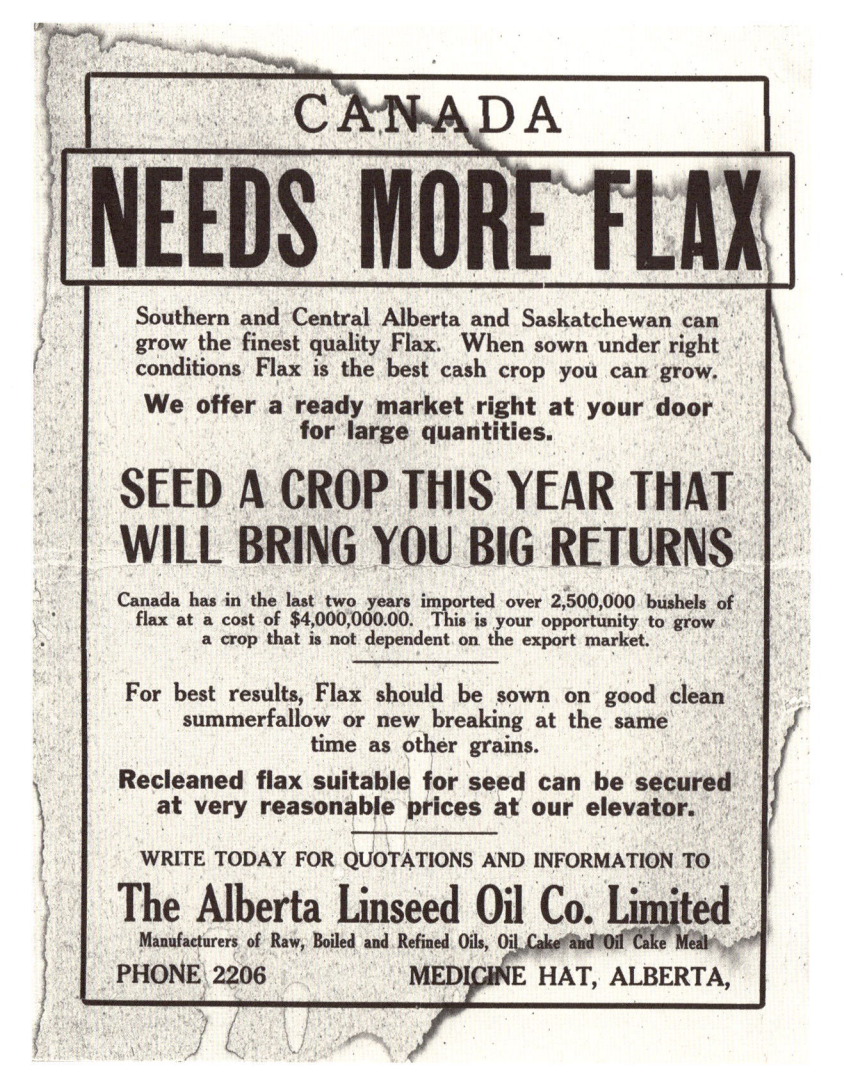

Figure 5.6   "Canada needs more flax"

was certainly in the latter category, but in some ways Livingston Linseed Oil bridged the two. When it came to raw materials, western millers hoped for a steady and protected domestic supply of flax seed, but others were content to buy seed from any market. James and John were initially dedicated to making the Canadian West a main source of their seed, but by the early twentieth century they imported large amounts from overseas, especially Argentina. At this point, Livingston used his political experience and

Figure 5.7 Minnesota Linseed Oil and Paint Company, Minneapolis, 1915

actively pressed for trade favours, as in 1907 when he petitioned Ottawa for drawbacks on exported linseed meal to offset the costs of importing seed.[41]

The US Urban Industrial Complex

The US linseed oil story developed along similar lines, but its industry grew faster and saw an even greater concentration in a few main businesses. In the mid-nineteenth century, when the American crop amounted to half a million bushels of seed, most seed was grown in Ohio, Kentucky, and New York. We saw in Figure 1.3 how important Ohio was as both an oil producer and a seed supplier in 1860. Ten years later, the crop was over seven million bushels, and two-thirds was grown in Illinois, Indiana, and Iowa. One decade later, the largest producer was Minnesota, and by the turn of the century, it was North Dakota. The westward movement of settlement and flax production left the linseed oil mills in Ohio and New York without a nearby supply of flax seed. Demand for linseed oil in Midwestern cities prompted many eastern crushers to expand operations there, especially in Chicago. In the process, the consolidation of linseed oil crushing began. During the 1850s and 1860s, over half the 168 mills present at mid-century disappeared. Situating linseed oil production in the frontier myth was how the industry made sense of itself. Some of the well-known crushers argued that their Midwestern businesses closed down and "followed the raising of flax" from Ohio to Indiana, Illinois, Missouri, and Kansas and on to Iowa, Minnesota, and the Dakotas.[42] William Archer, for example, was involved in one of the first linseed oil mills in the Miami River valley, Ohio, and in the 1880s, leaving his son George in Ohio, he went to operate mills with another son in St Paul and Yankton, Dakota Territory.[43]

The cost of flax seed represented between 75 and 85 per cent of the value of oil and cake – one of the highest ratios of material cost to output value of all industries in the United States.[44] After the seed itself, the freight on shipping seed, oil, and cake was a crusher's largest expense. Freight on seed from the West cost anywhere from 2.5 to 4.5 cents per bushel at docks on the Great Lakes, and it cost an additional 4.0 to 5.5 cents per bushel to have it hauled to seaboard cities by rail.[45] Shipping seed from the West was far too expensive for eastern crushers on the seaboard or away from lake ports.[46] Offsetting that in locational calculations was the cost of transporting oil and cake to consumers. Oil was usually sold in 50-gallon barrels.[47] Barrels and barrelling cost about twice as much as the cost of the labour and hydropower required to produce the oil. For crushers who shipped most of their product great distances (such as those in the Northwest), tank cars were used instead. Up to 21 per cent of the cost of freight could be saved this way, but cars required tank stations and barrelling operations in destination cities.[48] Oil was shipped to a variety of consumers, and as in Canada, most were urban.

The structure of linseed oil crushing changed dramatically in the late nineteenth century, and by 1910 production was concentrated in the Northwest, the east end of the Great Lakes, and the Eastern Seaboard. The millers that relocated because of this restructuring blamed the westward movement in flax production, but that was only one factor along with growing economies of scale and improvements in processing and refining technologies that enabled new establishments in any location to earn a profit at a lower price than older and smaller ones. It is true that many small mills in Ohio closed down in the late nineteenth century and that Minneapolis became a major centre after 1900. Even Chicago, which had the most mills of any city in 1900, lost its share to Minneapolis, which by 1907 was crushing over a quarter of the nation's linseed oil.[49]

However, as Table 5.1 shows, the number of mills declined in all states outside of Minnesota; New York State was always a leader; and none of

Table 5.1 Number and percentage of national output of linseed oil mills by state, 1880–1920

State	1880 No.	1880 Output	1890 No.	1890 Output	1900 No.	1900 Output	1910 No.	1910 Output	1920 No.	1920 Output
CA	1	3.0%	2	2.0%	1	0.8%				
IA	7	5.0%	7	9.0%	4	5.9%	2		2	
IL	12	20.1%	8	12.3%	9	25.5%	5	8.2%	2	
IN	8	8.3%	3	2.7%	2	1.7%				
KY	1	0.5%								
KS	3	0.7%	2	2.0%	1	0.8%	1		1	
MN	2	2.3%	3	6.6%	5	12.0%	6	30.0%	6	27.6%
MO			5	7.6%	2	1.7%	2			
NE	1	0.8%	1	1.0%	1	0.8%				
NJ	1	0.1%	1	1.0%	1	0.8%			2	12.4%
NY	9	36.9%	7	31.9%	9	31.0%	6	41.9%	6	29.5%
OH	21	15.5%	14	15.2%	7	11.5%	4	9.9%	3	7.8%
OR	2	0.3%			1	0.8%	1		1	
PA	9	4.5%	5	5.2%	4	5.7%	1			
SD			1	1.0%						
VA	1	0.0%								
WA									1	
WI	3	2.1%	3	2.7%	1	0.8%	1		2	
Total	81	100.0%	62	100.0%	48	100.0%	29	90.0%	26	77.3%

Note: Although total national output was given each census year, it was only broken down by state for the largest producers. Hence, figures in italics represent the average output multiplied by the number of mills in the state, when known. In 1910 and 1920, output was specified for only three states, and estimates are only given here for Illinois and New Jersey, the next largest producers in each year.

Sources: United States, Department of the Interior, *Seventh Census, 1850: Embracing a Statistical View of Each of the States and Territories, Arranged by Counties, Towns, etc. ...* (Washington: Robert Armstrong 1853), lxxii; *Eighth Census, 1860*, Vol. 3, Manufactures (Washington, DC: GPO 1865); for outputs in 1850, 1860, and 1870, see United States, Department of the Interior, *Ninth Census, 1870*, Vol. 3 (Washington, DC: GPO 1872), Table 10 – Oil Vegetable, 618–19; *Tenth Census, 1880*, Vol. 2, Manufactures (Washington, DC: GPO 1883); *Eleventh Census, 1890*, Vol. 6, Manufactures, 3 Parts (Washington, DC: GPO 1895); *Twelfth Census, 1900*, Vol. 7, Manufactures: United States by Industries (Washington: United States Census Office 1902), Part 1, 323; *Thirteenth Census, 1910*, Vol. 8, Manufactures: General Report and Analysis (Washington, DC: GPO 1913), Table 7, 68; United States, Bureau of the Census, *Fourteenth Census, 1920*, Vol. 10, Manufactures, 1919, Reports for selected industries (Washington, DC: GPO 1921), 748–52.

the main linseed oil–producing states lost their share of the country's total output between 1880 and 1920. Buffalo became a central city for linseed crushing despite its remoteness from the Northwest. Five of New York State's nine mills were located there in 1900, making linseed oil the fourth-largest industry by value of output in the state's second-largest industrial centre.[50] The closure of so many small mills in the Midwest reflected the industry's new structure and the benefits of scale. Between 1890 and 1910, the number of mills dropped by half, the capital invested per mill almost tripled, and outputs increased from $24 million to $37 million. The industry changed radically in scale and became an integrated enterprise between the mid-nineteenth century and the end of the First World War, but it did not move west in the same way as flax production.

Another element of the restructuring was horizontal integration. In January 1887 a group of smaller crushers merged to form a trust, which they named the National Linseed Oil Company. It consolidated over forty-nine mills, forty elevators, and a fleet of tank cars and tank stations.[51] One member of the trust claimed in 1889 that it had created "an absolute monopoly of the linseed oil trade of the country."[52] Naturally, he claimed that this was a necessary response to cutthroat competition and inefficiency, and his testimony helped situate linseed crushers in the larger trust-building movement of the late nineteenth century. The reasons he gave for the combination also reveal some of the industry's idiosyncrasies and its suitability for trade cooperation. A major patent-holder company had inflated the price of linseed oil–crushing equipment, and the trust was meant to force them to "come down from their high horse, and ... ask a reasonable price for their machines."[53] Once established, National Linseed formed a committee to assess new technologies and if it found a technology useful, it would be purchased for all members of the trust.

Cooperation such as this meant that crushers could buy seed directly from the growers at uniform prices and stop speculators from cornering the market. The trust enabled crushers with mortgaged mills to pay a uniform interest rate of 4 or 5 per cent, instead of the 8 to 10 per cent that some of the newer mills were paying in the West. Cooperation also helped crushers avoid excessive fire insurance fees by using mutual insurance and enforcing the most appropriate precautions in the trust's mills. Insurance was especially important for linseed oil manufacturers and consumers because drying oils are prone to spontaneous combustion.[54] In other words, the members of National Linseed sought to put their managers in a better position to influence the cost of technology, loans, and insurance in addition to the cost of seed and the finished oil.

The 1887 merger settled some "old feuds," and members thereafter claimed to be "managing all the works in the trust as if it were all a single property, controlled by different managers." Several branch offices were established, and anywhere from six to twenty-one small mills – which the trust claimed were inefficient and moribund anyway – were closed. The price of oil did increase, from 38 to 52 cents per gallon in 1887 and 1888, which the crushers claimed was corrective and a fair price.[55] Yet in the 1890s, National Linseed's share of the industry declined. In fact, it was the least successful of the major trusts in the late nineteenth century because the linseed oil trust had fewer products to sell, smaller overseas markets, and fewer sources of raw materials than the petroleum and cotton oil trusts.[56] Several of its members left to take advantage of the higher prices with new, more efficient mills; these included George Archer and his friend John W. Daniels, who had been directors.[57] Buffalo's Spencer Kellogg and Sons, the oldest family name in linseed oil, had sold its Buffalo mill to National Linseed in 1889 and built a new mill there with thirty-six presses in 1894. Daniels left in 1901 to help Cleveland's Sherwin-Williams become a large competitor in linseed oil. He then started his own oil mill in Minneapolis in 1902, and Archer joined him three years later to form Archer-Daniels Linseed Company.[58]

In 1898 the trust was reorganized as the American Linseed Company.[59] Most of the new company's mills were in Iowa, Illinois, and Ohio, reflecting a commitment to the Midwest and the Miami River valley.[60] This new merger gave American Linseed between two-thirds and 85 per cent of the country's oil production, and Frederick T. Gates, Jay Gould, and John D. Rockefeller Jr joined its board in 1901.[61] But it did not include National Lead, whose mills on the Eastern Seaboard crushed up to 15 per cent of the country's oil in the 1890s.[62] Sherwin-Williams in Cleveland and Archer-Daniels and other crushers in the twin cities were important start-ups in the early twentieth century. And in 1909, Spencer Kellogg, which now had mills in Minneapolis and on the Hudson River in Edgewater, built the world's largest oil mill in Buffalo, with 186 presses.[63] American Linseed owned up to 60 mills, but many were small and distant from lake ports and the new frontier of flax farming. The trust's small mills were used intermittently and often not at all, and one study estimated that it operated only 360 presses continuously whereas Spencer Kellogg's giant mill ran almost all of its presses without rest.[64]

In the late nineteenth century, domestic flax seed amply supplied all of the US linseed oil industry's output, and cultivation was encouraged by the demand for linseed oil during the population and construction booms in the middle states. By 1910, according to Table 6.1, imports suddenly accounted for up to half of the seed consumed in the United States. This was to the benefit of Eastern Seaboard mills in the large population centres.

They could access seed relatively easily through Atlantic shipping routes, and their finished product was immediately marketable in the country's largest cities. The United States' interwar protective policies, discussed in Chapter 7, only briefly increased domestic seed production's share of total consumption. American farmers simply found that flax seed was too problematic and that other crops paid a better rent for the land.

The linseed oil industry seldom operated under a monopoly, but its horizontal integration and centralized administration reflected the importance of growing economies of scale and the liquid capital needed for acquiring raw material in ever-larger quantities. In the normal process of US integrated enterprise, linseed oil crushers integrated horizontally with each other and sometimes vertically with paint manufacturers, and paint and lead manufacturers integrated backward into linseed oil. The linseed oil crushers did not simply transplant the small-mill structure of the 1870s to the Northwest – the scale of crushing had changed too much for that – but they did abandon or consolidate small mills and set up highly efficient operations in locations that minimized the transportation costs to flax seed supplies, urban oil markets, and the overseas market for linseed cake.

Flax Seed Enters the Food Chain

Besides oil, crushers produced linseed meal as a by-product. Linseed meal was the cake of seed left after the oil had been extracted in the press, and like many oilseed meals it provided a protein-rich feed supplement for livestock. Crushers could package and sell it directly as cake, or they could grind it to make oil meal. It was always a by-product to North American crushers, yet European sales were important enough that they tailored their products to the exact shape, packaging, and general standard of quality demanded by their customers overseas.[65] Northern European countries were major linseed meal consumers, using it as a fertilizer, feed, and sometimes as a jelly for calves made from boiling flax seed in water.[66] Britain was an early mover in linseed oil production, and its cattle were voracious consumers of linseed meal in the late nineteenth century. This is evident in the nation's raw flax seed imports (especially from Russia, India, and eventually Argentina) and its imports of oilseed cake from other oil producers like the United States and Germany. Before the First World War almost all North American linseed meal was exported to Europe, Canada's to Great Britain and the United States' to Holland, Belgium, and Germany. The United States exported an average of 300,000 tons of linseed meal to Europe each year.[67] Every bushel of flax seed yielded a 9:5 ratio of meal to oil, but the meal value in each bushel was only $0.43 for every $1.00 of oil.[68]

The North American meat-packing industry was established on the idea of making the continent's vegetable produce "incarnate."[69] Meat and dairy products were considered more highly processed than livestock fodder, and therefore a more profitable form of export. From the soapbox of the early Toronto Industrial Exhibitions, Elliot & Company attempted to persuade farmers to use linseed meal and thereby export Canada's flax "in the shape of cattle or sheep." The crushers claimed that the Dominion exported 3,500 tons of cake to Britain at a value of $105,000, but argued that meat from linseed-fed stock "would have netted the country double that amount."[70] Midwestern farmers were equally reluctant to feed linseed meal to livestock, even in Wisconsin's dairy heartland where one might have expected increased consumption based on the availability of meal from local linseed oil mills. "If farmers fully appreciated that in addition to the oil, the cake is of great value as a food for cattle and sheep," W.W. Daniells, a professor of agriculture at the University of Wisconsin, argued in 1879, then "doubtless much more flax seed would be raised."[71] Daniells and others emphasized the increased value of manure from linseed-fed animals, and Hugh G. Van Pelt complained that "the fertility of American farms has been and is being transferred across the ocean to enrich European farms."[72] Manitoba promoters blamed local farmers' low interest in "one of the most valuable agents that can be used in the fattening of livestock" for the slow development of the province's linseed oil–crushing industry.[73]

In Canada, farmers refused to feed linseed meal to their livestock when pasture, fodder crops, and distillery swill were abundant. Practically all the linseed meal from James Livingston's mill in Baden was exported to the United Kingdom in the 1890s, with the exception of an occasional carload to Quebec.[74] Experts asserted linseed meal's superiority for fattening livestock, but its absence from most beef farms suggests that even if it was the best feed, it was not worth the investment with so many alternatives nearby. "In nineteenth century Ontario," Ian MacLachlan argued, "an intensive feeding industry developed based primarily on by-products." Although linseed meal was technically a by-product, MacLachlan had in mind the lower grades of grain from flour mills and the swill from breweries and distilleries. Their knowledge of livestock feeding might have been another explanation for distillers Gooderham & Worts's brief interest in crushing linseed oil in the 1860s.[75]

Farm scientists promoted the use of linseed meal as feed, both for increased meat and dairy productivity as well as for improved digestion and coats.[76] William Brown from the Ontario Agricultural College in Guelph claimed that 100 pounds of linseed meal would add 81 pounds of flesh and fat to an animal.[77] One hundred pounds of meal would last an animal from twenty to fifty days and would mean a daily weight gain of 1.6 to 4 pounds – the high end of weight gain for young steers.[78] But, how many animals could actually be fed on US linseed meal in a given year? The average annual output of US crushers between 1920, when reliable data began, and 1936 was 556,000 tons of cake. If every pound had been consumed in the United States, it would have only amounted to 3 million pounds per day, or at William Brown's feeding rates, enough to supplement the diets of anywhere between 600,000 and 1.5 million head of beef cattle. The population of US herds in the first third of the twentieth century ranged from 60 to 75 million cattle, let alone the millions of other livestock.[79] And only about one-fifth of the linseed meal produced in the US was consumed there, hardly enough to fatten the cattle of a few major Midwestern feedlots.

Tariff drawbacks were important to crushers who imported seed, as a certain percentage of the tariff on imported seed would be returned with exported cake.[80] Whenever the drawback on exported cake was removed – for example, in the 1913 Underwood tariff – linseed crushers began to share Elliot & Company's enthusiasm for teaching North American farmers to use the cake.[81] Cake was the most important form of export revenue from flax because to some European farmers it filled an important niche. By the 1920s Canadians exported about as much linseed cake to the United Kingdom as they had in the 1880s, yet farmers in other destinations were buying far more of the feed.[82] Successful crushers therefore had to know the domestic market for oil and the foreign market for cake. They had to gauge domestic production of flax seed and be prepared to supplement it with imported seed while reconciling the idiosyncrasies of the seed with the preferences of consumers on both sides of the Atlantic. The linseed oil and cake industry was a truly global trade, connecting Argentine flax farmers with US paint makers and Canadian oil mills with European dairy consumers.

Crushing Competition

In the post-bellum rush of settlers to the northwestern plains, the linseed oil and paint manufacturers came to the forefront of the chemical and oil sectors. The geography of this industry was initially contingent on the sources of its raw material, flax seed. Small linseed-crushing operations sprang up in almost any area where flax seed could be found, but flax was never grown in one place for long and the small crushers were necessarily transient. Some mill complexes, such as Eby's mills in Waterloo County, Upper Canada, were able to run small linseed oil operations without a significant supply of local raw materials. But these mills operated sporadically and were prime examples of the multiple outputs of many small-town mill complexes. By the end of the nineteenth century, several major linseed oil companies had emerged and soon came to dominate the industry. The large crushers relied on getting inputs at a lower cost than less favourably situated rivals, and their locations reflected the two major sources of flax seed, the farms of the Northern Great Plains and Prairies and the Atlantic trade from Argentina, India, and Russia. At mid-century most linseed oil crushers were located in New York and Ohio, but flax seed production began to move northwestward, and mills were constructed in Illinois, Missouri, Kansas, Iowa, and Minnesota. By the 1890s the two metropoles of flax production were Minneapolis and Buffalo.

In Canada a similar industrial geography was in place. Initially, Montreal mills processed local and imported flax, but with the expansion of the flax economy in Ontario, Livingston's mill in Baden displaced Montreal as the eastern oil metropole. Baden did not have the advantage of cheap lake transport, at least for the journey from lake ports to Baden, and as the industry expanded, Montreal, with its excellent access to Canada's largest markets and overseas seed imports, once again became an important centre. In addition, the Canadian West hosted some flax and oil production in Manitoba, and over the next decade it appeared as if the Canadian prairies would be the ideal location for flax seed farming.

The earliest linseed crushing mill in Canada, Lyman & Company, was vertically integrated, at least in its Montreal operation. Family relations were critical to the establishment of the business, but the Lyman firms took on several other partners throughout the nineteenth century. The industry organized itself by family first and then sometimes vertically. In

Canada, linseed oil was characterized by two vertically integrated family empires that competed fiercely for customers. The first was made up of the Livingston and Liersch businesses, which were tightly intertwined through family marriages, and the second was in the form of the Canada Paint and Manitoba Linseed Oil family syndicate.

When economies of scale gained importance in the international linseed oil industry, Livingston lost the advantage of locally controlled supply chains. The Baden headquarters was well situated for running the Ontario-Michigan-Manitoba flax empire, both geographically and culturally, with James Livingston's connections to family and local Mennonites. The Livingstons reduced their reliance on volatile seed markets by procuring seed from flax fibre farmers and their own operations in Ontario, and as we will see in Chapter 6, they provided agents, warehouses, and contracts with Mennonite immigrants to acquire flax seed in Manitoba. When this system failed to produce a large enough seed supply, Livingston imported from Duluth and overseas markets, and Baden lost its edge. Livingston rather quickly sold off his fibre operations in the 1890s and expanded his linseed oil mills to other sites, especially Montreal, to take advantage of market proximity. Family remained important here, and for James Livingston it meant training sons and sons-in-law to manage the new urban outposts while he remained in the rural headquarters. For Livingston's competition it meant keeping family on the boards of paint companies and integrating backward into linseed oil, as Lyman & Clare had done a half-century earlier.

In 1923 a US Supreme Court decision dissolved American Linseed, the largest flax-based corporation in world history, and several other upheavals in the remainder of the decade demonstrated the strength of the competition in the US linseed oil industry. In the same year that American Linseed was defeated, Archer-Daniels acquired Midland Linseed Oil, and the consolidated company (ADM) began to expand and purchase smaller crushers such as William O. Gooderich in Milwaukee and Fredonia Linseed Oil in Fredonia, Kansas. By 1928 the Rockefellers offered the remaining American Linseed mills for sale, and the two main companies to pick up the pieces were Spencer Kellogg based in Buffalo and ADM based in Minneapolis.[83] The complexity of the linseed oil conglomerates is evident from the fact that the most important feature of American Linseed's sale in 1928 was its line of non-flax foodstuffs; these were bought by Gold Dust, the successor to the American Cotton Oil trust.[84] Flax seed enterprises like

American Linseed had become part of a larger oilseed-processing sector, and although it might appear that companies like ADM were lining up to become the newest linseed oil "giants," they too were undergoing a transition to multiple vegetable and synthetic products. Because the linseed oil companies were constantly in flux, the cost of oil and paint was impossible to raise through cartels, and crushers and paint makers had to increase efficiency in every way in their plants. One of the new marks of efficiency and modern technology was the research chemist employed in linseed oil and paint mill laboratories.

Barrelled Sunlight or Saving the Surface? Marketing Paint

The mainly urban world of craft-made and ready-mixed paint led to a large new array of coverings available to a growing segment of consumers. Paint manufacturers and linseed oil crushers had traditionally sold their goods to other industries, but with ongoing competition and a new luxury market emerging in the twentieth century they aggressively promoted their goods as both "pure" and essential for everyday consumers. Pittsburgh Plate Glass Company's paint catalogue in 1923 literally painted a picture of a new world of colour made possible by their line of varnishes, concrete paints, and ready-mixed colours. Vibrantly illustrated pictures displayed next to sample sheets of available colours were meant to instill pride of property in the consumer. An average two-storey country home covered completely with shingles would "last many years longer by warding off sun and rain" with the use of Tor-on Shingle Stain. Much larger but still inviting institutions were pictured as places where "the severe wear of scuffing feet" demanded a paint like Patton's Porchite. Ocean liners crowded with passengers and bridges carrying automobiles needed protection "against rust, the great red plague" so as to ensure human safety. In private worlds, farms, which included houses, barns, fences, and outbuildings as well as tractors and implements – virtually every surface of which was covered in paint – were protected by Patton's Industrial Building Paint and Wagon and Tractor Enamel Paint. Patton's Auto Gloss finish was applied to wicker and outdoor furniture, to appliances, and to motorcycles, trucks, and a car by their owners, creating a scene that would have seemed odd to carriage painters two generations earlier. Varnishes were branded "aged" under the head of a Sphinx. Their application saved a variety of wood surfaces, including elegant polished floors, doors, automobiles, and watercraft. In the most obvious demonstration of an appeal to lay-painters, "Colored Varnish" advertisements showed five different people staining furniture and floors around the home. Only one person was a professional painter, and three were women in everyday dresses and aprons, staining leisurely with a brush in one clean bare hand and a mug of stain in the other.[85] The marketing of these products intended consumers to think of paint products as a coating for everything and as useable by anyone.

Many consumers had to be convinced that the expense of paint would be worthwhile in more than just an aesthetic sense. Paint marketing began to take this approach in the 1910s, when manufacturers emphasized paint's clean, bright appearance. In 1908, the *American Paint and Oil Dealer* magazine exhorted dealers to "make yourselves missionaries of the gospel of good paint – of brightness and thrift," by bringing paint to the darkest and dirtiest neighbourhoods. In 1912 the same journal received funding from manufacturers to promote the use of paint through the "National Clean-up and Paint-up Campaign." By 1925 this group was spending $60,000 per year on propaganda, and this continued until at least the 1950s. By that point the manufacturers had taken paint testing and propaganda to the extreme, notably on the Nevada Proving Ground, Nye County, Nevada. There they painted fences, small houses, and other test surfaces to observe the effect of nuclear heat on the paints and the surfaces they protected.[86] In Canada, the federal Home Improvement Plan and various industry programs (such as the "Clean up, Paint up, Modernize" campaign in Vancouver) reflected similar cooperation between government and building-supply industries in the late 1930s. The Home Improvement Plan helped Canadians acquire small loans for home renovations, including painting and linoleum installation, and the plan effectively helped industry normalize these products as part of a middle-class standard of living. Other scholars have explored how cleanliness was used by the state in the larger context of social and urban reform.[87] Starting in the early twentieth century, paint was presented as a solution to urban degradation, and the linseed oil and paint manufacturers took advantage of society's elite and middle classes to identify new markets for paint.

Nature also became more important to marketers who presented consumers with a way to create sunlight in the "darkest and dirtiest neighbourhoods." Individual paint companies converted implicit brand names such as "Gloss-O-Lite" to "Barreled Sunlight" and "The paint that brims over with sunlight."[88] Not only was solar energy a product now available in paint, but paint was now necessary to protect materials from the sun. Manufacturers claimed that paint and varnish products were "indispensable alike to cottage, factory, and skyscraper; they protect and beautify our possessions."[89] They presented the case, specifically to farmers, that paint had many returns and was "a good investment anywhere; in the home, on the farm, or on commercial buildings. The banker more readily lends money on buildings which are kept well painted. It is his assurance that his security will remain constant."[90] Also in this period, Ontario farm journals like the *Farmer's Advocate* began to run articles on the importance of painting houses and barns. Between 1910 and 1920, farmers read about "Paint Value on the Farm," "High Interest on a Coat of Paint," and "Paint and Profits." They were asked, "Does It Pay to Paint?" and were instructed to "Paint and Keep Your Credit Good."[91] Pittsburgh Plate Glass Company asked, rhetorically, "how often a dollar's worth of paint would save the usefulness of a hundred-dollar [farm] implement!"[92] Everyday life had changed in the new world of industrial goods and urban lifestyles, and now paint was necessary in these spaces, both for bringing sunlight in and keeping it at a safe distance.

Colour was also used to reinforce notions of rural modernity in this period. One farm journal editor gloated over the number of painted barns that were visible from the windows of his car in one county near Toronto, emphasizing rural Ontario's investment in roads, automobility, and property improvement.[93] The *Farmer's Advocate* promised that using paint would add ten to fifteen years to the life of a new house, barn, fence, or wagon, and that painted buildings increased farm resale value dramatically. The addition of colour was not described as an aesthetic improvement in any way, but rather as "paint insurance" and the best way "for the farmer to protect and insure his 'wooden things.'"[94] Buying "paint is a dollars and cents proposition," claimed another paper, and "it costs much more not to buy it."[95] Worrying about the opportunity costs of unpainted buildings and implements was as novel as rural paint itself in the early twentieth century. Daniel Tiemann argues that since for most of its history paint was expensive and wood was cheap, applying paint to preserve wood "would have been indeed gilding the tinsel."[96]

Nevertheless, paint manufacturers launched a "Save the Surface and You Save All – Paint and Varnish" campaign in 1919, appealing "to the individual's natural interest in what he owns." Their advertisements for "sun-proof paint" were written in the rays of a scorching sun and many contained

"PROOF" is the general trade name applied to Paints, Varnishes, Enamels, and other allied products of the Pittsburgh Plate Glass Company. This trade name is a distinction of quality, and applies generally to these products, in addition to the official trade-marks of the individual lines.

Glass, Paints, Enamels, and Varnishes are indispensable alike to cottage, factory, and skyscraper; they protect and beautify our possessions.

These and innumerable other products—insecticides, disinfectants, and chemicals—comprising the entire line of Proof Products, are available always in dependable supply everywhere, at the command of architects, dealers, contractors, painters, building owners, food growers, and manufacturers.

From raw material to finished product, the manufacture and distribution of Proof Products are under one ownership, one organization, operating through specialized manufacturing divisions, effecting incalculable economies—both in manufacture and distribution—assuring dependability of supply and consistent maintenance of Highest Quality Standards.

PITTSBURGH PLATE GLASS COMPANY
GLASS　　　MANUFACTURERS　　　PAINT

PAINT AND VARNISH FACTORIES: MILWAUKEE, WIS.; NEWARK, N. J.; PORTLAND, ORE.

Figure 5.8　Proof Paints, Pittsburgh Plate Glass Company, 1923 subtexts such as "the life of your property is unlimited if you keep it well painted." Four years later Pittsburgh Plate Glass Company claimed that the campaign had improved "the public attitude toward paint and varnish, but there is still much to be done." Whereas the coverings were once a luxury, they were becoming "prime economic necessities for the preservation of property, for cleanliness, and for their influence on morale."[97] By the 1930s, paint companies were promoting paint as a "tonic for tired homes": "A few magic strokes brush away that tired look" on houses "weary of fighting the weather."[98] To early twentieth-century paint consumers the sun was depicted as an oddly benign but destructive force that was happy to do its solar work from behind the protective barriers established by paint consumers (Figure 5.8).[99]

Most rural consumers were reluctant to colour and protect every surface in their homes, at least until the 1920s. Even in 1936, when the Canadian National Parks Branch acquired the house that would be "Green Gables" in the Prince Edward Island National Park, they found that this ordinary Maritime dwelling had "no paint on the house, just whitewash, and no trim of any colour." The branch possessed extensive resources for property enhancement, but if there had been no reason to draw attention to the shade of this National Heritage Place's trim, they might have stood by their original assessment that "the exterior does not actually need painting."[100] L.M. Montgomery herself felt differently about the general appearance of Cavendish in the interwar period, noting in 1924 that "Cavendish is getting so shabby. Almost all the houses are unpainted and dowdy. Times are hard, of course, but I fear there are other reasons – indifference, the dying out of the old families."[101]

The flax propaganda of the 1860s demonstrated the very limited correlation between the activities promoted by farm journals and what actually happened. In the countryside, farmers were faced with a barrage of logic intended to convince them to apply paint as an investment. It is not clear when rural people began to pay much attention to campaigns like "Clean-up, Paint-up, Fix-up" and "Save the Surface," but before 1920 defending wood from the sun was a selective strategy at best.

Chemically Pure: Regulating Paint

Up until the twentieth century, the world of colour was limited to small objects, high-traffic surfaces, and vehicles. Nevertheless, multinational paint and linseed oil companies like Sherwin-Williams promised to "Cover the Earth," and in the large industrialized cities it would seem that their industry was on its way to doing just that. The paint industry underwent major transformations between 1850 and 1920, and the failure of price fixing and the increasing importance of scale in the paint and oil industries led to another development in the flax-paint side of the commodity web. Manufacturers began to employ chemical researchers and technicians to improve the performance and uniformity of their products and, more importantly, to minimize wasted batches. The new demand for scientists and other researchers in the web was not only about increasing efficiency but also about standardizing refined linseed oils for a growing number of applications and an expanding variety of paint products. Thus, the role of chemists in paint and oil research and regulation deserves special attention. The Progressive Era debates over product purity prompted some consumers to argue that integrated paint manufacturers had disconnected them from paint-making processes. Oil and paint companies responded with a range of marketing and regulation efforts of their own.

In the mid-nineteenth century, white lead varied considerably in quality, and when apothecaries prepared it ground in oil, customers had more difficulty determining the quality of the product until it was applied as paint. In 1869 Canadians could choose between thirteen varieties of "Colors in Oil," including five grades of white lead. The white lead ranged in quality and price from "general," at $2.35 for a 25-pound tin, to "common," at only $1.30 per tin.[102] Similarly, Toronto druggists and oil manufacturers such as Lyman, Elliot & Co. advertised "various qualities" of white lead.[103]

There was also variation in the quality of linseed oil; most nineteenth-century prices for oil specified a high and a low price for both raw and boiled linseed oil. The terms "pure" and "Crown Pure" were used in the linseed and paint industries sometime before 1890 as an indication of grade and price. In 1890 a Montreal paint manufacturer promised that its Johnson's Pure Paints were made with only "Pure Linseed Oil Paints, containing no cheap mixtures."[104] Mixed paints were sold as "pure liquid colors," and "second quality" paints were sold at around 15 per cent discount.[105] But with white lead, the term "pure" indicated an "Association guarantee," for "pure, ground in oil" lead packaged in small kegs. The lead contained about 8 per cent oil.[106] This wholesale product was purchased by painters and glaziers, and the term "pure" assured them that they could expect the paint to mix, cover, dry, and last. Paint's durability was not immediately evident, but customers were quick to complain to painters when a product as expensive as paint did not turn out as they wished. Mixing, covering, and drying were critical to the painter's operation and profits, and painters could quickly tell if their materials were poor.

Chemists saw a place for their analysis in the paint industry from very early on. Sometimes they examined the health concerns associated with painting, such as painter's colic, and ways to make paint "free from smell and ... those noxious vapours which are so prejudicial to health."[107] As paint became a prepared good, people like Victor Biart claimed the title of chemist and held that their craft was necessary for identifying the "adulterations of paints" to consumers. Evaluating paint had been a possibility for consumers when "the druggist bought [lead] in the state of powder," Biart claimed. But by 1870 white lead was "generally sold ground in linseed oil," and fairly elaborate methods were needed to check for impurities such as sulphate of lime (plaster of Paris), carbonate of lime (chalk), and the most common adulterant, sulphate of baryta. Manufacturers apparently mixed this less expensive white mineral – barium sulphate – with white lead because it did not affect the weight or hardness of the paint. The main drawback of baryta was that the adulterated paint did not cover as well as "pure" white lead.[108]

An adulterant in linseed oil was any substitute not derived from pure flax seed that had been intentionally added to dilute the oil. Some adulterants included rosin oil, corn oil, cottonseed oil, menhaden oil, hemp oil, and rape oil.[109] Petroleum oils could be mixed with these others and added to linseed oil, but the result was a product that dried slowly and made surfaces impossible to repaint without scraping. Adulterated oils must have still been mostly linseed oil, but they were sold as competitors of "pure linseed oil" and were mostly used for rough outdoor painting. Unlike adulterants, impurities came from foreign objects in the flax seed. Difficult to detect, they affected the drying time and hardness of the oil. There were also substitutes for linseed oil as a drying vehicle. The oils that competed with linseed as a drying oil were poppy, walnut, and China wood oil, or tung, although so few of these oils were available in North America that they would only be cheaper than flax seed if their production somehow increased. Importing products like tung proved so difficult that the cost difference between them and linseed oil was negated.

Paint and pigments were more frequently and easily adulterated, and according to an industrial manual by mechanical engineer William Ennis, "Unfortunately, no products are more widely adulterated than mixed paints." The substitutions in adulterated paint were usually in the lead or zinc pigments or in other dilutions added to the paint in the factory.[110] However, some believed that poor linseed oil was a major factor. One writer regretted "that so many manufacturers of white paint labelled their paint 'Strictly Pure'" when in fact most used varying degrees of adulterants in the lead and linseed oils. "Painters generally agree," Ennis argued, "that there is nothing like pure linseed oil for paint. All other oils are practically worse than nothing."[111] Slightly adulterated linseed oil was probably not so harsh an ingredient, and mixing these products could have different purposes. Theodore Zuk Penn argued that "[a]dulterants is perhaps a harsh word here, for it is quite possible the whiting was used as a filler and was added purposely to extend the bulk of the expensive white lead."[112]

Adulteration and substitution were important issues for manufacturers before they became the focus of state regulators, and like many industries in this period, some of the first to cry foul over adulteration in linseed oil were the crushers and industry types themselves. The US anti-adulteration movement of the 1880s was mainly a movement within established food and drug businesses that saw adulteration more as an unfair advantage to competitors than a health concern.[113] The linseed oil and paint manufacturer associations were also engaged in the process, branding, standardizing, and regulating adulterants in their own products. The rise and dissolution of associations and trusts gave rise to frequent discussions of paint quality and prices. As historians from James Harvey Young to Christian Warren have shown, the messages from Progressive Era chemists and regulators were often mixed. Sodium benzoate was declared both poisonous and safe in the space of two years. Lead paint was declared pure in one country and poisonous in another.[114]

The threat of adulterants was used in part to warn consumers that low prices for linseed oil would translate to poor coverings for their walls, floors, wood, and other goods. Often the warnings were most intense when trade associations or trusts were forming or dissolving. In 1889 one nostalgic member of the National Linseed Oil Company made his case for its recent business combination. Because of intense competition, linseed oil had been adulterated to cut costs and redundant marketers were the only ones benefiting from competition. "A good many firms in the trade, who used to be above any thing of the sort," he disparaged,

> have been marketing some goods in the past few years which were not exactly the "pure linseed oil" which they were labelled. It's a

mean business – adulteration – but not many of our customers ever test their purchases. The one thing they are apt to look at is price, for they are buying to sell again ... We have often discussed the possibility of stopping these adulterations, but it was a hard matter to cure by mere mutual agreement. How do I know what my competitor ... does with the vats in his cellar after working hours ...? For I must confess that there are a few men in our trade who are as tricky as horse jockeys.[115]

These arguments were fashioned by the trade to ensure stability and price, and not out of a concern for whether the consumer good dried quickly on walls or withstood floor traffic.

In Canada an industry that lacked an association or suffered from price cutting was often called "demoralized," a term that to contemporaries meant a collapse in both "the ethical as well as the material bases of business success."[116] Canadian paint manufacturers used this terminology when the White Lead Association and the Paint Grinders' Association dissolved and prices dropped in 1894; trade journals warned that adulterated products were sure to follow. "It is a well known fact," a *Toronto Hardware* writer stated in responding to declining prices in lead, "that the nearer the price of an article gets to the cost of producing it, the greater is the probability that the quality will be sacrificed to price. In other words, the article is likely to be adulterated."

The Canadian government had a lead purity law in place in this period, but hardware journals argued that "cheap white lead can be produced without resorting to adulteration, just as there are more ways of killing a dog than by hanging." The manufacturers would "simply use the poorest-quality lead and oil, resulting in a paint that required more coats and longer drying time."[117] By 1901 there still was no trade association, and competition in the Canadian paint and oil trades was intense. The "association guarantee" under which premium white lead products had been advertised was now replaced on all manufacturers' products with the label "best brands, government standard."[118]

In the United States there were no government standards for lead, but legislation emerged in several states at the outset of the twentieth century that forbade the sale of any adulterated oil under the pretense of it being pure linseed oil. Pennsylvania was at the cusp of this regulatory trend with an Act in 1901. By 1906 in New York, the law demanded that the product name "raw linseed oil" only be ascribed to oil made entirely from flax seed. It also reserved the name "boiled linseed oil" for oil from the same source as had been heated to at least 107°C.[119]

Scientists such as USDA chemist Harvey W. Wiley were driving forces behind the federal Pure Food and Drugs Act of 1906, and the paint purity campaign had its own chemist at the helm, Edwin F. Ladd of the North Dakota Agricultural Experimental Station (NDAES). Both the chemists and the manufacturers came from the context of the anti-adulteration movement of the 1880s. Almost two hundred bills to curb food and drug adulteration were considered by Congress between 1879 and 1906.[120] Historian Christian Warren argues that lobbyists for the powerful paint manufacturers' association managed to fight off pure paint legislation by using a "network of spies" operating in most state capitals. Such regulatory bills were often introduced at the state level and, if successful and popular, were repeated at the national level.[121]

Warren's theory bears out, and according to George Heckel, an association watchdog, "proposed paint laws appeared regularly in a dozen or more states" after 1900. The local paint manufacturers were usually able to lobby against or propose amendments to the bill, but only if they knew of them. Heckel was tracking the bills personally when "the board of directors ... suggest[ed] that I take charge of the service officially. They allowed me ten dollars per state for a selected list of states, which I in turn paid to selected individuals in these states for reporting service – but none of us thought of North Dakota ... until North Dakota stepped into the limelight not a single paint law in any State got past the Legislature."[122]

Heckel's job was to collect drafts of "paint bills" as they came forward and distribute them to industry members, who would lobby against them, but North Dakota's populists were particularly successful in passing pure food, drug, and seed laws. Ladd, who later became a US senator, was the main crusader through the NDAES, but he was part of a larger movement that received strong support from the state's commissioner of agriculture and women activists from the WCTU and other groups.[123] According to one historian, Ladd's early chemical analysis concluded that North Dakota was "a dumping ground for shoddy products" from outside the state.[124] Presumably these evils lurked not far from the border because Ladd also targeted Minneapolis paint and oil manufacturers. Ladd had been influential in North Dakota's 1901 pure food law and subsequently pursued a pure paint law; meanwhile, his colleague from the field of biology, Henry

Bolley, worked on pure seed legislation.[125] Both were enacted in 1906 in North Dakota, and from that point paint manufacturers had to specify the amount of "adulterants" the product contained or the ingredients there were apart from "pure linseed oil," lead, zinc, Japan dryer, and "pure colours."[126]

Two could play the purity game. US crushers reacted to crusaders like Ladd as Canadian manufacturers reacted to price "demoralization," with the language of purity. Linseed oil and paint makers tied their own professional character to the purity of their product.[127] The trouble with strict regulations from the crusher's perspective was not the fines (which ranged from $50 to $500) but the damage to the manufacturer's reputation. The Archer-Daniels Linseed Oil firm wrote to NDAES, arguing that "our oil purity is connected to our honourable character" and that, as manufacturers, they were simply "gentlemen trying to do what is right."[128]

Ladd saw his Act as a mandate to analyze the nation's paint. He built paint-test fences to observe paint performance in real world conditions, he published bulletins with test results, and his department eventually established a paint/polymer branch. Furious paint manufacturers responded, asking Ladd why he examined paint that "is not offered for sale, nor is ever likely to be offered for sale in North Dakota," but not all industrialists opposed the law.[129] Retailers and master painters favoured standardization and organized support for subsequent state paint laws. After all, they were not only the primary paint consumers, but they had to answer to clients if a product failed. In an attempt to regain some control over the legislation, the paint industries united briefly and proposed a federal regulatory bill similar to the Federal Pure Food and Drugs Act. A federal paint Act would alter state Acts like North Dakota's in the industries' favour, but their efforts fizzled out by 1910.[130] Painted fences were meant to bring wide consumption experiences in multiple regions under the microscope. Here, everyday colour consumption was recreated in the middle of a field, and Ladd presented himself as both someone who understood and improved paint and someone who painted. The flax-paint web demonstrated the new scale of the urban industrial complex, and it shows how people like Ladd imagined their role within it.

A New Old-Fashioned Product: Responding to Regulation

The linseed oil and paint industries responded to these highly public experiments by publishing their own chemical research and marketing their products as old-fashioned everyday goods. In a direct response to the pure paint law in North Dakota, the National Lead Company adopted the "Old Dutch Process" marketing brand in 1907. This reminded and reassured consumers that their white lead was manufactured in a simple and centuries-old technique tracing back to the Netherlands.[131] The image of old-fashioned production was really about knowledge. The processes of linseed oil refining and paint manufacturing required special knowledge in the early twentieth century, and processors wanted to assure consumers and regulators that the products were not compromised by new processes. In this image, paint came to the customer through new business structures, ones that made the public more than suspicious, but the processes or knowledge of preparing oil and pigments were the same.

James Livingston marketed his linseed oils as "the kind your grandfather used."[132] Why would he market his product in that way and to whom was he advertising? It was more than a simple turn of phrase. Livingston collected his flax seed from suppliers around the world, processed it in unprecedented quantities in mills running around the clock, and sold it to manufacturers, who brought it to consumers in a new form: ready-mixed paint. He apparently believed that manufacturers and other people who bought linseed oil needed to be assured that they were buying a quality product. On the other hand, manufacturers like Livingston wondered how to expose a luxury product to a larger market. He produced an urban commodity from the middle of rural Ontario, and his home was a showpiece for visitors, vibrantly painted and decorated with rare Victorian mural work. His advertising was recalled by local residents, people who saw it on packaging and vehicles, so he may even have been anticipating an increase in paint consumption among rural clients.[133] If linseed oil was "an abstraction" to ordinary people, as Ennis argued, then why did linseed oil crushers like James Livingston and Archer-Daniels worry about the public perception of their product? Product purity was paramount to these crushers, at least in theory. Not only did crushers bank on the reputation of their product, but according to Ennis, they limited their involvement in other oil works so that their "consumer [could] remain satisfied as to the integrity of the product." He identified "an unwritten law of the industry that a crusher must not be interested in the manufacture or sale of other oils."[134]

The flax-paint commodity web, however, did overlap with other oil products. Although the exclusive commitment to linseed oil may have seemed true to Ennis in 1909, when he wrote about it, the cottonseed oil industry depended heavily on linseed oil technologies for crushing and industrial organization for refining the oil. The most important machinists to both industries were located in Ohio, especially the French Oil Mill Machinery Company in Piqua, the V.D. Anderson Company in Cleveland, and the Buckeye Iron and Brass Works in Dayton.[135] A large proportion of oil from the south made its way to northern urban refineries, and most of it had been crushed with machinery developed and manufactured by the linseed oil industry. Lynnette Boney Wrenn suspected that the side-by-side production of cottonseed oil and of lead and varnish meant that these businesses tried to use the southern by-product as an additive in their drying oils, but it is just as likely that they recognized the potential for producing multiple products with similar inputs and some of the same technologies.[136] Furthermore, linseed oil crushers were the first to study and work with alternative vegetable drying oils. American Linseed focused mainly on vegetable oil food products in the 1920s, and Archer-Daniels-Midland Linseed Company became a soybean manufacturer and then branched out into other seed and vegetable oil products in the mid-twentieth century.[137] Given the many movements of actors within and beyond flax's wider commodity web, the image of linseed oil crushing as a process with no connections to similar products and substitutes seems unlikely. It is therefore unlikely that crushers were as committed to product purity in the mills as they were in their marketing and media appearances. They were concerned that their products performed better than the competition's, but to oil and paint manufacturers the concept of purity was primarily about grades, and the notion of old-fashioned production was about cloaking relatively new products in tradition and trade knowledge.

In order to remain competitive, it helped to know something about chemistry. In the early twentieth century, the paint manufacturer and industry advocate George Heckel (himself an industrial chemist) attributed the perfection of consumer goods in paint and varnish products to the inventions of scientists.[138] Linseed oil and paint manufacturers did employ chemists, and in 1920 the chief chemist at Spencer Kellogg began publishing a series of letters in the trade journals that were later compiled as *Laboratory Letters*. The letters were mostly a description of and an advertisement for the "scientifically prepared Boiled Oil," Kellogg's Improved Boiled Linseed Oil. The manufacturers defended improvements that might have been considered "impurities" as ultimately the best way to standardize the performance of the drying oil. The critical element of refining boiled oil was the "minute quantity of metal ... which stimulates the drying reaction without becoming an integral part of the finished product." Spencer Kellogg used chemistry to understand and demonstrate the limits of purity. Improved Boiled was "a hundred per cent pure Linseed Oil minus two-tenths of one percent of metal, or 99.8% of pure Oil." The refined product, the company claimed, was a "Boiled Oil in every sense of the word, [with] practically the normal anatomy of pure Raw Linseed Oil." In other words, the improvements were additives but did not detract from the chemical structure of oil; they added value and were no threat to customers.[139]

Manufacturers had always been able to measure the important aspects of refined oils, such as their drying times, opacity, and evenness of film. Raw linseed oil dried in from three to five days, and the best boiled oils, by 1921, would dry in fifteen to twenty hours.[140] However, Spencer Kellogg argued that the "average manufacturer or superintendent has little time to experiment in the laboratory, or to study much of the technical literature of Linseed Oil in order to keep informed."[141] Part of the work of a linseed oil crusher was detecting adulterated oils, which required trained smell, taste, touch, and sight. A crusher could often detect oil that had as little as 10 per cent tung by smelling it. Specialists could also see mustard seed contamination in linseed oil, or identify mixtures of pure and inferior linseed oil undetectable by chemical analysis by looking for their greenish colour before boiling, or by observing their drying time.[142] Many of the tasks of testing and refining oils were performed by business managers and other non-scientists, but at some point in the early twentieth century, the very large mills incorporated educated staff who could monitor and experiment with quality processing more systematically and regularly.[143]

Chemists helped linseed oil manufacturers remain competitive by improving and monitoring the refining process. One method of processing linseed oil that eventually fell out of favour was to percolate the crushed seed in naphtha. Only about a tenth of US linseed oil was produced this way in 1909, but it would have provided work for chemists in oil mills. Even with regularly pressed oil, refining required careful monitoring. Since boiling often caused discolouration and burning, the refined product could vary from batch to batch, regardless of the purity of the oil. For painters and other paint manufacturers, a consistent and fast-drying oil was important for consistent and fast-drying paint. Various problems in the oil introduced by poor refining affected the consumer good. For example, paint's consistency was influenced by the average acid value in linseed oil, and paint blistered when its oils contained mineral oil or moisture.[144] Chemists helped provide this edge to companies like Spencer Kellogg, whose marketing logo, "The Test Tells," appeared on a distillation beaker printed on their barrels and tank cars.[145]

In Canada there is little evidence that the major crushers employed chemists. One of the few worlds that James Livingston was relatively removed from was professional science. He was well connected in the corridors of power in Ottawa, but he apparently solicited neither the knowledge of farm scientists nor the work of industrial chemists. As a fibre producer, James had access to a large body of literature and kept books such as Sharp's *Flax, Tow, and Jute Spinning* in his library.[146] However, flax seed industrialists and intellectuals produced research at a growing pace, especially in the United States, at the start of the twentieth century. The US linseed oil millers were eager to connect with and learn from flax scientists and other specialists, but Livingston was from a different place and an earlier era. In 1917 the elderly and ailing man's company continued to advertise his products as "old process."[147]

Why was there a continued emphasis on old process among some crushers, including Midland Linseed Oil, which in 1910 was only a decade old?[148] In some cases, "old process" was directly tied to the business's genealogy, and it appears to have been popular among the long-established mills in Baden and Piqua, Ohio.[149] However, to Spencer Kellogg, one of the oldest millers in North America, the difference between their refined oils and "haphazard Boiled Linseed Oil" was "the difference between scientific and empirical methods. Old processes often embrace the right philosophy, but suffer from lack of scientific understanding."[150] To Livingston, the difference between these two oils might have seemed small and not a major factor in competition. When Alberta Linseed Oil opened in 1913, it invested in all of the space, equipment, and likely the personnel necessary for industrial research. Images show that their test laboratories were similar to some of the large US mills.[151] However, their failed example suggests that industrial research was less important in Canadian mills, or at least that chemists alone could not guarantee success in the industry. Frank T. Shutt, Ottawa's "Dominion Chemist," addressed the American Chemical Society's first meeting in Canada, where he argued that these professionals recognized "no political boundary lines in the giving out of the results of their labor." However, the Canadian oil industry was a late adapter of the profession's knowledge.[152]

Manufacturers of paint and linseed oil experienced even more changes in their own roles than they witnessed in their product development. In the 1920s, paint and oil manufacturers continued to develop an image of themselves as scientific practitioners of ancient crafts. Pittsburgh Plate Glass Company showed images of Paleolithic painters and claimed that the only major change in process was that chemistry had replaced foraging for naturally available pigments. Even then, some of the purest colours came only from natural pigments, such as ultramarine of "azurite" from hand-crushed lapis lazuli. "Although ultramarine is manufactured today by chemical processes," Pittsburgh Plate Glass Company admitted, "modern artists sometimes prefer the old-time color because of its superior quality."[153] This was light lip-service indeed compared to "ancient" methods, because the paint that was coloured with hand-crushed pigments in the early twentieth century would have been an extremely small fraction of the new world of colour.

Not only designer colours but everyday milling as well was romanticized, and according to Pittsburgh Plate Glass, "the essential operations of early paint manufacturers continue to this day." Paint was still ground between stones and mixed in containers as it had been for millennia, but the processes were now mechanized "and operations ranging from the simplest and most elementary to the most scientific and complex have their part in the process."[154] White lead, for example, was "the oldest white pigment known" and was apparently "made by a process not widely different from the present-day Old Dutch Process, the method now most widely used." On the other hand, Pittsburgh Plate Glass treated what regulators were calling adulterants, such as basic sulphate of lead and blue lead, as recent discoveries with new applications that were "produced by a similar method" and were in many ways superior to the "pure" white lead.[155] Images of the complicated steel roller grinding mills were juxtaposed with an elderly "stone-dresser" wielding a pneumatic hammer on a stone that resembled the painter's slab. This craftworker's "[l]ong experience makes [him] an adept and in recent years the automatic hammer has lightened his task considerably."[156]

From Biart's distillations, to the refined oils offered by Elliot and Company in the 1880s, to Ennis's record of chemical developments, to Spencer Kellogg's publication of *Laboratory Letters*, and to Pittsburgh Plate Glass's catalogue and trade history, chemists certainly had an important role in refined oils and were even more critical in the paint industry.[157] It is possible

Elevator—Capacity, 1,000,000 bushels Flaxseed

Research Laboratory

Figure 5.9   Spencer Kellogg, laboratory, c. 1921

that their role was overplayed in some publications, especially where photographs of laboratory scenes figured heavily. As Philip Scranton and Walter Licht found in their study of industrial images in Philadelphia, the oily presses and dirty workers did not instinctively draw the photographer's or enthusiast's eye.[158] Most collections of images of a linseed oil mill include one of a single worker in filthy clothing loading presses with seed and removing cake, as well as a few photos of the press floor layout. The collections almost always include a series of images of the laboratory equipment. One exception is Ennis's industrial manual, which is richly illustrated with over seventy sketches of machinery and photographs of

Figure 5.10 Battery of twenty French heavy-duty screw presses. Installed by Archer-Daniels-Midland Company in Minneapolis, MN

Spencer Kellogg's recently constructed mill, but the images are completely devoid of either human or laboratory subjects.[159] The Buffalo miller's own publications were just the opposite, with workers visible in almost every image, including chemists in the company's lab (Figure 5.9).[160] Until the installation of monstrous "super-duo expellers" and other automated systems, linseed oil presses and paint mills were mundane and unchanging photographic subjects.[161] It is no surprise then that the chemist and the laboratory tools became popular features in industry publications. They proved the manufacturer's commitment to "purity" and progress, and they offered clean and interesting perspectives of the mill.

The contrasting images in Whitney Eastman's history are subtle indications of the former manager's assessment of change.[162] A "battery" of twenty old-style screw presses at Archer-Daniels-Midland represented the height of the late nineteenth-century process (Figure 5.10). It drew on

Figure 5.11   Rotocel extractor

the image of military machines and simple but efficient manufacturing. It was followed by the massive linseed oil expellers representing the new industry, highly complex black boxes. The first process was also mostly enclosed, but there was a certain simplicity and transparency to the system that allowed customers to understand the production process if they cared to know about it. In mid-twentieth-century processing, it was no longer easy to tell from looking at the machines how flax seed entered and oil and cake exited. The Blaw-Knox rotocel replaced the industrial design of the press altogether, bringing the seed into a circular rotating system for cleaning, crushing, heating, pressing, and expressing as oil through a series of pipes. A passive observer gazes through the device's one small window, and readers are able to see the complex non-linear system at work in the cutaway.[163] Here the face of linseed oil mill photography merged the sterility and cleanliness of the laboratory with the continuous processing and refining of a messy but confined commodity. The manufacturer is depicted as an actor who once led the mechanical troops with military precision but now stares through a peephole in the rotocel (Figure 5.11).

There is an argument that general consumers became very concerned with the origins and quality of the ingredients in their consumer goods during the progressive pure food and drug movements of the 1890s and 1900s. But this was as much about the reputation and competition of manufacturers as it was about the interests of consumers. The main actors in the paint purity movement were not concerned citizens who bemoaned a disconnection between the flax farmers of the Northwest and the ersatz on their paint brushes. Rather, the debates over product purity were between the various branches of the oil trade in a time when the industry was undergoing rapid organizational change and painters were recognizing the convenience of, but also the threat to, their craft posed by ready-mixed paint. The great irony of paint purity laws is that their promoters did not care as much about the well-known toxic qualities of white lead as they did about food's purity and healthfulness.[164] The main concern was that the paint would be quick drying and free of adulterated ingredients, including "water for oil, and cheap minerals for lead" even if the "pure" mineral was poisonous. Paint purity laws actually ensured that the product would be more dangerous.[165]

Because they manufactured such a novel product, linseed oil and paint manufacturers designed marketing strategies with at least three approaches. The first, oddly enough, was a claim that the industrial processes and products were *old-fashioned* and had been in use for centuries. Second, they presented the mass-produced commodity as a necessity for protecting human goods from the oppression of natural and unnatural elements, and the commodity web that began with solar energy ended with "sun-proof paint."[166] A third message communicated through marketing was meant for the scientific world, and it was to assure chemists, especially those involved in regulation, that the manufacturers were capable of designing their own products using their own selection of scientists.

The world of linseed oil had been turned upside down, and it was now the primary ingredient in a mass-produced luxury good that most people previously could not afford. Oil and paint industries were expanding at exactly the same time as trade associations were, as well as the anti-adulteration and pure food movements, and trade knowledge was deeply intertwined with the standard trends in business organization. Manufacturers blamed excessive competition for the rumours of adulteration, and reformers blamed the associations for conspiring to adulterate consumer goods. The work of the chemist was important for discerning impurities, and North Dakota's head chemist and pure-food law crusader Edwin Ladd blew whistles whenever he suspected adulterated oil. Linseed oil crushers reacted in different ways to regulation. Firms like Spencer Kellogg and Alberta Linseed Oil emphasized transparency in their processes and the role of their industrial chemists in bringing the best standard of linseed oil to the covering industries. Other companies, like National Lead and Livingston's Dominion Linseed Oil, reassured customers that they used time-honoured methods of production and therefore produced the kind of reliable oil and paint that one's grandfather might have used.

Chemistry was not the only or even the primary way linseed oil was made and monitored. Consumers had not asked that paint be added to the list of regulated goods, and they likely did not feel in any way disconnected from the old-fashioned processes of paint making. On the contrary, when the value and complexity of paint increased and demanded inputs from chemists, consumers actually became more closely connected to the various strands of the commodity web, in the sense that many could now buy and apply significant quantities of paint. However, the process of painting was connected to the process of farming, and specifically to the farms of the Northern Great Plains and Prairies. There, flax became one of the tools in the hands of sodbusters. At first it was used cautiously, it was only grown with the assurance of a strong market price and on good agricultural land, but eventually farmers took greater risks both financial and environmental.

CHAPTER SIX

Cover Crop

*Growing Flax for Linseed
Oil and Paint*

Beginning in 1878 and especially by the start of the twentieth century, the Northern Great Plains and Prairies became North America's flax belt. US crushers and farmers took the business of flax westward with settlement in a simultaneous and symbiotic process of supply and demand, but in Canada crushers brought flax to the West with a single journey and a contract. Unlike its sister oilseed crop – cottonseed – flax cultivation moved west and did not expand near its original industrial heartland.[1] If grain farming was – after ranching – the second spear thrust of the Anthropocene into North America's grasslands, then this oilseed irruption was the thin edge of the wedge. Not only were the Northern Great Plains well suited to flax cultivation in terms of land use and growing conditions, but flax seemed best suited to new land. For various reasons flax was thought to outperform other crops on new breaking, preparing the soil and eventually conceding its space to wheat and corn. The plant grew quickly on new breaking and best on land that was free of disease, and therefore it became a significant first crop on some of the most fragile ecosystems and unforgiving northern grasslands. The active relocation of flax seed production to areas of extensive new breaking was both a response to the demand-side shift in flax's commodity web and a biological innovation that was not found in other producing regions, such as Argentina and India. The innovators included producers who combined cultural experience learned both locally and in their places of origin, and processors who, when the industry was nascent and smaller in scale, knew more about the idiosyncrasies and opportunities in flax than many flax farmers.

The change was an example of market-responsive and biological innovation – that is, it was an attempt to increase value and output per acre with non-mechanical adaptations to new markets. The flax story contributes, in this way, to a Canadian historiography of western environmental adaptations, including dry farming and new plant varieties, and to Alan Olmstead and Paul Rhode's more recent revisionist approach to the emphasis on mechanical innovation and labour productivity in US agriculture. To their list of "nonmechanical innovations – new plant varieties, fertilizers, pesticides, irrigation or drainage systems, improved cultural practices," we should add farmers' mobility in marginal and agriculturally hazardous environments and in less familiar "first" practices such as sowing flax on new breaking. To the now standard Canadian story of scientific farm improvements and prairie super-staples we can add regional specialization and the influence of industrial knowledge on agricultural expansion.[2] It is likely that the first North Americans to prefer planting flax on new breaking were the flax millers themselves, and the first farmers to record other reasons for the practice were the Mennonites of Manitoba, where flax had been introduced by James Livingston and other Ontario flax millers.

Beginning with the trade relationship formed between James Livingston and the Mennonites of the Manitoba West Reserve in 1878, the Northern Great Plains soon became a major source of the world's linseed oil and paint. For Livingston and other millers with backgrounds in the fibre industry, a specific consideration was the effort to minimize their costs of raw materials by maintaining close connections with their suppliers. In Salem, Oregon, a linseed oil company with the name Pioneer Oil used contracts with local farmers to secure the product of about 6,000 acres in the early 1870s.[3] Back in the Great Lakes region, as the centre of US flax seed production moved from Ohio to Illinois, the much smaller Canadian crop was declining in Quebec and rising only slowly among Ontario's fibre farmers.

Two Thousand Kilometres to Market: Establishing a Western Connection

It is not surprising that Canadians thought that the new Canadian West had the potential to be a profitable and possibly more predictable source of raw material. James Livingston targeted Mennonites to supply his raw material, and he quickly coordinated the necessary infrastructure to supply seed to Mennonite farmers and to buy their harvests under contract. Industrialists like Livingston believed the heavy prairie soil was good for flax cultivation, and once distribution and marketing systems were established in the following decades, this part of the prairie did become the ideal flax-growing region. This North American flax zone developed under near perfect conditions and provided much of the raw material for the exploding markets for oil and paint in midwestern and eastern cities. If the rate of western settlement disappointed railways and wheat merchants, it at least satisfied the Canadian market for linseed oil rather well.

As we have seen, mid-nineteenth-century millers adapted to local markets by retooling and refitting their mills to include flax. Practically simultaneously, other manufacturers as large as the Lymans and as small as Eby's oil mill were moving into linseed oil, and again, it was because linseed oil lined up with the other commodities they were producing. Although the manufacturers did not have to take control of production, they stimulated a demand for seed that prompted a reaction among flax producers. Within a generation of the turning to specialized flax production in Waterloo, flax seed producers in the Great Lakes region no longer satisfied the industrial demand for linseed oil. Although they roughly doubled Ontario's annual seed output by 1890 (to over 70,000 bushels), that fell far short of the rate of increase of demand. In the 1890s the largest oil business in the province, Livingston Linseed Oil, crushed half a million bushels per year and claimed to sit inactive for weeks at a time because of seed shortages.[4] For the first time the region's manufacturers followed new producers rather than create and coordinate production from local resources.[5]

Flax was eventually on the minds of many settlers heading to the North American West. To farmers, it was a now-valuable commodity to mix with wheat and other cash crops. To processors, western development and flax's suitability to the region represented limitless sources of raw material. Both concepts were as novel as the mechanical fibre processing had been in the mid-nineteenth century. Farmers previously considered the seed a by-product, and extensive production was out of the question without an industrial framework for labour and other inputs. Flax industrialists had been accustomed to investing in capital frameworks and then seeking raw material. Now the raw material was best produced in a region far from linseed oil's industrial establishments and consumers.

Livingston likely noticed the conspicuous presence of Russian Mennonites who sojourned in Berlin (Kitchener) and Waterloo on their way to Winnipeg in 1874. They followed the news of these immigrants in subsequent years, and at some point they planned to survey the land themselves and look for potential flax producers in Manitoba. Jacob Y. Shantz, a Waterloo Mennonite community leader and businessman, has been called a "vicarious pioneer" for his work helping these immigrants establish themselves in Manitoba.[6] Mennonites in Waterloo County were interested and deeply invested in the Manitoba settlement schemes for Russian Mennonites, but the connection was purely ethnic and not related to the flax they began growing for Waterloo millers. Several years after Elias Eby sold his linseed oil mills in Bridgeport, he commented on the immigrants as they passed through Waterloo "to the so called grassland" and hosted one family of ten in his home in 1874.[7]

Eby did not mention Livingston's journey to Manitoba in his diary. There is no cultural reason to think that local Mennonites like Shantz or Eby would have considered their co-religionists likely candidates for flax cultivation. By 1871 there was no more preference for flax-growing among Waterloo Mennonites than there was among any other farmers, and in fact, most of the flax was grown in townships further north and west. Of the forty flax farmers in Livingston's own Wilmot Township, half were German and three were Pennsylvania Dutch by origin. However, only eight flax growers were Mennonite by religion, and the cultural connection to flax was tenuous indeed.[8]

Initially, Livingston dealt with Mennonites in the East Reserve, but he soon concluded that not much flax would be grown east of the Red River, and he moved to the West Reserve along the Canadian–American border.[9] There, the company established seed warehouses and hired German-speaking agents in at least five locations. The first of these agents was Philip Erbach, a former cooper from Baden who operated out of Emerson and serviced the West Reserve.[10] Later, James Livingston's own nephew Peter Livingston occupied this role. Peter was living in Milverton, Mornington Township, when he married Eliza Ann West in 1874. He initially worked in the family business as a flax dresser.[11] When he took the job in Manitoba, he and Eliza were living in a company-owned house in Morden, on the West Reserve. The second Livingston depot was built in 1883 in Gretna where, the *Winnipeg Times* stated, "Messrs. Livingston Bros., flax merchants, have erected a large warehouse ... for the purpose of buying about

25,000 bushels for shipment to Ontario."[12] Ten years later the Livingstons had contracted with the Northwest Transportation Company to move 100,000 bushels, all purchased through their agents in southern Manitoba.[13] By the end of the century, the company's real estate in Manitoba included warehouses in Gretna, Altona, Rosenfeldt, Plum Centre, and Winkler, and the dwelling in Morden.[14]

The company's role in the reserve was an important link in bringing large quantities of flax seed from previously uninterested farmers to a distant Canadian mill. In Erbach's home town of Baden, he was recorded as a "commercial traveller for flax mills."[15] He wintered in Ontario and spent most of the remaining seasons in Emerson (West Lynn, more specifically). Work as a commercial traveller involved much more than buying seed. Erbach's role was to bring flax seed for sowing from Ontario and elsewhere, buy the flax seed crop from farmers in the fall, select and store the best seed for sowing, and then clean and distribute it in the spring. It appears that Erbach's major occupation in the spring was to operate a seed cleaner that he had brought from Ontario and to produce "pure clean flax of a very superior quality, free from all impurities and other seeds of any kind whatever after passing through the machine. This is the seed," according to the *Times*, "very much required in Manitoba." Farmers were no doubt cautious about the purity of their sowing seeds, but they were primarily interested in Erbach's seed-loaning role. Erbach loaned seed to farmers on the condition that they sow it and sell their product back to Livingston. It is not clear from the *Times* how the contracts were arranged and at what prices, but we do know that farmers were offered the seed in March and that "applications for seed [came] in so fast that Mr. Erbach found it necessary to have his machine for cleaning the seed running night and day to supply the demand called for by the farmers."[16] The early date of this article, 23 March, suggests that Mennonites knew well in advance of the growing season how much flax they would plant. Sowing took place from about 25 May to 7 June, two months after they contracted with agents like Erbach.[17] As the urban industrial complex began to spread its influence over the prairie grasslands, farmers like the West Reserve Mennonites entered into a relationship with the millers that bore some resemblance to fibre factorship in the East. However, the urban complex was not yet complete, and these Mennonite farmers continued to exercise a great deal of choice with respect to the planting and harvesting seasons.

Planting the Mennonite West Reserve

Consistent with the nineteenth-century flax fibre promoters, Canadian observers such as the *Winnipeg Times*, the Department of Agriculture, and other government agencies often expressed bewilderment when they heard that Mennonites and other Prairie flax farmers discarded the fibre. They hoped that flax would be grown and processed for fibre to compete in the northern European market.[18] E.B. Biggar of Montreal presented a paper at the Colonial and Indian Exhibition in 1886, promoting flax fibre production on the Prairies and noting that Manitoba's Mennonites were burning the fibre for lack of a market.[19] For most of linseed oil's history as an industrial commodity, very few except for farmers and millers sensed that flax seed was the only major industry offered by the crop. Western flax was almost always harvested by the blade, and even if it was hand pulled, the fibres were coarser than flax grown in the East. The original expedition to the West had included fibre millers like Brown and Bauer, two fibre millers from Hay Township, Huron County, who actually claimed and purchased a section and a half with the intention of establishing a scutch mill and a "'thousand acre' flax farm" near Emerson.[20] However, these millers failed to "prove up," and the farm was abandoned. Livingston soon realized that fibre was unprofitable in this context. Manitoba's farmers were not interested in the risks associated with using scarce labour for pulling flax.

The Livingston seed contracts made it possible for Mennonite farmers to produce increasing amounts of flax seed and eventually supply the greater part of the domestic raw materials required for Canadian-made linseed oil. The Mennonite immigrants had difficulty adjusting to the Canadian winters and to having to break the prairie sod so that it would be ready for crops if needed quickly.[21] They did not grow flax in any systematic way before they were approached by Livingston through his German-speaking agents in 1878, and they would not have tried to grow flax prior to having an assured market. In 1877, according to the Manitoba Department of Agriculture, Mennonites produced only 280 bushels of flax seed, likely in conjunction with Brown and Bauer in Emerson. The following year, production had increased to over 5,000.[22] The first full census of agricultural production in the Mennonite West Reserve counted almost 7,000 bushels in 1879, and as we have seen, the Livingstons hoped to buy at least three times that amount in Gretna alone in 1883. In 1891 the District of

Selkirk, where most Mennonites settled, reported 32,000 bushels of flax seed (92 per cent of the province's crop and therefore most of the flax grown in the Canadian West), and in 1895 production approached 100,000 bushels, all sold to Livingston for one dollar per bushel.[23] By the turn of the century, farmers in Lisgar and Provencher produced 77,844 bushels of flax seed, 95 per cent of the province's crop, and in 1906 they harvested 148,000 bushels. No other district in Manitoba grew more than these two areas from 1890 to 1906. Mennonites relied mainly on Livingston and their own local seed supplies for spring seeding. They evidently explored other sources of supply, such as the Traill, Maulson, & Clark Company, grain exporters and agents in Winnipeg. In 1886, G.J. Maulson planned to import "several thousand bushels" from the United States specifically to offer the new Mennonite flax growers "a change of seed." However, imported seed was not duty-free. Maulson applied to the minister of customs for an exemption, but it is not clear whether it was given or whether his company ever sold flax seed to the Mennonites.[24] Meanwhile, the Livingstons enjoyed many years doing business with Mennonite flax growers in the West Reserve.

The participation of Mennonites in world markets seemed to transgress their commitment to separation in every way from worldly behaviour, and many have thought of Mennonite culture as completely autarkic. Mennonites, however, not only sold flax outside of the West Reserve, they also were never entirely self-sufficient in the nineteenth century. Prairie farmers used some seed medicinally, and some women used boiled flax seed as a wave-set for hair styling, but of all the grains, flax seed was probably the least suitable for on-farm consumption.[25] Linseed meal was a popular feed for livestock in Europe, but even it had to be milled first. Clearly, flax could not be considered a feed crop in the same way as oats and corn. Although it was once commonly believed that Mennonites in the West clung to European-style peasant economies and resisted the market-based commodity production of their neighbours, a judicious involvement in commodity markets was actually what allowed Mennonites to remain culturally separate in their new country.[26] Even before emigrating from Russia, the Mennonites traded and exported most of their agricultural produce, and from the earliest days of their settlements in Manitoba, they participated regularly in markets as consumers and produced commodities like flax seed that were not for consumption in local households, or even in distant homes, without an elaborate and specialized manufacturing process. In these communities, where some held that it was worldly to own a bicycle or to attach bells to horse harnesses, there were apparently no arguments made against producing commodities for a luxury trade – paint.[27]

Mennonite communities in the Prairie West have long been viewed as a curiosity by outsiders, especially in their unusual settlement patterns based on villages rather than on individual farms. When one Dominion land surveyor travelled to the West Reserve in the fall of 1881, he judged that there was little sign of "agricultural improvements." Like many visitors to the Plains, he experienced the disorientation and vastness of the landscape. However, when he approached one of the villages and saw the concentration of grain and haystacks, he felt he "had to consider the matter of improvements in a different light."[28] To many visitors, these signs of work were signs of life, settlement, and "improvement," but they appeared differently in the West, and especially in Mennonite communities. An examination of Mennonites through their use of flax reveals other signs of work and environmental adaptation, as well as distinct forms of human and natural ecology. Manitoba's early flax culture differed from Ontario's and reflected the ethnic and economic dissimilarities between the two provinces. Mennonite villages either produced large per capita quantities of flax seed or none at all. Just a few villages in the Mennonite West Reserve together produced up to 100,000 bushels per year, or more than all of Ontario's annual crop combined.[29] Within each flax-producing village, most if not all of the farmers grew some flax. The village of Rosenort produced over 1,000 bushels of flax seed in 1881. All but three of the twenty families grew flax (two of the non-producers were likely herders or employed outside of agriculture). However, there were very clear frontrunners, and over half of the crop was produced by four farmers, quite unlike the equal spread of wheat, barley, and oats among the villagers.

The major producers in Rosenort tended to be close neighbours, and most of the flax was produced by two groups of neighbours. The advantage of living beside other flax-producing families was unrelated to any shared home production of textiles or oil, as the flax was entirely sold off the farm and the fibre was burned. It was likely useful to have a neighbour in flax who could share information, implements, transportation, and perhaps even labour. Mennonite historians have shown that the typical Mennonite *Kagel* was ordered so that each villager had several long, narrow plots of land that surrounded the village, and each had relatively the same distance to travel to each plot.[30] The village system indicates that the advantage of living next to a flax farmer was simply to have access to knowledge and experience and not necessarily to have shared production in the field.

Rosenort was an unusual case in that it was home to flax farmers yet was relatively distant from the flax depots established by the major buyer, Livingston Linseed Oil. Other flax-producing villages, such as Silberfeld and NeuAnlage, were located much closer to Livingston's operations in Gretna and Altona. These farmers cultivated less land on average than farmers in Rosenort, and they produced much smaller quantities of flax and other crops. In Silberfeld and NeuAnlage, the flax crop represented a significant proportion of each village's total wealth. In NeuAnlage the year's flax was worth about $600, or the value of half of the village's wheat crop, but in Silberfeld the wheat crop was smaller and the $520 flax harvest was worth over two-thirds of the market value of wheat. Most villagers earned a surprising amount from growing flax, and their willingness to farm what so many other Manitobans did not suggests that they were confident in their knowledge of the crop and of the strength of the market provided by Livingston and others. Silberfeld and NeuAnlage also differed from Rosenort in that only one farmer in each village stood out from the others as a major producer. In Silberfeld, Jacob Spenst harvested 135 bushels, or a quarter of the village's production, and in NeuAnlage, David Schellenberg harvested the same proportion at 150 bushels, or the product of about 15 acres. It is not clear what might have made individuals like Spenst and Schellenberg more interested in flax than their neighbours, but their crop choices reveal concentrations within concentrations.[31]

Flax seed production was just as concentrated and specialized as the fibre crop in Ontario. It would take about 700 David Schellenbergs to cultivate a crop like the West Reserve's 10,000 acres in 1883. There *were* roughly that many farmers in the West Reserve (769 families in 1880), but since several villages had no flax at all, there must have been a few places where production dwarfed all the others. The reserve and its contiguous townships produced 97 per cent of the flax reported by the Manitoba Department of Agriculture. Mennonite farmers sowed flax on about 8 per cent of their cropland, a little more than other Manitoba townships reporting flax. Again, production tended to cluster, and in one township adjacent to Mennonite settlements almost a quarter of the cropland was in flax. Historians have pointed to flax cultivation as one of the Mennonites' contributions to Manitoba settlement; this kind of flax culture did not come to the neighbouring North Dakota and Minnesota counties for at least another decade.[32] In 1889 the entire span of counties bordering Manitoba produced less than 1,000 acres of flax seed, and the largest flax-farming county within 300 kilometres of the West Reserve was Walsh, North Dakota, which cultivated less than 5,000 acres. The enclave of flax producers in Manitoba brought flax to the northern plains long before the supposed march of this frontier commodity reached the US Great Plains.[33]

The State's Sodbusting Plant

Flax earned its reputation as a frontier crop when experts began to notice that it was a popular first crop on new land. From the outset of flax production in Manitoba, it seems that farmers reserved it for their newly broken land. Local newspapers, essayists, and the Department of Agriculture noted that flax performed well on first breaking, often when sown at half the normal rate. In 1878 the *Manitoba Weekly Free Press* noted that Mennonite farmers used flax as a "catch" crop on new breaking.[34] In 1881 M. Cooke, a farmer from Norway House, claimed that the crop grew best on newly broken sod according to Mennonite practice, and five years later J.Y. Shantz told the Committee on Immigration and Colonization that this was indeed how most Mennonites farmed flax.[35] However, these observations were based specifically on the practices of a concentrated group of farmers with ready access to a strong flax seed market. It is unclear whether the idea of seeding flax on the most recently ploughed prairie was one they had taken with them from eastern Europe or whether it was a concept suggested by Livingston and his agents. One Mennonite in Waterloo County recalled that flax was "well adapted to freshly broken land" because it could be planted later and allowed time for spring breaking to dry. There are very few references to this practice in Ontario, and this one in particular may be presentist.[36] Whatever its origin, the practice of sowing flax on new breaking served mainly to make the best use of land and labour inputs. Mennonite flax producers found they could produce "quite as much wheat as their neighbours ... [because] the seed was only put in after wheat sowing was done, or on new breakings on which wheat could not be sown."[37] In addition, John Lowe, deputy minister of agriculture, wrote an article on the success of the Mennonite flax crop in an interview with the *Daily Mail and Empire* in 1895. He promoted flax by pointing to the Mennonite example and claiming that the plant "can be grown with success on the first breaking, and sown after all other grains are put in." Lowe reported a flax crop in Morris County of 160 acres, "the whole of which last fall was unturned Prairie sod." The yield from this new land was expected to be an impressive 22 bushels per acre, and for extra authority he cited "a Mennonite accustomed to looking at flax fields."[38] So, flax's life as a frontier crop was likely as much about the innovative ordering of time and land resources as it was about any other botanical cause.

It does not appear that sowing flax on new breaking was recommended by other Manitoba farmers or experts. The Department of Agriculture circulated a questionnaire on first crops, and most correspondents argued that nothing should be sown on first breaking for twelve months. Others said that two years were required to prepare the soil for planting, and only a couple of farmers suggested planting wheat or oats on new soil. Flax was not discussed because most Manitobans had never grown it and had no reason to think that it would pay for its stay on the land. Cropland was not scarce but cash was, and growing even a thin amount of a potentially high-yielding crop did not make sense if it did not pay for the expense of buying and farming the seed. Nevertheless, what we do learn from the Mennonites' and other farmers' early experience with flax in Manitoba is that it was only grown where there were expectations of a strong and accessible market, and that what appears to be their practice of using flax on new land advised the next generation of prairie farmers and helped form the opinions of flax experts and other observers.

The salient features of the production chain were its temporal instability and its geographic fugacity. To the nation's top farm officials, mobility was symptomatic of flax's role as a sodbusting crop. State officials initially held that flax grown on freshly broken sod helped prepare the land for subsequent crops. Thus, in 1896 William Saunders, the nation's top farm scientist, claimed that flax grew well on first breaking in the Northwest and also on fallowed land. He endorsed its role in the prairie plough-up, claiming that when "flax is sown on first breaking, a seed bed ... is provided, the farmer derives a revenue from his land the first year, and the crop effectually rots the sod so as to admit of ploughing to the ordinary depth in the autumn."[39] State farm officials discussed flax's movements in the United States during the 1890s, and in 1902 the USDA claimed that "[t]he flaxseed crop of the West has been distinctively a 'pioneer' crop ... [and] insofar as the reproduction of flaxseed is concerned, quickly exhaustive of the soil." The rhetoric intensified in later years. In 1907 a US farm scientist considered flax "valuable [for] subduing the virgin sod" and noted that "since the opening of the new lands in the West ... it has been a crop with which to reclaim the native sod."[40] In 1910 students at the Ontario Agricultural College observed that "on the prairies of the West flax is often sown on freshly-broken sod," although one study warned that flax was injurious to the wheat crops that followed.[41]

The notion of flax as a "nomadic" crop became part of frontier mythology for the better part of the twentieth century. An important manual for the linseed oil industry explained, in 1909, that flax's frontier nature drove production westward from Ohio to the virgin soils, after 1865.[42] In 1917 the Harvard economist Walter Barker claimed that "after a few failures the standard grain crops tend to displace flax seed and drive it on to another frontier." By 1890 it had moved into the Dakotas, and by 1900 Montana, "where it is now held up in its march by the mountains." According to this rhetoric, ordinary farmers could only use flax for a limited period on their newly broken land. But, with the help of agricultural science, "flaxseed may settle down so that a thoroughly scientific culture may be developed."[43]

The first prairie flax farmers in Canada, together with the linseed oil millers who bought their seed, did not think of the crop as particularly nomadic. Their experience with flax seed shows, rather, that it was *only* grown with expectations of or even contracts for a good market. James Livingston's initial journey and Erbach's subsequent work in the West Reserve are early examples of eastern markets established in the West. The story of flax seed as a first crop in the West represented innovation and adaptation and spoke to environmental stresses and human survival on marginal land, or to the process of learning what land can or cannot grow and what should and should not be farmed. Planting flax on breaking was another small form of biological innovation and an example of non-mechanical technology used to survive the difficult first years in the new West.[44] The flax-paint web also shows the reciprocal influences of market demand and productive ecosystems on the actors who produced and consumed flax. As markets for seed increased and political and economic expansion allowed the resettlement of new agro-ecosystems in the West, farmers and millers responded and adapted by converting fibre to seed production.

Some Mennonite groups have been considered the epitome of the myth of a people resisting industrial capitalism and embracing a simple life on the land. Yet months after migrating to the West Reserve, Manitoba's Russian Mennonites responded to demand-side market shifts and soon became the largest producers of Canadian flax seed. Linseed oil crushers turned to western sources of supply, and seed growers adapted their agricultural practices to flax's short calendar and suitability for new breaking work in the northern grasslands. In the Canadian example this transition began in Ontario in the 1860s and was complete by the early twentieth century when plentiful seed was supplied from the West.

Biological innovation helps to account for the use of flax on new breaking, but it does not explain why some practised it and many others did not. Earlier chapters addressed why flax fibre operations might have appeared in clustered geographies, but even western Mennonites grew flax in concentrations in certain villages, and across the rest of the extensively cultivated West, flax usually appeared in a concentrated area. Soon farmers in the Northern Great Plains and Prairies became globally competitive flax seed producers, but trends that began with a small group of Mennonites continued as the crop entered the rotation of tens of thousands of farmers in the region each year.

The Edge of Expansion: Flax on New Land

Flax seed was the third most valuable cultivated crop in the three Prairie provinces in 1910, and afterwards it was briefly the second most valuable grain in Canada.[45] A similar acreage of flax had been cultivated each year in the United States for over three decades, and after 1910 the two countries' outputs followed similar trends. All of this acreage was for seed, and most fibre was simply destroyed. Unlike wheat, flax was grown mostly on new land and then abandoned for other cereals. Historians have noted this practice, but the land-use strategies have not been explored.[46] Through a spatial analysis of data from the Canadian and US Censuses of Agriculture, one can see the constant migration of the centre of flax production and its relationship to newly ploughed land.

The semi-arid regions of the Dakotas, Saskatchewan, and Alberta were initially considered unproductive. Early Canadian observers of this area, later dubbed Palliser's Triangle, declared it too arid for cultivation. However, in 1880 John Macoun's second western expedition for the Canadian government took him through the dry belt of the Prairies. There he decided that the dry grasslands could be farmed profitably and would benefit from breaking and prolonged cultivation – exactly what his expansionist sponsors wanted to hear.[47] He also considered flax fibre "one of the best products of the Northwest" and reported seeing some crops grow twelve feet high.[48] If this apocryphal tale was true, the crop was almost certainly hemp, not flax. Bill Waiser has argued that Macoun's "'rains follow the plough' notion was fraught with difficulties [and] created the false impression that bringing the shallow and light soils of the second and third prairie steppe into agricultural production was a simple matter of cultivation." In fact, these drought-prone areas would require "special techniques" for prolonged cultivation, techniques now canonized in the story of environmental hazards, technological and scientific triumphs over nature, and eventual abandonment of areas too arid for cultivation.[49] Along with the development of dry farming and early-maturing and drought-resistant varieties, flax cultivation was another element of adaptation to grassland ecosystems. Canadian fibre promoters such as John Castell Hopkins used Macoun's reports to encourage hemp production in the Prairies, and Hopkins declared the grassland "lying beyond the line of safe wheat growing," an ideal environment for up to one hundred million acres of flax.[50]

Flax cultivation was part environmental adaptation, part business strategy. A farmer's land-use system operated like a differential equation with environmental, economic, cultural, technological, and demographic variables. Seeding flax on new breaking, which seemed to contemporaries and some recent scholars like a transplanted American activity, could not have been a predictable cultural trend. The culture argument rested on the assumption that settlers moved from one agricultural region to another and their practices represented some homogeneous culture in their places of origin. But historians have shown that Prairie settlement was part of a series of individual and multi-generational migrations, Plains farmers were surprisingly mobile, and these producers were sometimes away from their farms more than they were on them.[51] It is impossible to gauge the extent to which background practices were valued and implemented, although economists and geographers have used place of origin as a measurement for Plains migration and land use.[52] It is well known that expert discourse encouraged Prairie settlers to transplant an eastern-style agricultural system, but it is unlikely that farmers naively assumed new wine would keep in old wineskins – or that traditional land use would work in new environments.[53] People brought practices with them, but they also evolved with every stop in their journey.

A better explanation for the practice of allotting new land for flax includes a range of factors found on the commodity web, from the idiosyncrasies of the crop to the value of the finished object. State officials, farm

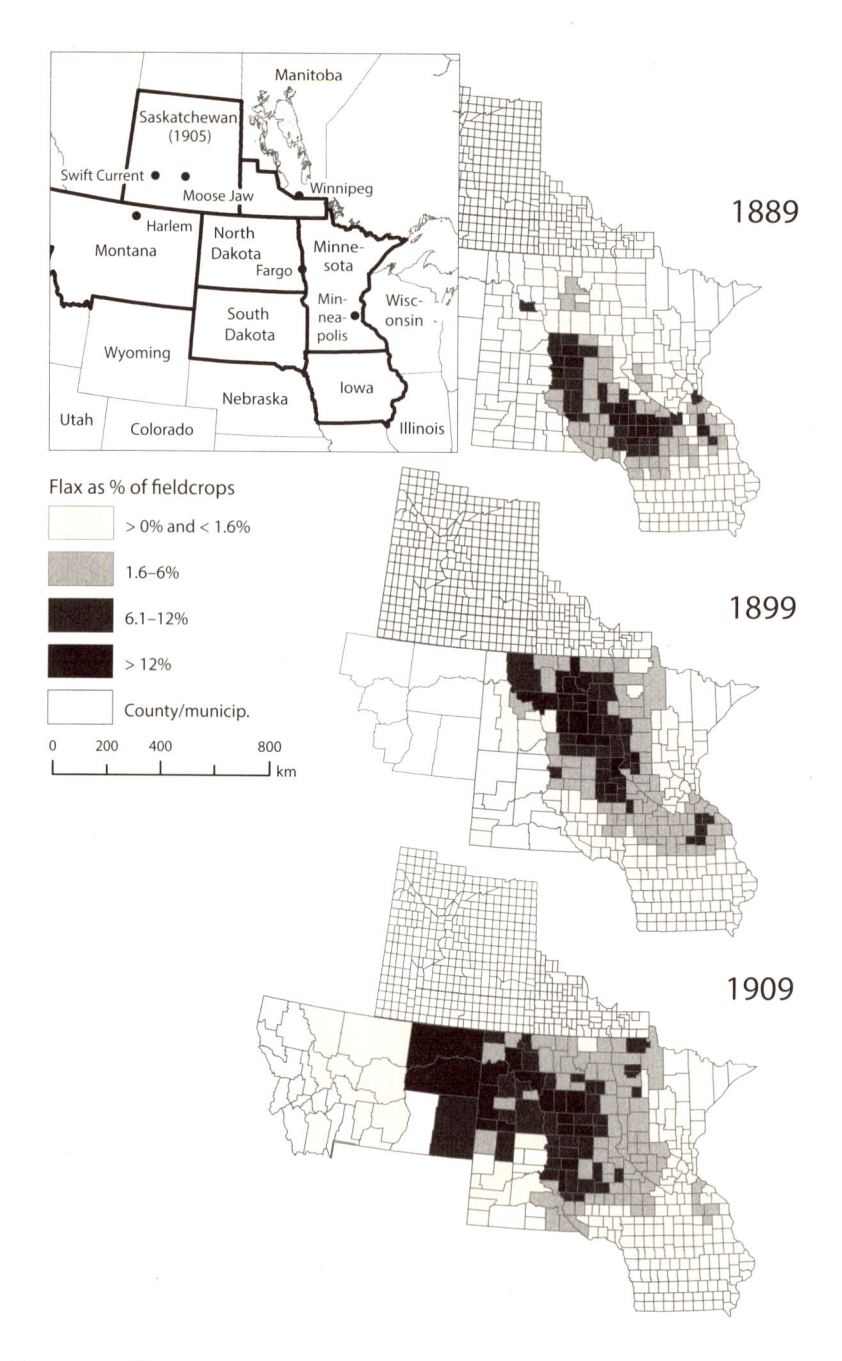

Figure 6.1 Flax as a percentage of all field crops sown on the Northern Great Plains and Prairies, 1889–1926

Sources: Canada, Dominion Bureau of Statistics, *Census of Canada 1921* (Ottawa: Dominion Bureau of Statistics, 1924); *Census of the Prairie Provinces,* 1916 (Ottawa: J. de L. Tache 1918); *Census of the Prairie Provinces* (Ottawa: Dominion Bureau of Statistics 1926); United States,

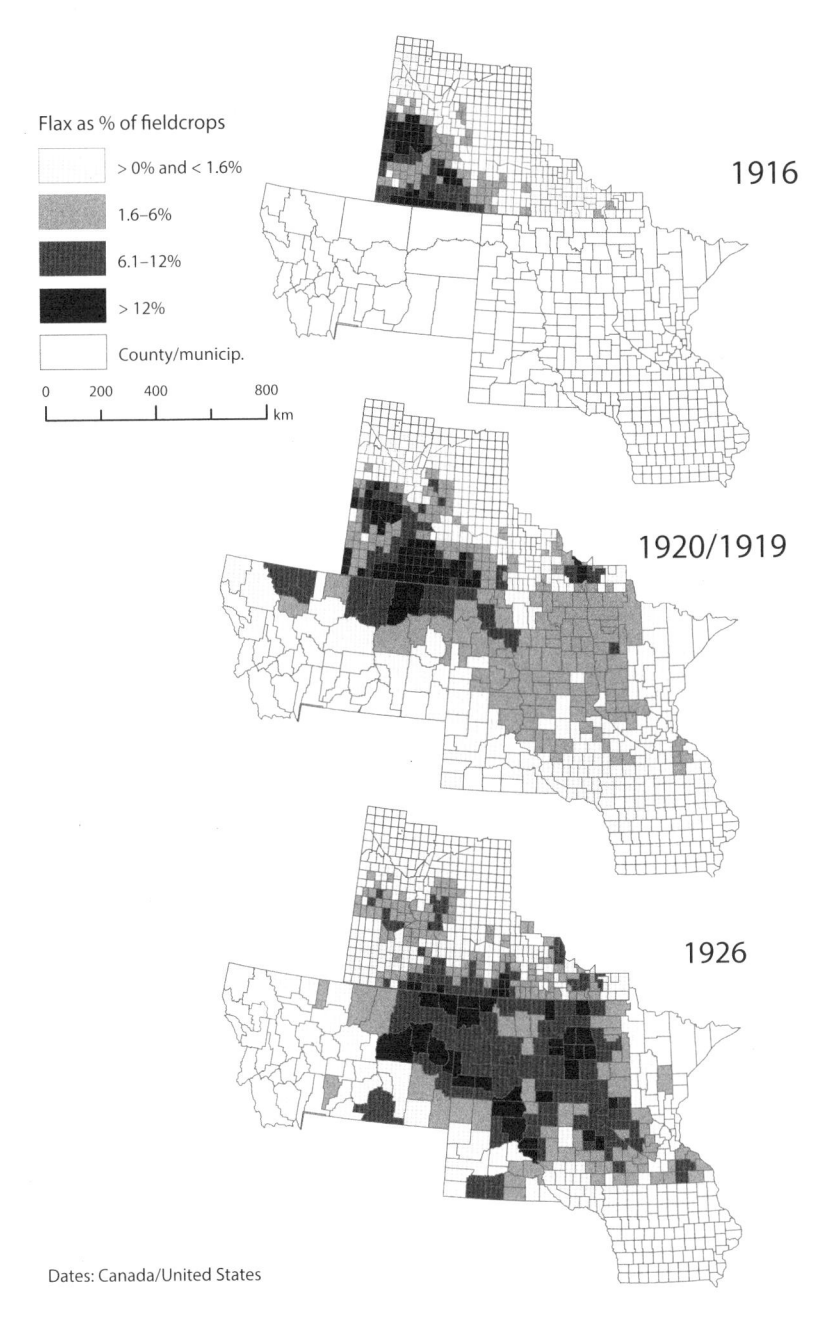

Flax as % of fieldcrops
- > 0% and < 1.6%
- 1.6–6%
- 6.1–12%
- > 12%
- County/municip.

0　200　400　800 km

1916

1920/1919

1926

Dates: Canada/United States

Department of the Interior, *Eleventh Census, 1890* (Washington, DC: GPO 1895); *Twelfth Census, 1900* (Washington, DC: GPO 1901); *Thirteenth Census, 1910* (Washington, DC: GPO 1913); United States, Bureau of the Census, *Fourteenth Census, 1920* (Washington, DC: GPO 1921); *Census of Agriculture, 1925* (Washington, DC: GPO 1927); *Fifteenth Census, 1930* (Washington, DC: GPO 1931). US spatial boundary files are used courtesy of Historical United States County (HUSCO) Boundary Files, Louisiana State University.

scientists, capitalist commodity producers, and the oilseed industrialists all had grand explanations for why flax appeared mostly on new breaking, but in reality farmers would only put flax on new – or any – land when a range of economic and environmental conditions were in place. It was never sown without the promise of adequately high prices or in the absence of affordable seed and other inputs. When price allowed, it tended to appear on new breaking, as it could be planted later and transported further without upsetting the balance of other activities and without the farmer having to learn many new techniques. Government officials argued that flax was an agent of pioneer agriculture; to farmers, flax on new breaking was something so ordinary and routine it was rarely discussed, merely practised.

Officials based their theories of flax's westward movement not on experimentation but on the pattern of flax production visible in aggregate census data. The centre of flax production shifted several times between 1889 and 1930 in the Northwest states, Manitoba, and Saskatchewan (Figure 6.1). In fact, the nexus was never in the same place twice over any given quinquennial or decennial period before 1925. Beginning in Iowa, the location of farmers who chose to grow flax clearly moved north and west between 1889 and 1916. This was followed by a slight southeastward movement in Saskatchewan until the Great Depression. Then Canadian flax virtually disappeared from production, until wartime prices encouraged a gradual recovery. In all of flax's early history in Saskatchewan, the farmers who made it a significant part of their field crops usually lived in districts along the US border.

Despite its apparent presence as a frontier crop, flax did not suddenly become a major component of land-settlement strategies in Canada. New breaking increased quickly in Canada after 1900, but it was only in 1909 that the Saskatchewan Department of Agriculture noticed an increased interest in flax.[54] Then the Canadian flax crop, located almost entirely in the Prairie provinces, doubled its 1910 acreage to 1.4 million in 1911 and averaged about 1.1 million acres per year over the next decade (Table 6.1). The Canada census attributed the unprecedented flax production to "the opening up of the Prairie provinces where flax has been sown to good advantage on new breaking." Even as the federal report was being compiled, provincial officials were calling 1911 "a real flax year in Saskatchewan." They connected it to the mass immigration of settlers from the western United States and the "consequent increase in new breaking."[55] American settlers were considered "desirable citizens" in many ways, especially because they

Table 6.1 North American flax seed production and price, 1901–1931

	United States				Canada			
Year	Production, mil bu.	US, % production of total consumption	Price, $ per bu.	Tariff, $ per bu.	Production, mil bu.	Acres '000	Price, West, $ per bu.	Acre Value, $ per ac.
1900	16.0				0.2	23		
1901	27.6	116%		0.25				
1902	36.1	116%	1.05	0.25				
1903	25.4	103%	0.82	0.25				
1904	22.6	99%	0.99	0.25				
1905	28.7	127%	0.84	0.25				
1906	27.6	136%	1.01	0.25				
1907	23.8	120%	0.96	0.25				
1908	20.6	102%	1.18	0.25	1.5	139	1.07	11.50
1909	19.5	79%	1.53	0.25	2.2	139	1.48	23.53
1910	11.4	51%	2.31	0.25	4.2	582	2.17	15.86
1911	18.5	73%	1.82	0.25	15.4	1,351	1.83	20.88
1912	28.1	86%	1.29	0.25	26.1	2,022	1.05	13.57
1913	15.1	68%	1.23	0.25	17.5	1,553	1.04	11.70
1914	12.9	57%	1.31	0.20	7.2	1,084	1.14	7.57
1915	11.3	49%	1.68	0.20	6.1	463	1.68	22.21
1916	11.9	55%	2.31	0.20	8.3	658	2.41	30.20
1917	8.4	42%	3.11	0.20	5.9	920	2.97	19.15
1918	12.8	62%	3.58	0.20	6.1	1,068	3.24	18.37
1919	6.8	23%	4.42	0.20	5.5	1,093	4.55	22.79

/continued

arrived with dry-farming experience as well as animals, implements, furniture, and capital. Some were what Paul Voisey has called "professional pioneers" who had bought and improved successive frontier farms in many parts of the United States and Canada.[56]

Government officials saw a role for themselves in advising newcomers, and in 1911, A.F. Mantle, deputy minister of the Saskatchewan Department of Agriculture, feared that thousands of farmers were homesteading "with little or no knowledge of their work, the soil, the climate or those laws of nature and plant nature with which they must work in harmony." "The first crop that most of these men will attempt to grow," he asserted, "is flax." Farmers used it to "subdue new land" and for "reducing the soil to such a condition that it would grow wheat." Perhaps even more important than its

Table 6.1 /continued

Year	Production, mil bu.	US, % production of total consumption	Price, $ per bu.	Tariff, $ per bu.	Production, mil bu.	Acres '000	Price, West, $ per bu.	Acre Value, $ per ac.
	United States				**Canada**			
1920	10.9	39%	2.33	0.20	8.0	1,428	2.07	11.60
1921	8.1	26%	1.65	0.30	4.1	533	1.54	11.87
1922	10.5	31%	2.08	0.40	5.0	566	1.95	17.21
1923	16.6	49%	2.12	0.40	7.1	630	1.92	21.77
1924	31.2	77%	2.18	0.40	9.7	1,277	2.28	17.32
1925	22.3	57%	2.26	0.40	6.2	843	2.10	15.54
1926	18.5	46%	2.03	0.40	6.0	738	1.74	14.14
1927	25.2	58%	1.92	0.40	4.9	476	1.90	19.50
1928	19.1	47%	1.94	0.40	3.6	378	2.02	19.31
1929	15.9	40%	2.81	0.40	2.1	382	2.48	13.37
1930	21.7	74%	1.61	0.65	5.1	582	1.14	9.93
1931	11.8	45%	1.17	0.65	2.5	648	0.94	3.98

Sources: United States, Department of the Interior, *Seventh Census, 1850: Embracing a Statistical View of Each of the States and Territories, Arranged by Counties, Towns, etc. ...* (Washington, DC: Robert Armstrong 1853), lxxii; *Eighth Census, 1860*, Vol. 3, Manufactures (Washington, DC: GPO 1865); for outputs in 1850, 1860, and 1870, see United States, Department of the Interior, *Ninth Census, 1870*, Vol. 3 (Washington, DC: GPO 1872), Table 10 – Oil Vegetable, 618–19; *Tenth Census, 1880*, Vol. 2, Manufactures (Washington, DC: GPO 1883); *Eleventh Census, 1890*, Vol. 6, Manufactures, 3 Parts (Washington, DC: GPO 1895); *Twelfth Census, 1900*, Vol. 7, Manufactures: United States by Industries (Washington: United States Census Office 1902), Part 1, 323; *Thirteenth Census, 1910*, Vol. 8, Manufactures: General Report and Analysis (Washington, DC: GPO 1913), Table 7, 68; United States, Bureau of the Census, *Fourteenth Census, 1920*, Vol. 10, Manufactures, 1919, Reports for selected industries (Washington, DC: GPO 1921), 748–52.

sodbusting strength was flax seed's high price, and he recommended flax as a commodity that would provide homesteaders with much-needed capital in their first years of prairie farming. On the other hand, he wondered, "is it worth the while of large companies and farmers having capital – men who need not sacrifice the future to immediate returns – to sow even clean flax seed upon their newly ploughed breaking?" Some claimed that the crop introduced noxious weeds, and its soil was said to become "flax-sick" or useless for growing flax in short rotations. Mantle warned that as a result "the crop has become nomadic in character, constantly seeking new localities, and has been used almost exclusively as the first crop sown on virgin lands that were being brought under cultivation."[57] The majority of linseed oil crushers in this period also thought of their raw material as a "nomadic" crop.[58]

By the 1920s when sodbusting had attenuated, the role of flax in breaking was clear to agricultural and other scientists. "Flax," John Bracken argued in *Dry Farming in Western Canada*, is "a dry land crop [that] deserves greater consideration than it has received in the past, particularly on breaking or sod land, and on heavy land that does not blow." It is "very popular" he claimed, "as a first crop after breaking," and by that point it was confined to marginal lands.[59] Historians and other social scientists in the 1920s affirmed that flax had been an important crop in the prairie plough-up and was therefore a frontier commodity. Victor Clark described "the settlement of the Prairie state, where flax proved an excellent crop for subduing wild land."[60] In 1931 a study of Prairie agricultural history found that flax seed production peaked "when much new land was being brought under cultivation." It concluded that flax was most common on newly broken land, and "its acreage has consequently fluctuated" to correspond with settlement.[61]

In the late nineteenth century, John Murch moved from England to North Dakota and then north to Saskatchewan in a series of farm migrations. He moved back to North Dakota in 1906 and then filed on a quarter section in 1909, near his brother's homestead, in Shackleton, Saskatchewan. Their homesteading experiences reflect standard operation for that period and place. John and Alfred dug a well on John's land and built shacks on each property to begin proving up. Breaking was a slow process, beginning in 1910 with ten acres under a walking plough pulled by oxen. In 1911 John planted oats on his new land, threshed them in 1912, and in the next year sowed a modest crop of flax that yielded nine bushels per acre. A promissory note dated 6 August 1913 shows that John agreed to pay the International Harvester Company of Canada $60 for a "McCormick Binder with flax carriage" before October.

Alfred's progress was practically identical and suggests why these farms developed so slowly. Alfred was actually a labourer on the railway, hoping to raise the funds necessary to operate his and his brother's farms. They lived in the same house on John's land and stored supplies in both houses to meet homesteading requirements. The middle-aged John married Rachel Vistnes in 1913 and began to operate his farm separately from Alfred's. Alfred recalled the small flax crop of 1913 and described his next flax crop,

Figure 6.2   Flax at Joe Bellas's homestead shack, Carlstadt (later Alderson), AB, 1911

in 1915, which yielded sixty bushels per acre and earned enough for them to build a house.[62]

Various new settlers grew flax on the Prairies in different quantities. A photograph, dated 1911, of Joe Bellas's homestead shack near Alderson, Alberta, shows that he grew a small plot of flax and that it ripened shortly after his wheat crop was harvested and stooked (Figure 6.2). Bellas was a forty-two-year-old bachelor who had immigrated from the United States in 1909.[63] Judging from his house, he had not been on the land for very long and was probably homesteading a single quarter section. The native short-grass prairie in the foreground suggests that Bellas's breaking strategy was to sow flax on the new land furthest from his home and adjacent to the property he planned to break next. The farmer's flax land acted as a moving buffer between the unbroken prairie and his wheat or other crops. The late maturation of the crop confirms that this field was the last to be sown, probably on sod broken that spring or in the previous fall. In a collection of Prairie documents including photographs focusing on machinery and texts telling a narrative of numbers, this image of place and work stands apart. Although black-and-white images could not fully capture it, this is also a photograph of a flax crop in bloom, its blue flowers catching the photographer's eye and prompting a shot at this angle.

Figure 6.3   Cutting flax on Canadian Wheatlands fields, Suffield, AB, c. 1912, Suffield, AB, c. 1912

Standing in stark contrast to Joe Bellas and John Murch were several operators who grew flax on new land on a large scale. A photograph of the Canadian Wheatlands Company in 1912 or shortly thereafter shows a dozen horse- and mule-drawn binders and ten binders pulled by a tractor, all harvesting an immense field of flax near Suffield, Alberta (Figure 6.3). O.S. Longman's first field work at this company included preparing an entire section for flax that had been ploughed the previous fall. The disking and packing took the better part of three weeks. Repairs demanded additional time, since the many exposed parts of the machinery were constantly being coated with dry prairie dust. In that same year Sir John Strutt grew flax on several sections of new breaking in Flaxcombe, near Kindersley, Saskatchewan. The niece of one of Strutt's hundred-person workforce recalled that the first "year's breaking was seeded to flax" and yielded "some 65,000 to 70,000 bushels."[64] Suddenly a few large producers were each cultivating as much flax as the Mennonites on Manitoba's West Reserve ever grew in one season. This was not a sustainable practice; it was not intended to be.

Large producers invested heavily in a rapid transformation of the landscape. Mechanical innovations such as steam-powered breaking outfits helped maximize the use of capital, and biological innovations such as

Figure 6.4   Wilcox-area steam tractor ploughing and burning flax, 1907

planting flax on new breaking helped producers maximize the use of time. Figure 6.4 shows how, in 1907, mechanical and biological innovations in the production of flax oil were combined in a single outfit fueled by flax. Flax straw was usually a waste product of prairie seed production, but large producers might have been able to organize its collection and storage for this purpose. Most people could not have accessed these operations, because, as this image demonstrates, a single plough could have required up to seven workers and a few horses to support it. Risk was enormous in a market as unpredictable as flax seed's, and this may explain why some sodbusters, such as Charles Noble, had difficulty attracting capital. In the vicinity of what became Nobleford, Alberta, this large producer sowed 3,000 acres of crops on new breaking. Most of these acres, in 1911 and 1912, were in flax and yielded a total of 48,000 bushels of seed. "He was trying out new ideas," wrote an excited local newspaper writer, "and at prevailing prices it does not take a savant to figure out what Mr. Noble is going to make on this year's operations." Noble had tried unsuccessfully to attract investors to his project, and after initial rejections he wrote of his high yields to a friend in Edinburgh. "I am sure this would have been very interesting indeed," he gloated, "to those who might have joined in the purchase of this property." Noble continued to expand his holdings and sow flax on new breaking, even reporting a crop worth between one-third and one-half of a million dollars in 1919. In this year of intensive sodbusting, wheat was only the second most valuable crop, worth between 67 and 80 per cent of the value of flax.[65]

Most farmers were eventually discouraged by flax seed's problematic markets and the process of cultivating it in new and highly unpredictable environments. Some lost their shirts on the crop and most probably only made modest gains before abandoning flax for other crops.[66] However, whenever farmers experienced just the right combination of high prices and good growing conditions, it was possible for them to make a small fortune from flax. Tom Goldsmith, who farmed near Marengo, Saskatchewan, bought a section in 1914, and recalled breaking "it all that season and [harvesting] a bumper crop of flax in 1915."[67] Emmett Cronan, an elevator operator who bought the seed, illustrated what a dramatic success it was for the small farmer. "In the spring of 1915 [Goldsmith] put those 600 acres of rough plowing into flax … and yielded 15 bushels per acre when harvested. At the selling price of $1.80 a bushel it amounted to approximately $17,000.00." Goldsmith finished delivering his flax seed on a Friday and demanded that the operator pay him in one cash payment. Cronan could only secure cash in $10 bills, and when Goldsmith refused to accept the money before his bank opened the following week, the operator hid the 1,700 bills in his elevator's living quarters and guarded them nervously with a revolver for three days.[68] In a smaller operation in North Dakota, African-American Era Bell Thompson recalled when her father sold his 1,000-bushel flax crop at the unusually high price of $4.08 per bushel, only one day before the price dropped dramatically. He had accepted the price when his crop was still flowering, and "[t]hat year we had a bumper crop, the new land yielding more than the thousand bushels."[69] What stood out to those retelling these stories were the unusually large profits. The practice of growing the crop on new land and in harsh conditions was an unexceptional part of rural life on the Plains.

Oil from the Northern Plains

The increasing popularity of flax growing in the far Northern Plains is evident in the maps of production from 1889 to 1919. In 1909 we see the importance of flax to farmers on the border of Montana and North Dakota.

Several counties there devoted over 12 per cent of their cropland to flax, and Billings, North Dakota, grew flax on almost 25 per cent of its field crop area. In 1916, 1919, and 1920, districts on the Saskatchewan–United States border devoted more of their cropland to flax than any other areas. This demonstrates the importance of studying transborder industries. The crop grew in both countries in similar quantities and was constrained geographically in the same ways north and south of the border. Knowledge, producers, and practices traveled relatively freely across the border, but it posed other significant restrictions. For example, both governments developed elaborate regulations to keep agricultural products thought to contain contagious diseases or pests from crossing the border, and worked to quarantine suspect cases.[70] Like most crops, the most commonly feared contamination in flax was weed seed, a prejudice held since the 1860s and probably as a result of the introduction of Russian thistle through a shipment of contaminated seed in 1877.[71]

In 1909 Ennis's industrial manual projected that the "migration" of the flax crop would plant it firmly on the Canadian side of the border. In fact, the ideal growing region, according to a recent taxonomy of flax seed, is the brown, dark brown, black, and dark grey chernozemic soils of the Canadian prairies and the most northerly Great Plains.[72] For a dozen years after 1910, flax production followed strikingly similar trends in Canada and the United States. The initial surge in 1911 and 1912 was followed by a period of low production and then an apparent increase in the 1920s. Major flax growers such as Noble continued to cultivate it on newly broken land, and large US outfits did the same. In 1918 and the following spring, the Campbell Farming Corporation broke and sowed 7,000 acres of Montana grassland to winter wheat and flax.[73] It continued breaking rapidly in 1919 and presumably included flax in its sodbusting strategy. A USDA scientist claimed at the end of the decade that "fully 80 per cent of the [flax] crop was produced on breaking," and "in practically all the area lying west of the ninety-eighth meridian flax is grown as the first crop on breaking." Although flax's relationship to new breaking was never systematically examined, it was widely embraced. If by now the crop was most favoured by farmers in the dry borderlands of Saskatchewan, Alberta, and Montana, it was, one geographer argued, because "there is little other virgin land in this section of the Great Plains."[74] However, something changed after 1920 that prevented flax from disappearing altogether on farms in Montana and North Dakota. Canadians grew their third-largest area of flax in 1920, and then suddenly production dropped by almost two-thirds in 1921. Conversely, the northwest states experienced a major resurgence in production after 1922, and the map for 1924 indicates that most counties in North Dakota and a few in South Dakota and Montana devoted at least a tenth of their cropland to flax.

The trend had nothing to do with declining consumer demand in either country. Painting continued apace in this period, with new marketing campaigns (such as Pittsburgh Plate Glass's images of bright white kitchens) encouraging consumption. Rather, the turnaround in production trends represented the earliest days of the retreat from Palliser's Triangle and the partial abandonment of the dry belt. Western Canadian rates of new breaking dropped sharply between 1921 and 1926, especially in Saskatchewan and Alberta. Many of the farmers who had made flax a part of their entry strategy in southwest Saskatchewan in the 1910s had now either retreated from the semi-arid region or had ceased to bring new land into production.

Even in this period of decline, the spatial patterns of flax production in Saskatchewan reveal much about the plant's place in grassland agroecosystems. Most flax production in Saskatchewan took place in two semi-arid areas: one near Kindersley and the other forming a large triangle with a base stretched along the US border and a peak near Moose Jaw. These were the last truly uncultivated lands in southern Saskatchewan in the 1910s, and a prime location for sowing flax on new breaking. In the decade after 1915, this triangle in southern Saskatchewan reveals something about the differential equation of the Canadian flax industry. Most important, flax production never disappeared in the way that areas of concentration left certain parts of the northwest states before 1920. This remained true even when the total acreage of flax plummeted in the 1920s. The 60 per cent drop between 1920 and 1921 did not change the geography of flax growing significantly. Instead, it seems most farmers simply made flax a smaller proportion of their cropping. Farmers were evidently able to scale their investment and land-use decisions. They were less constrained by the supposed environmental idiosyncrasies of flax and were quite able to think on their feet and use their knowledge of prices, climate, and factor availability to determine the best allocation of their resources.

During the 1920s the centre of flax production ceased to move north and west, and it seemed as if flax would return to the Great Plains. A rising price for seed and a tariff that favoured domestic production briefly encouraged flax cultivation in the northwest United States. The tariff on imported flax seed since 1913 had been 20 cents per bushel, but in peacetime protectionists claimed that a dependency on foreign supply could mean disaster for the linseed oil industry if another war cut off the source of seed. This reasoning lay behind the Emergency Tariff of 1921, which increased the rate to 30 cents. The following year, the tariff was increased by another 10 cents, and in 1930 a completely prohibitive tariff of 65 cents per bushel was introduced for imported flax seed. In 1934 the Kansas Agricultural Experiment Station felt that this tariff was "effective," but in Canada the effect of these policies on production was small. Canadian flax seed exports to the United States had continued despite the tariffs of the 1920s, and during the Great Depression both countries produced very little flax seed.[75]

The return of flax to the northwest states also indicated something of a "second frontier" opening in southwest North Dakota and Montana in the 1920s. This belt of counties with relatively high proportions of cropland in flax was especially visible in 1924 and had expanded into western South Dakota by 1929. Geoff Cunfer has shown that this area was probably quite recently ploughed, in 1920. It experienced a moderate increase in cultivation over the next fifteen years at which point the forbidding topography, temperature, and aridity "stopped sodbusters in their tracks."[76] Before these limits were reached, farmers employed flax as part of their sodbusting strategy. Yet flax had not been at all important here in 1919. Prior to the imposition of a protective tariff on flax seed, these farmers clearly grew other crops on their new land for some years.

Just as with flax fibre experts, the wisdom of the seed specialists should be taken lightly.[77] Those who imagined flax on every acre of newly ploughed land present only a very partial picture. If flax was grown mostly on new land, we could expect to find something like a one-to-one ratio between acres "improved" each year and acres of flax sown. Even more, this model would be consistent with the period of sodbusting, which in Saskatchewan peaked around 1908. But instead, Canadian flax peaked a few years later, and even then, thousands of acres were improved across Saskatchewan every year without the help of flax cultivation. Other crops were seeded on new lands in the larger grasslands region. Charles Noble found that oats had a slightly higher value per acre than his flax, even in the first years of breaking. The *Farm and Ranch Review* claimed that growing wheat was the fastest way to make money during "the early years of settlement."[78] John Bracken demonstrated that wheat grown after flax yielded better than wheat grown after wheat, but not nearly as well as wheat grown after corn and other fodder crops. Donald Worster originally argued that most new land under cultivation was sown to wheat, and Cunfer's more detailed crop analysis suggests that corn was considered a breaking crop and followed a westward movement in the Central Plains similar to flax's northwest migration.[79] Thus wheat, corn, oats, and flax were all considered "breaking crops" in different locations, but sowing flax on breaking was a form of biological innovation on the frontier because of the increased feasibility of ploughing and planting in a single season.

Flax's relationship to newly broken land gave the oilseed a distinctive signature. Flax was another breaking crop – appearing in an area complementary to and north of the corn belt. Using census data from the Great Plains Population and Environment Project in a Geographic Information System, it is possible to compare US flax production with the rates of grassland ploughing during the first major increase in flax cultivation, from 1899 to 1909.[80] In 1899 we see the first major crop of flax recorded in the United States, and we are able to compare it on a county level with the amount of new land ploughed in the previous decade (Figure 6.5). For the sake of comparison, the breaking data for counties whose total area changed by more than 200 square miles in that decade are excluded. There is a clear correlation in this map between the areas of highest flax acreage and the areas with the most new land improved in the previous decade. By 1909 the crop was slightly more dispersed and its centre of production had shifted westward again throughout North and South Dakota. Its movement followed another shift – the frontier of land under improvement. Again, land-use data are not displayed for counties whose boundaries changed to any great degree, but we can extrapolate from the trends in Figure 6.5 to suggest that the areas in north-central North Dakota were also undergoing significant breaking.

In Canada, the absence of crop data for standard census districts before 1916 precludes a study of the relationship between flax farming and breaking in the peak years of either activity. However, early records suggest that flax was most popular in the areas undergoing the most new breaking. Qu'Appelle, the Saskatchewan district with the most new land improved between 1905 and 1906, was by far the nation's largest producer of flax seed. In fact, 3.3 per cent of the district's cropland was in flax in 1905 and almost twice that percentage was flax acreage in 1906. We get an excellent picture of which rural municipalities in Saskatchewan improved the most

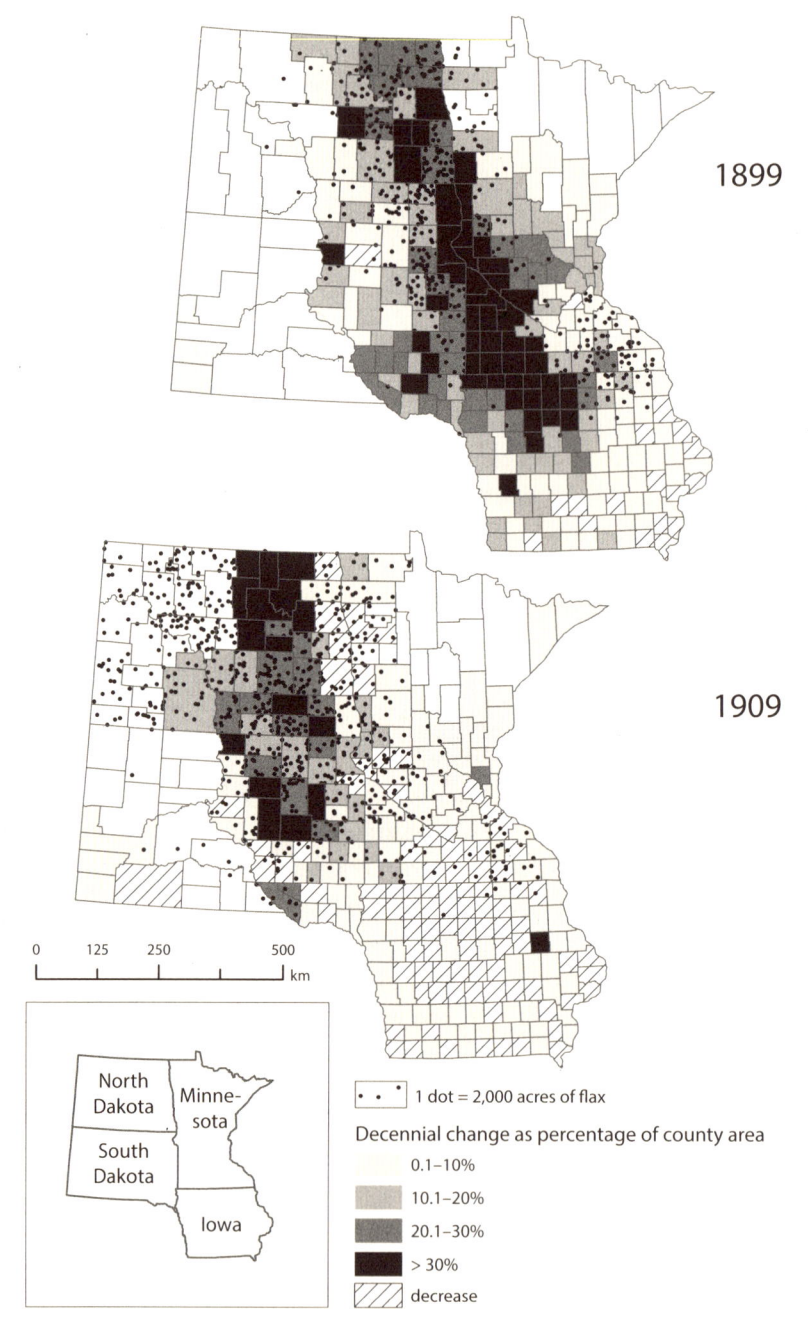

Figure 6.5 Flax acres seeded and change in improved acreage, 1889–1899, 1899–1909

Note: There are several complicating factors with the acreage of new land under improvement. The category does not necessarily indicate that all land under improvement was ploughed. In Canada it included land in field crops as well as idle land, fallow land, and land in pasture. In the United States, discrepancies between the questions enumerators asked about improved land in 1910 and 1920 mean that more pasture was likely included in the latter year's data. Fortunately, earlier data are more comparable in the United States, and excellent records for the Prairie provinces in the 1920s enable research in that decade. Specific to the flax issue, the crop may have been more attractive to sow on land that had been broken in the spring of that growing season than on new land that had been broken the previous summer. The census does not allow us to differentiate between the two, and we are left to examine more anecdotal evidence.

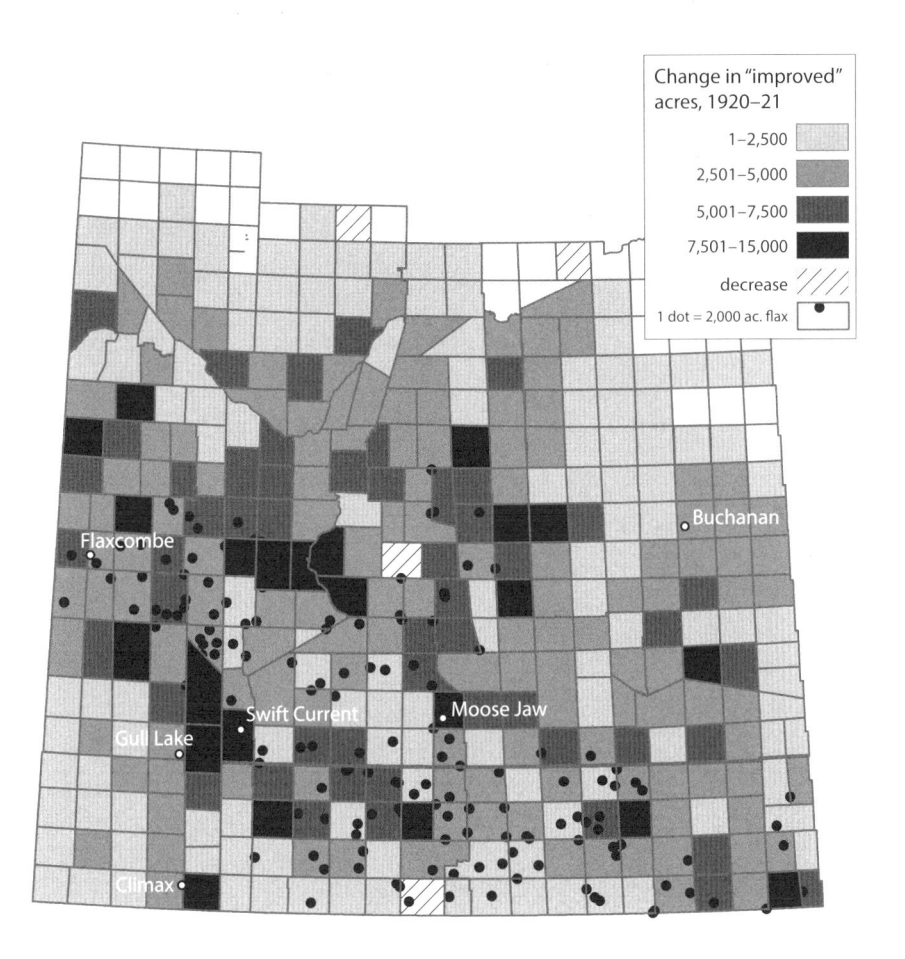

Source: US Census Office, *Eleventh Census, 1890*; US Census Office, *Twelfth Census, 1900*; US Bureau of the Census, *Thirteenth Census, 1910*. US spatial boundary files are used courtesy of HUSCO.

Figure 6.6 Flax acres seeded, 1921, and change in improved acreage in south Saskatchewan, 1920–1921

Source: Canada, Dominion Bureau of Statistics, *Census of Canada 1921* (Ottawa: Dominion Bureau of Statistics 1924).

land between 1920 and 1921, and we are able to compare that with flax production in 1921 (Figure 6.6). Many areas, especially in the northern half of the map, experienced significant breaking, quite apart from any serious flax production, but in the areas where flax was grown, there was a potential relationship to the areas of most intensive sodbusting. Saskatchewan's rural municipalities did not reflect a perfect ratio of one acre of flax for every acre of new land. By far, most of the municipalities reporting any flax at all reported small amounts and were home to more recently improved acres rather than to acres of flax. Nevertheless, these data suggest that new breaking and semi-arid grassland were fundamental aspects of growing flax.

Flax experts had several explanations for why farmers grew flax on new land and seemed unable to grow it on old land. Flax's impact on the soil was often discussed – especially as Prairie farmers grew it in unprecedented quantities – and these discussions shed light on what farmers, promoters, and scientists thought of the environment. When state officials observed flax seed production moving between states, they were less concerned with where flax moved than with where it disappeared. All the major cereals appeared in newly settled areas, but unlike flax they were grown for extended periods. So, flax was no more a crop that did well on new land than it was a crop that grew poorly on land previously cropped to flax. Flax's constant relocation convinced scientists that it was detrimental to the land and could only be grown once or twice before it exhausted the soil of the essential elements needed for flax cultivation. Soil exhaustion was a handy excuse for farm scientists in the late nineteenth and early twentieth century: it allowed them to measure the effects of supposedly slovenly and short-sighted husbandry and to demonstrate better practices to the public.[81] It is not a surprise, therefore, that scientists who observed the disappearance of flax from old land pointed to soil exhaustion as a reason.

The problem with the supposed threat of soil exhaustion was that farmers around the world learned how to farm soils for centuries without "exhausting" them. Producers in Argentina, India, and the fibre-producing regions of Northern Ireland and the Baltic countries all farmed flax extensively for long periods. Bracken realized this and claimed that "it should be possible for us to do what older agricultural countries have done, viz., conserve our soil resources while still developing them."[82] Conversely, flax was far from a prerequisite crop on new breaking. There is no evidence that flax was used on new soil in areas geographically similar to the Northern

Great Plains. Even in Argentina, where grain cultivation and settlement schemes developed in ways similar to Canada's, farmers produced flax seed in short rotations (flax occupied over half of all cropland in some counties) for generations.[83] However, in a relatively short period of new breaking, North American flax production soared and flax's role on pioneer farms became part of standard knowledge.

By favouring new breaking, flax growers hastened the cultivation of grasslands that ultimately could not sustain extensive agriculture in dry periods. This was one of the effects of increased demand for linseed oil and the low cost of acquiring land in some of North America's most ecologically sensitive grasslands. One of the reasons flax was grown at all, given the agricultural economy generally known for wheat monoculture, was that it required few additional capital inputs. Almost anything a farmer invested in to grow flax could be used later or sometimes even in the same season for cultivating other grains. Flax farming occasionally demanded granaries for long-term crop storage or leak-proof wagons for hauling, but these were rather general improvements that a farmer could justify even if he or she never intended to grow flax again.[84] The only investment a farmer was unlikely to make use of after flax growing was a special binder attachment that facilitated mechanical harvesting. Flax straw was much stronger than the straw on other cereal crops, but even so, most regular equipment would harvest it under careful operation. Many farmers grew such small amounts of flax that they could harvest their crops without employing the more efficient modifications.

There must have been other reasons for growing flax apart from soil preparation or cultural preference and for not growing it apart from soil exhaustion. The more likely explanations for flax's distinctive geography are found in the spatial patterns of grassland resettlement and the political economy of the flax-paint commodity web. The production chain's output – seed – had a new and growing market, and demand for linseed oil in the paint and varnish industries led to high prices that coincided with the period of grassland settlement. Transporting those outputs from newly broken land distant from railways or across international borders was possible with this low-bulk and high-value commodity. The major input necessary for flax production – land free of flax diseases and generic weeds – was plentiful on the edges of cultivation. The crop performed well on rough breaking, often the only kind accomplished by new settlers, and had a season short enough to allow planting directly after spring breaking.

Breaking Sod or Breaking Even? Alternate Explanations

When Mantle encouraged flax production as an early cash crop for small farmers, it was likely because of reports such as this one from the *Lakota Herald*: a North Dakotan farmer in Walsh County, it claimed, "raised 2,500 bushels of flax from 100 acres of a $750 farm, and is still selling it at home for $1.75 per bushel for seed. A $4,375 crop off a $750 farm is pretty swift farming."[85] The yield this farmer reported in 1900 was over twice the previous year's national average and seems unlikely. However, Walsh County was one of several important flax-producing districts in the Red River valley, and its flax farmers averaged over 15 bushels per acre in 1899. At that rate, the farmer would still have earned $2,625 for his crop.

In 1910, at the cusp of the first major flax boom in Canada, the Saskatchewan Department of Agriculture claimed that farmers were earning similar prices. Well over a dozen correspondents who wrote to the department querying the increase in various crops made mention that farmers grew flax because the high price made it worthwhile. Others gave reasons for growing flax that did not include breaking new sod, such as its low bulk, profitability for distant transport, and the low price of other commodities, such as oats. Farmers reported an increase in flax in many areas and for several reasons – mainly its price.

Flax generally had a much higher price per bushel than wheat (between 1908 and 1931 the median price per bushel in Canada was 90 cents higher), but to the farmer the relevant consideration was predicted return in terms of value per acre. Here, flax seed tended to be much closer to wheat, although fluctuations in the market and crop yields produced wide variations. Over three decades, these data showed a positive correlation between the direction of flax price movements in one year and the area of land seeded to flax in the next. For example, regular data on annual flax acreage and prices were available in Canada beginning in 1908, and in the first four years the value per acre of flax surged upward (Table 6.1). The acres seeded to flax increased as well, but they lagged by one year, between 1909 and 1912. The value per acre dropped from 1911 to 1914, and flax acreage followed, 1912–15. Similar trends occurred after the war – for instance, in the first half of the 1920s. Tracking these prices was critical to the equation of land use, but other factors, such as drought and the general rates of new breaking, had an effect.

The Canadian linseed oil industry could only use a small amount of the country's seed – between 500,000 and 1 million bushels – so shortly after 1900 the country became a net exporter of seed. In a letter to A.F. Mantle in 1910, Thomas Thompson, an influential grain merchant in Winnipeg, estimated that the domestic market for flax seed would never demand more than a million bushels.[86] Indeed at the time of writing, Canada produced just over 2 million bushels annually, but in a matter of two years its production would increase to 26 million bushels of seed. The largest markets were in Great Britain and the United States. Along the Canadian–US border, where Saskatchewan dry belt farmers often seeded over a quarter of their cropland to flax, the oilseed was hauled to US elevators, apparently without interference from customs officials. Climax area farmers like Bud English hauled the short distance to Harlem, Montana, not just for flax markets but for a variety of services and destinations. Emil Kluzak hauled flax to Harlem in 1912, starting from Gull Lake, a depot on the Canadian Pacific Railway's main (and only nearby) line. "There were no customs on the border then," Emil claimed, "and people went back and forth as they pleased. We hauled about 65 bushels on each team. Flax was worth from 95¢ to $1.00 a bushel, but you could buy a whole load of groceries for a load of flax."[87]

Many farmers along the international border homesteaded in Canada but lived, worked, and sold to some extent in the United States. Flax farming fit the social and economic situation of these cross-border farmers well. The commodity was ideal for hauling longer distances, and the loose border controls meant that it could be brought to the best market at either side. The land on the Canadian side was uncultivated and free of the weeds and diseases that plagued the crop. Farming it required few inputs or skills unfamiliar to Great Plains farmers, and by taking free homesteads and seeding them to flax, these farmers were investing in a promising product – land. Any inputs they put toward flax farming added value to their investment. Seed, implements, and other settlers' effects were available in familiar contexts, and hauling flax to US markets was easily combined with family visits or short-term work trips.

Those who grew flax on new land found that ideal conditions included a suitable market for their seed, access to breaking outfits, and early, dry spring weather. In the *Lakota Herald* of 12 June 1903, a farmer in Nelson, North Dakota, claimed good weather "made it possible to do much breaking and ... much flax has been sown on the new breaking." In the spring of 1910 the Saskatchewan Department of Agriculture offered some explanations for the increased amount of flax it observed under cultivation that year. The early spring allowed more land to be broken and sown in the same season, and flax was the crop of choice for such a quick use of new land. Another early spring in 1911 meant that "new settlers continued breaking until late" and planting the land to flax in the "new districts" north and south of the Canadian Pacific main line west of Moose Jaw. These farmers gambled with nature by seeding late and were struck by misfortune when early snows caught and ruined their standing flax. However, the flax crop continued to increase, "especially in newly opened districts," and the department began to encourage a flax-breaking rotation. It recommended seeding new land to flax, followed by wheat, hay, another cereal, and then flax again.[88]

Sowing any crop on new breaking in the Prairies required an intricate ordering of time and resources with the crop's calendar. Shipping and time constraints meant that most flax growers would have had to secure seed and break land before anyone knew when the land would dry, but good weather and the fact that flax was the fastest-maturing grain at that time may have given them more flexibility and enough time to order seed. One linseed oil crusher distributed literature to Canadian farmers as late as 25 May, because it was assured "there is yet time for considerable planting of flax, and the bulletins may do good."[89]

One possible solution for last-minute flax production may have been breaking outfits. John Bracken described a tractor outfit that ploughed, disced, seeded, and harrowed flax in a single operation, but this presumably could not be afforded by ordinary homesteaders. In four of Saskatchewan's nine crop districts, provincial correspondents reported gas- and steam-powered outfits breaking land and sowing it to flax. In Buchanan, Saskatchewan, a correspondent to the Department of Agriculture reported seven such outfits breaking flax land in his township. "As near I can tell there were 640 acres broken this last week for flax," he wrote on the eighth of May. O.S. Longman broke land for flax on a steam ploughing outfit that could break an entire section of Alberta prairie in less than a week.[90] Outfits need not have been large; smaller producers hired a variety of neighbours to prepare land and sow it to flax. This was probably popular among single women farmers, according to Glenda Riley. For example, "Bess Corey hired men to break sod, put up fencing," and a "Montana woman similarly hired neighboring homesteaders to break her land, remove the rocks, and seed forty acres in flax."[91]

Some oil crushers tried to reduce the costs of a farmer's inputs by providing seed through local agents, much as the American Linen Company had once done in Massachusetts and as James Livingston had once done in Manitoba. E.H. Smith, of American Linseed Company, attempted to have local banks and land companies finance and distribute flax seed to sodbusters for him. Smith sent a request to a county commissioner in Scranton, North Dakota, to have flax seed distributed in counties "with so much rich unbroken land."[92] The Johnston Elevator Company sold land on the crop payment plan and felt that farmers "would break up additional land upon their holdings if they could get [flax] seed" subsidies from crushers.[93] American Linseed approached banks such as First State Bank in Mott, North Dakota, asking them to finance and distribute the flax seed the company provided. The crushers knew where to find likely flax farmers, since, according to the bank manager, "there are 125 mammoth steam and gasoline breaking out-fits tearing up the sod in the County this spring as well as all that can be broken by horses. This will be planted to flax."[94] Paint manufacturers contemplated assisting local banks to finance seed for farmers, but ultimately, "they … decided to let the farmer finance his own seed."[95]

The argument that flax yielded better than other crops on new land may also explain why farmers grew flax on breaking when prices made it worthwhile. In 1916, J.H. Grisdale of the Canadian Department of Agriculture argued that flax was suitable for the new settler because it outperformed other crops on new land.[96] US economists such as Walter Barker expressed similar optimism in 1917. In his opinion, flax was "the only grain which can be grown to full yield the first season on land ploughed and planted in the spring. Many farming lands have been paid for with the flaxseed crop obtained from the first breaking of the soil." Consequently, "the bulk of the flaxseed crop has been produced by the frontier farmer."[97] According to Cora Hind, a leading Prairie agriculture columnist, "[B]y far the safest crop for the beginner or on freshly broken land is oats, or flax put in in the spring on ground broken the previous summer," but many experts, including Hind, were reluctant to endorse flax in Canada for reasons ranging from weed control and market instability to soil exhaustion.[98] For example, on 13 June 1914, Archer-Daniels Linseed read an article in the *Winnipeg Free Press* entitled "Flax on Spring Breaking," and they felt that Canadian flax seed was under enough of a threat that they should rebuke the editors for their fear mongering. In a separate letter to plant pathologist Henry Bolley of NDAEC, Archer-Daniels complained that "some woman in Winnipeg," undoubtedly Cora Hind, had attacked the practice of using flax on new breaking.[99] Bolley commented on the company's reply to the article, claiming that Hind argued from an English tradition of summer fallow and the practice of taking a full year to break sod, whereas Americans were "used to plowing over the ground and then putting in the crop" or, in other words sowing flax on breaking.[100]

These incentives for flax production do not necessarily explain why flax was often abandoned when new breaking ceased. Farmers on established farms found that whenever the spread of acre values between flax and other crops was very narrow, the risks of growing flax quickly outweighed the incentives. One of the major risks was a disease that Bolley called *Fusarium lini*, but that was more commonly known as flax wilt. In 1911, when US flax crops were mostly concentrated in the Dakotas, Mantle claimed that flax wilt was the real cause of "the disappearance of flax as a general crop in all but a few of the north-western states."[101] When established farmers abandoned flax, they often considered their land flax-sick, and some North Dakotan neologists even claimed that their land had been "flaxed to death." At the turn of the twentieth century, Bolley discovered that the problem was not the result of soil exhaustion but was caused by a fungus deposited by the plant itself. When the crop was grown in a short rotation, it became prone to disease, especially flax wilt.[102] Bolley's perspective on flax cultivation and on the linseed oil industry's campaign to "save the flax crop" is discussed further in Chapter 7. Here, I have examined flax from the farmer's perspective, and what was most crucial to farmers was obtaining the best possible outputs from whatever they chose to plant. In the first years this was about managing limited resources and adapting to new environments.

Thus, flax gives us a new glimpse of the prairie "plow-up," and this glimpse suggests that when prices were encouraging, farmers used the crop in some parts of the Northern Great Plains for quick and effective use of newly broken land. It is unlikely that the economic allure of flax induced many farmers to break new land to accommodate it, and even more unlikely that farmers believed flax would help prepare their new soils for other cropping. However, this research affirms the probability that when, for a variety of reasons, farmers chose to grow flax, they grew it in areas with more new land under the plough. The main factor influencing farmers' decisions to grow flax was its price and the acre value spread between it and other crops. A secondary factor was the suitability of flax for their businesses or the way flax fit with their available inputs. Without a high enough price as incentive, no amount of appropriate land could entice farmers to grow flax. There were many uncertainties in these two factors. Flax seed prices were even less predictable than those of the major cereals, and the most fertile flax land required certain growing conditions to secure a good yield. Diversifying production helped minimize risk, but it was only possible if the spread between commodity prices was relatively narrow.[103] In certain years and places, the acre value of flax seed came close to that of wheat and oats, and this coincided with a period of rapid settlement and available breaking on which to sow flax.

Flax was not the critical factor in sodbusting; rather, its growing importance in a period of major settlement and rising demand for linseed oil meant that Prairie people thought of new ways to make flax work in their businesses and their environment. A new thirst for colour and the political economy of the paint and varnish industries determined when flax was grown and the cost of exporting it to the United States. Within national boundaries and during years of high prices, the environment and the land-use practices of farmers encouraged flax growing in the Northern Great Plains and semi-arid prairie. The environmental constraints within this region were overcome by a variety of technological improvements, so long as they were affordable, such as dry-farming and early maturing varieties. Crop diseases were especially crippling for flax yields, and although disease-resistant varieties were developed, the most effective adaptation was long crop rotations or moving production to land that had never been in flax. The changing environment and economy of the flax boom meant that innovations were necessary to find the best areas for flax and the best botanical strains for those areas. Flax did not create the agricultural prairie landscape, but the people who did often grew flax on new breaking.

CHAPTER SEVEN

Saving Flax

*Industry, Science,
and the Tariff*

In the first decade of the twentieth century, a wonderful circle of influence in the flax-paint commodity web swirled around the North Dakota Agricultural Experiment Station (NDAES) in Fargo. As we have seen, the station's chemist, Edwin Ladd, pressured the linseed oil and paint manufacturers by calling foul, reporting on paint ingredients and performance, and setting up paint-test fences on station soil. On a nearby experimental plot, NDAES plant pathologist Henry Bolley had been cultivating and selecting disease-resistant flax varieties, and in this chapter we see how the crushers organized themselves to support Bolley's work under the mantra "Save the flax crop." The crushers found it a useful way to get information and to perhaps convince farmers of their good intentions and commitment to the Great Plains flax growers. Bolley and Ladd received the recognition as professionals they wanted for their science, the manufacturers found valuable information and another excuse for strengthening their association, and farmers were able to sell flax, paint country homes, and feel that their voices were being heard through regulators and farm scientists.

Linseed oil crushers in the United States believed that the American Linseed trust had failed because of the nature of the flax crop and because its mills in states such as Ohio and Iowa were too far from the new source of seed. Although distance alone was not the issue (we have seen that Montana's and Saskatchewan's flax seed was profitably milled as far away as Buffalo and Montreal), crushers were genuinely worried that US and even Canadian flax production (the only justification for milling linseed oil west of the seaboard) would cease with the closing frontier. Since "we have now gotten down to the last ditch in this country," George Archer warned, the industry would crumble.[1] Crushers turned to farm science for a solution, particularly the flax research of the NDAES. Henry Bolley, the station's botanist, had been testing flax for disease on Plot 30, a small piece of land that is now on the National Register of Historic Places because of the discoveries he made there. In 1910 the industry looked to this soil and this institution for some way to bring flax seed cultivation back to the Red River Valley and perhaps even the Midwest. A Flax Development Committee (FDC) was formed to encourage flax cultivation and flax research. The mantra "Save the flax crop" was repeated in Bolley's letters, plastered on signs in country stores, and mailed to farmers in thousands of station bulletins.

Diseased Soil: Plot 30 and Plant Pathology

At the close of the nineteenth century the North Dakota Agricultural Experiment Station devoted significant attention to flax. Production had been expanding so much in the region that it seemed clear that flax offered a promising use of the station's resources. John Shepperd, a professor of agriculture who had observed flax experiments where the same variety grew well on some plots and poorly on others of similar soil,[2] suggested that Bolley perform some experiments shortly after he arrived in 1890. They decided to dedicate a small piece of land in the experimental plots west of campus to studying flax. "Plot 30," as it was called, became famous as the longest continuous flax field in the United States.[3] The plot is historically significant not because it was a space devoted to continuous flax, but because it remains a few square metres of diseased soil, cultivated for the purpose of saving the Great Plains from further infection. Plot 30 was a microcosm of what Brian Black calls sacrificial landscapes – a deliberate and even cultivated miasma in one place considered acceptable so that humans in other places could experience greater prosperity.[4] Growing flax year after year on the experimental plot had two purposes. The first was to identify which characteristics of the soil or seed made subsequent crops fail, and the second was to select and reproduce any varieties that thrived among the obstacles of repetitive cropping (Figure 7.1).[5]

Figure 7.1   "Survival of the Fittest Garden," Plot 30, NDAES, Fargo, ND, 1921

Alan Olmstead and Paul Rhode suggested that in 1895 United States Department of Agriculture (USDA) scientists began breeding wilt-resistant cotton varieties. The cotton experiments were not on dedicated plots, but clearly plant pathology was in a ripe environment for such new directions.[6] Bolley had begun flax research when he first arrived in North Dakota, and it was his major scientific activity between 1900 and the early 1920s. Several versions of the discovery of flax wilt exist. All place Henry Bolley at the centre, although some paint the picture with Bolley as the sole driving force behind flax research and others describe a variety of factors leading to and sustaining the scientific discoveries and resources for the flax industry. The latter is closest to the truth. Bolley was initially intrigued by flax's tendency to fail after two or three years of cultivation on the same soil, a problem that farmers reportedly attributed to soil exhaustion.[7] Bolley suspected a communicable disease specific to flax, since his early experiments showed that continuous flax crops did poorly even on the best soil, whereas the supposedly exhausted soil accommodated non-flax crops and weeds without trouble. Furthermore, flax planted on land that had come into contact with runoff or dust from nearby flax crops afflicted by the condition also began to fail.

Bolley declared the soil on old flax land to be flax-sick. "[F]ew men," he would later argue, "realize how far the damage to the soil has already occurred in the northwest."[8] In July 1900 he collected pure cultures from young wilting flax plants on experimental plots sown continuously to flax and from spores on flax seed samples from around the world. The most important of Bolley's discoveries in flax research was a common organism in the diseased plants, the fungus *Fusarium lini*, or flax wilt. The disease infected both the soils of his experimental plots and the new lands of the Northwest being sown to flax, and it remained a threat to any subsequent flax crop sown in the next five to seven years. One diseased plant was capable of infecting an entire rod (25 square metres) in four years. Bolley claimed that the *Fusarium* infected flax crops in Europe, but that it had never been understood there; rather, it was usually avoided through rotation, a solution not always available in the extensive flax farming of the Northwest.[9] A continuous flax plot was never ideal; something like it had been attempted by some fibre manufacturers (see Chapter 2), but they could obtain sufficient raw material in a small-enough acreage that they seldom had made continuous use of a single plot of land.

Exposing a continuous crop to disease for the purpose of selecting resistant strains was a little-known practice, and some reports claim that Bolley was the first to attempt it.[10] In effect, he was cultivating not a crop but a disease. His aim was to identify diseases (flax wilt was only the most serious of several), to isolate plants that might be able to resist the disease, and to develop chemical treatments to eliminate it. Bolley was trained in evolutionary theory and was particularly influenced by J.C. Arthur at Purdue University, who had studied Darwinian theory under such "new botanists" as Charles E. Bessey, William G. Farlow, and, most importantly, Asa Gray.[11] He described his plot as a "garden for the propagation of disease to eliminate the weak." Any plants that could survive flax wilt could then be selected and developed as resistant strains. In 1902 the NDAC was able to collect moderate amounts of seed from about sixty flax plants that had survived the highly concentrated diseases left by a decade of flax on Plot 30. The following season Bolley repeated his experiments on the diseased soil with hundreds of samples of seed from around the world, in the hope of isolating more plants that could resist the effects of wilt. Bolley also traveled to northern Europe and Russia that year with the support of the college and the USDA to search for varieties that had survived on continuous flax land.

At this point Bolley had demonstrated that flax did not exhaust the soil inordinately, and he had identified the disease that caused flax to fail. And so, "it was with great hopes that in 1904," Bolley later recounted, he "commenced the planting of seed samples ... to observe the life and death struggle of the different flax varieties and types against the organisms in the infected soil." Some plots were treated with fertilizers that proved ineffective against flax diseases, although one soil treatment – steaming the soil to 82°C one-foot deep – was successful if impractical.[12] Other treatments included spraying sowing seed with formaldehyde to exterminate any residue of wilt or rust. The experiments yielded plants that could resist flax wilt. Between 1908 and 1911 the wilt-resistant varieties NDR-52, NDR-73, and NDR-114 were released to farmers for commercial use. The news of Bolley's discoveries spread quickly through the agricultural science community and beyond. Not many were concerned about flax or aware of the problem of flax migration that Bolley was trying to solve. But many, like Canada's Dominion chemist, Frank Shutt, were impressed with Bolley's challenge to widespread nineteenth-century notions of certain crops causing soil exhaustion.[13]

Bolley's publications notwithstanding, there was disagreement or confusion among farmers, scientists, and industrialists over just what afflicted the flax crop in the Northwest. To the farm press and scientists, the northwestward movement of production meant that farmers in old areas abandoned the crop and therefore must have considered it exhausted. In reality, farmers abandoned flax for a variety of reasons: some thought that flax exhausted their land of the nutrients required to grow flax and others passed over it for more lucrative commodities. Some farmers found other causes for poor flax performance. In 1902, for example, a farmer from Murray County, Minnesota, wrote to Bolley claiming that "there is no such thing as a flax disease." He declared ants the real culprit and explained that virgin land was amenable to flax cultivation because its hard soil did not allow ants to attack the plant's roots.[14]

Many processors and seed dealers in the flax-paint commodity web did take farm science seriously and eagerly sought officials' information on the supply end of the web. Several of these were Canadian flax seed and grain merchants who wrote to the Saskatchewan Department of Agriculture for advice on the seed crop there.[15] Canada had not dedicated any of its own farm scientists to the problems of the flax crop in this period, mainly because flax seed production had previously been limited to a small area in southern Manitoba. A new market for flax knowledge emerged as the crop began to expand along with the unprecedented influx of farmers into the Canadian prairie. However, it did not translate quickly into scientific research in flax, probably because the region's few crop scientists were preoccupied with creating a knowledge base for wheat cultivation and dry-farming techniques in relatively unfamiliar topographies such as Palliser's Triangle. The massive increases in Canadian flax cultivation in 1911 and 1912 were centred in the newly farmed grassland of southern Saskatchewan, and they created enormous interest in the crop among that province's state officials and media. The provincial deputy minister of agriculture, A.F. Mantle, amassed a large file of flax correspondence in this decade with letters from US, Canadian, and Irish flax millers and other flax experts from Europe and North America. Almost all correspondence with the department was in regard to the massive amounts of flax straw that were burned or discarded as a waste product of the new Canadian flax crop. Many people claiming to be fibre specialists wrote to Mantle asking about the feasibility of starting linen and twine plants in Saskatchewan; one of the more promising relationships he developed was with E.H. Smith, a manager of the American Linseed Oil branch in Duluth and president of a new fibre company called Western Linen Mills, in the same city.[16] It appears that the straw from prairie flax crops was always too short, coarse, and poorly handled by threshers for the purpose of linen production. Large flax twine factories were established in Minnesota (Figure 7.2), but they eventually either closed or shifted to other fibre inputs such as sisal and henequen. Both Canadian and US state officials (including Mantle and Bolley), as well as municipal elites (such as Charles Roland of the Winnipeg Industrial Bureau), expended a tremendous amount of effort on promoting a fibre industry in the region. But ultimately the seed was the only product of any value from prairie flax. Several attempts were made to use tow from prairie flax straw for paper manufacturing, but the tow from flax seed crops was significantly lower in quality than tow derived from fibre flax.[17]

Canadian farm scientists were not interested in studying flax seed problems, but they were aware of Bolley's work and directed many inquiries south of the 49th parallel. However, when Canadians wrote to Fargo, they were often as interested in issues of trade policy as they were in crop science.[18] By far the most frequent queries to flax scientists like Bolley came from the large US crushers, and besides requesting literature, they regularly

Figure 7.2   "Flax Twine Plant," St Paul, MN, International Harvester Co., c. 1907

asked for evaluations of crop conditions and predictions of crop yields. In 1903 J.W. Hirst, of American Linseed in Chicago, wrote to inquire if Bolley knew how to arrest flax wilt, and if he did whether he would send them some publications to put "into the hands of our farmer friends." Later that year Bolley was also asked for literature by Pittsburgh Plate Glass, who considered themselves "among the largest consumers of linseed oil in this country." Bolley readily agreed to send literature, but he was particularly opposed to estimating outputs, claiming such prophesies were imprecise and financially detrimental to the farmer.[19] Bolley knew that farmers were not friends but suppliers in this context; crushers determined the price farmers would receive for flax seed and they based the price in large part on global crop estimates.

Saving the Flax Crop, and the Industry

Over the last half of the first decade of the twentieth century, American flax yields began to drop, and crushers in the Northwest who had few other options for their raw material turned to farm science for solutions. In 1906 George A. Archer wrote to Bolley to introduce his company, Archer-Daniels Linseed, and to report their experiences with flax wilt. Archer had been in the business since 1868 in Ohio, where his father had crushed flax for fourteen years previously.[20] They noticed that flax was never grown in less than a seven-year rotation to avoid what they were convinced was soil exhaustion. Since then, his business followed the production of flax from Ohio to Indiana, Illinois, Missouri, and Kansas and on to Iowa, Minnesota, and the Dakotas. After reading Bolley's work, he realized that he had seen flax wilt when the crop was grown too frequently, but he believed that if farmers practised good rotation, they should not have any problems with wilt and should "raise all the flax that [they] should raise in order to get a good price for it and raise it for many generations to come."[21] C.M. Durbin of Sherwin-Williams wrote late in 1907 to say he had read the USDA's bulletins on flax wilt carefully. Noting that the disease was hastened by moisture, Durbin worried that wet weather during the year's harvest could have fostered the growth of wilt spores and that the crop of 1908 would be harmed more than usual by wilt. He suggested that farmers be reminded to use the formaldehyde treatment in the spring. Bolley was so impressed with Durbin's inquiry that he made note to "publish this letter."[22] Such letters were unusual; most crushers appeared to believe that low flax yields were only temporary, the result of a series of dry years in the Dakotas. Their inquiries were intended to mine for data on crop conditions and on speculations about yields. Such information was critical to an economy of scale where raw material costs represented up to 85 per cent of the value of outputs. As Ennis's 1909 industrial manual put it:

> No crusher can succeed unless he is a shrewd buyer and seller of seed. The fluctuations of the market are always his first concern. A few seconds of time, when prices are varying widely, may wipe out his profits. In the absence of control of these fluctuations, the most essential thing is to provide for their close observation and for a constant alignment of selling policy with them.[23]

In January 1910 Fredonia Linseed Oil wrote Bolley to ask if he thought Northwest farmers would continue growing flax seed and to see if there were some way they could partner with him to help make it happen. They claimed that farmers in their county paid for their land with one crop of flax, yet were still so reluctant to grow it that the company had to contract with farmers, selling them seed at $2.00 per bushel and buying the harvest for $1.35 per bushel. Bolley's reply to Fredonia was discouraging; he lamented that keeping "North Dakota farmers in the business of growing flax" was a losing battle without an "enormous educational campaign" and unless his varieties did better than even he expected.[24] His reply to Fredonia was prophetic, because within months his work gained momentum and the dialogue between farm scientists and business increased rapidly. There were at least three reasons for these developments. First, for two decades Bolley had been adding significant knowledge to the scientific community and he wanted to see farmers make use of it in the form of pure and resistant flax seed, new seed treatments, and increased acreage on old flax land. He was not content to rely passively on his publications and was fiercely committed to practical science and the economic improvement of farmers in North Dakota and surrounding states. Second, flax production declined steadily in the United States from 1906 to 1910 (Table 6.1), and Bolley sensed it was something his science could cure. Third, Northwest linseed oil crushers were having difficulty obtaining flax seed, a direct result of the drop in production.

Bolley's correspondence contained a sense of excitement and new energy. With the new varieties created at NDSU and the heavy demand for linseed oil in the United States, he believed that a long and lucrative industry could be created with the flax of the Great Plains. Thus began an intensive educational campaign to "save the flax crop" (Figure 7.3).[25] To the campaigners, the fundamental problem was the disappearance of flax from the Northern Plains. However, if flax was indeed following its course to the newly broken lands, then bringing it back required real justification. After all, farmers in Ohio, Illinois, and then Iowa abandoned flax without much complaint from local flax experts, and Bolley had to convince his contemporaries that what seemed like an artificial measure would be worthwhile. Flax researchers expected to win the battle for flax using the science itself. Now flax could be grown on old land just as productively as on new land and probably more so, as the frontier was encroaching on marginally productive crop lands. If former flax growers used Bolley's wilt-resistant varieties and followed his instructions for seed selection and cleaning, they could once again make lucrative gains from the crop.[26]

The demand for seed in Northwest linseed oil mills had increased to the point that crushers needed fresh ideas for procuring it. In the Duluth branch of Sherwin-Williams's linseed operations, W.H. Kiichli considered taking control of primary seed production. He wrote Bolley to ask if "constant flax" was ever an agricultural possibility should a company attempt to cultivate it on its own land. Bolley replied that he would never advise planting flax after flax, but he had to phrase this recommendation carefully. The premise of the scientist's experiments was that flax could be grown on

Figure 7.3 "Help save the flax crop, ask us how," L. Simmons Hardware, Foxholm, ND

old flax land with certain treatments, but it required a minimum five-year rotation even when the treatments were applied. Bolley had grown flax on Plot 30 for two decades when he gave this advice, and he found that wilt was only one of the many diseases that would attack a continuous crop of flax. He cautioned that his work on the plot was not a precedent to show that it could be done, but a tool for identifying flax diseases and developing treatments and resistant strains. Bolley's suggestion to Sherwin-Williams and to other crushers was that they should rent enough land to devote a fifth of it to flax in rotation or they could rent land that had been flax-free for at least five years.[27]

Although the notion of crushers cultivating their own flax seed came up from time to time, it seemed more feasible to encourage farmers to raise the seed for them. In January of 1910 Bolley received a letter from Charles Thornton, an editor of the *Commercial Record* in Duluth and Bolley's liaison for the linseed oil–crushing and paint-manufacturing trades, who asked if "any assistance could be rendered you in furthering your investigations" and if there were specific ways of growing flax if the scientist would write them down so that Thornton could "take the matter up with the trade." In his reply (on which he later scrawled the words "The Beginning of the F.D.C."), Bolley outlined the need for "a Bureau for the special purpose of propagating proper information with regard to the culture of the flax plant in the northwest if we wish to save the crop. Thousands of farmers do not get the Experiment Station literature."[28] By the end of the year, this brainstorming had led to an offer to the scientist and his department to promote good flax cultivation through various competitions and educational campaigns. Both Bolley and the crushers were aware that the funding provided by private companies would appear suspect to farmers. Again, Durbin's perspective is illuminating. He wrote in 1910 that "if this industry is to be saved," the companies must act and they would cooperate with flax researchers. However, he warned, the "movement on foot" must be a small crusher's initiative or else farmers would suspect big business.[29] The fact that business expected farmers to suspect corporate collusion says something about the strength of anti-monopolism in the Progressive Era and the fine line Bolley had to walk when working with both farmers and crushers. More important, it shows that flax was only one choice among other crops and that there had to be a very good reason to invest in it if its acre value was not significantly higher than that of other crops.

The Flax Development Committee

Saving, or at least strengthening, the flax crop of the Northwest had long been a goal of the North Dakota Agricultural Experiment Station, but the full campaign began when the various linseed oil industries took interest. In the spring and summer of 1910, Bolley and Thornton quickly attracted the support of five industrial associations (crushers and the manufacturers of paint, lead, varnish, and linoleum) and dozens of individual companies representing almost all linseed oil production and much of its consumption. In 1917 the contributors to the FDC included four of the twenty-one largest US chemical companies, including the largest, E.I. Du Pont de Nemours. The members included American Linseed (formerly National Linseed) and National Lead, two of the eight major late nineteenth-century trusts; Sherwin-Williams, which by 1914 was considered one of the few multinationals in the chemical and oil industries;[30] and the Northwest giants, Midland Linseed and Archer-Daniels Linseed companies. Several other small crushers contributed to the FDC, at least initially, and the organization accounted for virtually all of the country's oil production.

In the early days of the FDC's conception, the linseed oil industry organized support in a matter of weeks. The linseed oil crushers were led by Archer-Daniels in Minneapolis, who learned the skill of agricultural promotion from the railroad companies.[31] Bolley's plan for a bureau of flax propaganda was brought to crushers in February; crushers and paint manufacturers spoke to Bolley and to each other in March, hoping it was not "too late to accomplish anything for this [1910's] crop"; Bolley suggested a meeting with the crushers in a letter dated 22 March, and a reply from Archer-Daniels the following day announced that a meeting had been arranged, including a date, place, and assurance that all the crushers of Minneapolis would attend.[32] After several meetings, twelve crushers contributed $7,000 to the college for Bolley's work, and the paint manufacturers secured an additional $10,500.[33] By August the crushers had agreed on their contribution, and the first cheques arrived the next week and were forwarded to Bolley's department.[34] Bolley replied, saying he had commenced the work of the FDC. His first tasks would begin immediately as he challenged Senator McCumber to raise the tariff on flax seed and used his influence as the state's pure seed commissioner to persuade elevator companies to store and sell resistant flax seed separately from regular grades. In an effort to win support from elevator companies, he warned them that the use "of mixed seed on the lands of the Northwest is destroying them for flax." The disease spread more quickly than Bolley's ability to educate farmers, and he needed the help of the elevator companies, which "are in direct touch with the farmer."[35] Here, just fifty years after the Perine brothers established mill complexes and thereby entered into relationships with potential flax producers, millers in the urban industrial complex found the best way to get "in touch" with farmers was through elevators via scientists. The FDC's official purpose was to "develop and encourage the cultivation of flaxseed," and for twenty years it reiterated the same goal: "to improve and to enlarge the production of flaxseed in the United States so that the industries consuming linseed oil might have a dependable domestic supply of the raw material ... so that at least 80 percent of our country's requirements are taken care of by flaxseed grown in our northwest."[36]

From the crusher's perspective, Bolley's most important role was as crop informant, and early exchanges reflect the industry's expectation that he was a source of, and took interest in, the kind of information on crop conditions and yields that would influence market prices. Archer-Daniels regularly sent Bolley their circulars on market conditions and occasionally notified him of when trading would begin or how world supplies of flax seed compared to the oil industry's demand.[37] When the first cheques arrived in August 1910, Bolley was instructed to "send advice to" the contributors, and in reply he promised to "keep them informed."[38] During the first spring season of the FDC's funding, in 1911, Bolley could only suggest that many more farmers were growing flax that year than previously.[39]

Crop estimates were critical in a market as unpredictable as flax seed. Flax seed prices often fluctuated widely in a single season (for example, from $2.14 to $1.28 in 1915), not only because of war but also from "surprising" rates of consumption in the United States and "impressions" formed about future markets based on Argentinian imports.[40] Research from the Kansas Agricultural Experiment Station in 1934 found that over the previous thirty-two years the market price dropped substantially every year from September to December as the market became saturated with the new crop of flax seed. Scientists at the Kansas station printed information from Fredonia Linseed Oil that showed that almost 80 per cent of the crop was sold in July and August, "before the bulk of the northern crop has depressed the price."[41] Furthermore, the entire world crop was not so large that its market could not be influenced by major purchasers. There was risk for everyone with a share in the flax-paint web. The large integrated companies such as American Linseed sometimes ran several years without profits and were specifically injured by two speculative purchases in the 1890s.[42] The market was cornered several times in the early twentieth century.[43] Nevertheless, the large businesses were less likely to collapse than smaller crushers or primary producers.

Most major crushers employed their own crop reporters. Midland Linseed Oil found that their reporters did not always agree with Bolley in his reports, and they refused to increase their contribution to the FDC because they did not have a clear sense of what Bolley did for them.[44] Some of the crushers in search of crop statistics approached Bolley with tact, others simply offered cash for any crop information he could give them.[45] At first Bolley refused to give the FDC subscribers certain kinds of information, claiming that it hurt the farmer and was unfair to other businesses. Eventually he began to print letters and farmer testimonials in his reports to the FDC and to release his field agents' reports of crop size and yields, but these were only for the small number of resistant flax plots that they inspected.[46] In 1918 Pittsburgh Plate Glass requested "telegraphic information daily" on probable acreage on the Northern Great Plains and Prairies, on the estimated time of planting, on the amounts of flax seed stored on hand, on the effects of weather, on the movements of seed to major centres, and on any other factors concerning the "flax seed and linseed oil situation."[47] The company was not even part of the FDC, and so Bolley reminded them that such information was imprecise and not something he gave up easily.

Several stakeholders were interested in an accounting of how the $17,000 annual FDC fund was used, but Bolley's answers usually depended on who was asking. To the crushers and linseed oil consumers, the scientist described his primary role as making "every effort ... to collect a fund of information regarding the crop." Secondary tasks included performing scientific experiments to stay abreast of "the latest available information as to how to better the results of flax cropping," as well as preserving better seed to sow, forming a seed breeders' association, and performing field demonstrations. Lastly, and then only "if there were any remaining funds available," Bolley hoped to increase the distribution of his station's flax propaganda. The emphasis on collecting data and creating information was certainly what appealed to the fund's contributors – the first official history of the FDC actually left out Bolley's interest in distributing literature and in forming a seed breeders' association – but in reality most of the fund was spent on putting literature together and distributing information that Bolley already knew.[48] To other scientists in farm experiment stations and the USDA, Bolley's description of the FDC was more cryptic. For example, the director of the agriculture extension, at the University of Idaho, wrote Bolley asking for help getting some resistant flax seed for Idaho farmers. When Bolley merely sent him a list of growers, the director replied, "[W]hat about getting this seed from the flax propaganda?" His own extension fund was inadequate, he claimed. Bolley tersely replied, "There is no 'Flax Propaganda Fund' for the purchase and distribution of seed." The director thought Bolley meant that there was no FDC at all: "I happen to know," he demurred, "that the Fl. Dev. Com. are annually giving into your hands several thousand dollars for flax development work," and he simply wanted "a little help."

In September 1913 Bolley reported to C.T. Wetherill, chair of the FDC, that scientists in other states had been advocating for their own support from FDC subscribers. They were asking each of the subscribers to consider dividing the fund between different states according to the number of members in each state. Bolley claimed that he was in favour of a decentralized network, in principle. "I am willing to co-operate," he pronounced, "but I do not like to be dictated to by those who have not yet shown anything accomplished for the flax crop."[49] When writing to other flax scientists, however, especially those who did not ask for money, Bolley was more supportive (and romantic), claiming that "the flax crop needs all the helpers it can get, because the world at large is against it."[50]

Bolley considered the FDC's service purely educational, a new way to teach farmers "soil and seed sanitation."[51] He believed that convincing local farmers was the same as convincing business people; North Dakota farmers were intelligent, and with only a couple of years of reiteration they should learn "to take up a new method of doing business."[52] Bolley's strategy for increasing flax seed propaganda was to print and distribute as many of his pamphlets on flax cultivation as possible. The station did not distribute free seed, wilt-resistant or otherwise, but merely instructed farmers where they could buy such seed. In 1911 he mailed out close to half a million circulars to farmers, railroad managers, and other business people; in 1912–13 that amount increased to 750,000 and his calculations did not include his publications in farm journals and other media.[53] Bolley claimed that he wrote about one article per week for college publications, farm journals, and local newspapers, but he qualified this official activity with an image of himself as farmer of the state, assuring industrialists that he would write summaries "as soon as I am through with my spring seeding."[54]

The photograph of Bolley's office in Figure 7.4 demonstrates rather well the level of activity during his tenure as a FDC scientist and propagandist. In the left foreground, Bolley reclines at his desk with a document, and behind him four women and two men work on other documents and a seed sample. Another typewriter sits on the desk in front of the photographer. What the photograph does not show are any signs of farming or experimental science. It is not clear how the other staff in his office were employed or what they were employed to write about, but in this pose anyway it appears that Bolley's funding paid for writers, reporters, postage, stenographers, and printers. Neither is it clear who took this photograph or who Bolley was posing for, but he certainly wanted to appear equipped and productive in his research, propaganda, or both.

Bolley's initial reports to the crushers and the larger FDC were, not surprisingly, claims of early success. More farmers than ever before were interested in flax and were practising good flax-farming methods.[55] Still, a constant complaint by contributors who wanted to see some real returns for their investment was that the progress of the FDC was difficult to

Figure 7.4 Bolley's office, Propaganda Headquarters, c. 1915

measure. Further routine criticism followed, but he discounted it as being misleading reports by false experts; he threatened to gather enough letters from farmers in his support to keep doubting crushers reading for weeks.[56]

Bolley tried to get flax grown on old land, but any land would do if it meant the crop would continue to supply his industrial employers. In fact, there was no shortage of "new" land in 1910. Bolley argued that some counties in western North Dakota "have sufficient new land to duplicate the entire crop of the state ... Morton County alone could almost do that."[57] In the second year of the FDC, Archer-Daniels sought to have Bolley's funding renewed among the crushers. They affirmed that the scientist's efforts to have flax grown routinely and profitably on old, wilt-infected land were critical to the industry's success; yet even then they reported resistance from some crushers to calls to renew subscriptions, and they had little confidence in the permanence of the funds.[58] Archer-Daniels asked for Bolley's estimation of how much of the new land remaining could be sown to flax. The scientist replied that a "tremendous" amount was available for flax farmers, most of it new land and some of it land that had been ploughed but never sown to flax. He estimated that a third or half of North Dakota's arable land was cultivated. "All of the remaining lands are awaiting only the turning of the plow," Bolley assured the company, and "all the land we would ever need for flax" was available to US growers.[59] Yet he was discouraged about the flax campaign in 1912. He said privately that he could not be expected to "change a whole farming system in a year or two." If wilt was thriving, it was only because North Dakota produced so much flax and by so many inexperienced and uneducated farmers.[60] By the summer of 1914 it became clear that the flax crop was smaller and the yield per acre lower than earlier levels. To Bolley, this was the result of farmers reducing their flax acreage because the price had been dropping since 1910 and they feared "the price would be lower than they would like to grow flax for."[61]

The partnership between state science and industry was already beginning to falter. Spencer Kellogg withdrew its support in 1913, and Midland Linseed followed in 1914, making many of the small crushers reluctant to "pay into something that the big crushers take advantage of" without paying. By the summer of 1915 the fund had dropped to $5,280 from ten contributors. Archer-Daniels claimed that the decline was due to many linseed crushers moving their plants east, as they planned "for the disappearance of the flax crop" in the Great Plains.[62] They feared that a third big crusher, American Linseed, might abandon the campaign. If it pulled out, then Archer-Daniels would not support the fund and "did not feel like asking the 8–10 small crushers" to continue.[63] Even Durbin, the Sherwin-Williams manager, admitted he was hearing reports that all "old land" was plagued with wilt and that "as soon as the wild land is all broken and has raised one crop of flax, flax raising will practically end" in North Dakota.[64] Durbin was a genuine believer in the science of flax disease; his comments reflect the fact that by now even the most dedicated crushers doubted the influence of farm science. Archer-Daniels continued to hear that wilt-resistant flax was a failure and that farmers were abandoning the crop. "You are the only one that gives us any encouragement about raising flax on old land," they despaired, and "We sincerely hope that your experiments and theories will prove to be a lasting benefit to the crop."[65] The myth of a frontier commodity continued to give the linseed oil industry a framework for understanding their own connection to place in the region.

By 1919 Bolley began to think that saving flax was futile. His correspondence with Archer-Daniels lost its previous business and logistical approach. Farmers do what they want, he claimed; "we just talk," and when the industry made sensational claims of flax's potential "it makes [farmers] unrestful on such matters, makes [them] grasp after something that only gets them into deep water." Ultimately it made farmers lose faith in farm science.[66] Bolley earnestly believed that his allegiance was to the farmer, and this partly explains his bewilderment here and in later years when farmers would mistrust him or simply ignore his findings and advice.[67] Always some form of flood or drought prevented the crop from recovering, but he continued to hope that farmers would try again. Bolley never stopped promoting flax as a good crop when markets allowed, but he did appear to give up hope that farmers would cultivate it in such a way as to eliminate wilt. In reply to an industry circular listing "30 years of linseed oil prices," Bolley argued that he had worked for a decade to increase flax output and lower the price of oil for "all the people of this country."[68] However, presumably lowering the price of oil would also lower the price of seed earned by the farmers he represented. What he was really coming to terms with was the uncomfortable alliance he shared with the corporate side of the flax-paint web. His work was meant to benefit the farmers of North Dakota, and he believed they would only prosper from flax farming if they received a fair price from the crushers. He believed it was up to the linseed oil crushers and consumers "to stand solidly behind the farmers" and up to the government to "protect [farmers] against the foreign crop" through the tariff.[69]

The FDC continued as an active force until 1931, but its funding for farm science in the 1920s was intermittent and usually for defined events, such as Bolley's research trip to Argentina and Uruguay in 1930. L.P. Nemzek of Dupont penned a condensed history of the FDC, and his annual crop summaries always glowed, always argued that when flax did not increase, it decreased less because of the work of the FDC and never mentioned that the support of major contributors, and therefore the faith of the small contributors, was constantly in doubt.[70] The FDC's primary goal of establishing 80 per cent of the industry's raw-material needs in the Northwest was not achieved, but the committee continued to develop flax in other ways. Those who wanted to see flax stay in the region worked on alternate solutions: Bolley continued to develop and distribute his resistant varieties, crushers experimented with distributing seed to farmers, and US farmers fought for a protective tariff against imported flax seed.

Even if corporate funding had been strong and consistent, it is unlikely that the state's scientific infrastructure could have created capacity for fully wilt-resistant flax farming. As previously mentioned, the flax work of NDAES involved selecting and producing wilt-resistant varieties. Their two most successful varieties in the 1910s were NDR 52 and especially NDR 114. In five years of testing, the average wilt resistance of the latter was 75 per cent, as compared to NDR 52, which was effective in only 13 per cent of tests, and "common flax," which resisted at a rate of 0.7 per cent.[71] In 1914 Bolley's agents inspected 14,000 acres in 269 fields. Most were seed plots of from 5 to 25 acres sown with carefully selected seed and weeded by hand to remove impurities from the finished product. Bolley estimated that at 8 bushels per acre these fields would yield over 111,000 bushels of "well bred seed."[72] If all the resistant seed was saved for sowing in 1915, which was unlikely (evidently much was sold to crushers), it would be enough to plant 110,000 to 160,000 acres. North Dakotan farmers planted from four to six times those amounts the following spring.[73] It is unlikely that Bolley was able to get his publications out to many farmers before 1905, when US production began its first major decline. After receiving assistance from the FDC and distributing countless papers, posters, and news briefings on flax, he seems to have found that his scientific advances made little difference to overall production levels.

By 1926 NDAES found that 62 per cent of farmers who submitted seed samples to the station possessed resistant flax for sowing, a "much higher percentage than usual." This was probably not representative of "the general run of growers," but it seems possible that more farmers were using resistant flax than had done so earlier. The rate of compliance was far slower than Bolley had hoped for and less effective than in other disease-resistant crops.[74] As we saw in Chapter 6, the acre value of flax was simply too close to that of other commodities to justify paying extra for seed that might only resist the effects of wilt in 75 per cent of cases. Argentine farmers produced more flax per acre and by this time had presumably perfected the art of extensive flax rotations on old land. North American farmers had other options, and it was hard to justify the risk and inputs involved in reintroducing flax. They had used flax creatively in the early years of settlement, but with increasing droughts and a declining rate of new breaking in the interwar period, it would only make sense to experiment with bringing flax back into rotations if its acre value was higher than that of competing crops.

Industrialists cared about encouraging and preserving the Northwest flax crop because they had invested in the region, and some crushers, such as Archer-Daniels, had strong connections to place. Ironically, Archer and Daniels were two of the major startups that left the American Linseed trust to take advantage of the new sources of flax seed. They recognized the limitations and legacy costs of the integrated enterprises based in Ohio and the Midwest and chose to leave for the Twin Cities. Now, they were the established order with substantial fixed costs, and in years of low production in the Great Plains, they found themselves very far from the world's flax seed. Before 1930 their circulars only show seed receipts in Minneapolis, Duluth, and Winnipeg.[75] It was this industrial geography that limited some crushers' access to overseas markets. There were ways for linseed oil crushers to stay in the region, and these strategies included applying existing human and capital resources to agricultural commodities that *could* be locally produced. Archer-Daniels Midland (ADM) later turned its attention toward soybeans, corn, and other commodities available in the Northern Great Plains.

Protecting Flax

Before they abandoned flax seed altogether, the crushers of the Northwest turned to tariffs. When strategies like endowing science, providing seed, and promoting cultivation failed to reverse the global shift in seed supply, some western crushers supported the seed tariffs of the 1920s provided there were compensatory tariffs on linseed oil. All of these business decisions occurred within a political economy and in the rapidly globalizing linseed oil trade of the 1920s, which included economic protectionism. Restricting trade at the border did not immediately sit well with producers and farm scientists whose families and colleagues worked on the other side. The work of Bolley and the FDC was initially meant to benefit farmers across the Great Plains, including Canada, and in other countries as well. Bolley had a long relationship with Canadian flax seed experts and Canadian branches of linseed oil businesses. American Linseed ensured that Bolley included Canadian farm scientists like John Bracken on his mailing list, and they even requested that Bolley send bulletins to their Argentinian representatives.[76] However, when it came to encouraging increased flax production through tariff protection, Bolley drew the line at the border.

Flax seed was one of the many US agricultural products given increased protection under the Emergency Act and the Fordney McCumber Act (Table 6.1). By 1922 a tariff of 40 cents per bushel was charged on imported seed. The policies were victories for US farmers who felt they needed protection against postwar European and other agricultural imports. Flax promoters such as Bolley were committed to increased protection. Like many of his contemporaries, he claimed to "have been pretty much a free trader all my life in spirit."[77] But, for most of his career he argued that, for better or for worse, the country was now on "a tariff basis," meaning that farmers paid duties on all manufactured products and should therefore be able to produce protected commodities. The industry needed tariffs on imported flax seed as well as "compensating duty on linseed oil and on all substitutes."[78] Once again Bolley found a friend in Archer-Daniels, who believed that only southern governors, vegetable oil importers, and other "enemies of our farmers" protested the duty in 1922.[79] For Archer-Daniels, the tariff provided a competitive edge against processors like Spencer Kellogg and others based in the East. So long as the price of foreign seed increased and US farmers planted a large crop, western mills could run longer and with lower costs than eastern mills, which suffered the opposite effect.

The root of Bolley's argument for protection was rather standard. If farmers were not protected, the crop would disappear from the US Plains. If the crop disappeared from the region, the country's entire linseed oil supply would be dependent on foreign raw materials. Foreign suppliers would then have the power to set prices and could not be trusted to play fair. The *Paint, Oil, and Drug Review*, an industry publication that favoured protectionism and wanted American farmers to grow more flax for linseed oil, warned that if US farmers "won't grow flax the farmers of Canada, Argentine India and Russia will, and the American farmer must pay tribute to them on every printed book or newspaper, every gallon of paint or varnish, every foot of linoleum or oilcloth used by him."[80] Although a variety of factors were at work, the crushers and Bolley attributed the massive increase in the US flax crop in 1924–25 to the tariff. Protection did not raise prices dramatically, but they increased consistently. Strong prices and good yields per acre in the early 1920s meant flax's value per acre was on the rise, and some farmers began to think of acre value as a predictable part of crop rotations. With a strong record of acre value, a steady market for paint in construction, and a protective tariff, US farmers gave flax another try. Archer-Daniels was not radically committed to the tariff, but they accepted it so long as seed flowed through Minneapolis and linseed oil received protection as well. They certainly would have preferred reciprocity with Canada, and they lobbied for it in 1911.[81]

Figure 0.2 ("Global flax seed production") shows that the competition from flax producers outside of the Northwest and Canada was strong and growing in the interwar period. The data also suggest that the benefits of protection were short-lived for US producers. We have seen that Canadian flax production was fairly resilient despite US protectionism. Prices for flax seed were not drastically different from US prices, and exports did not decline suddenly. Instead, Archer-Daniels predicted that "Canadian surplus would go to Europe," and the company's buyers in Winnipeg shipped very small quantities to Minneapolis in 1923 and 1924.[82] Companies like American Linseed took advantage of Canadian seed production by building a branch mill in Winnipeg, and flax specialists like Charles Thornton moved to Manitoba and continued to advise the industry from there.[83] The early prairie flax belt had also produced a transnational knowledge network based in the Northern Plains. Ralph Estey has shown that scientific research on flax diseases in Canada was insignificant before the 1930s.[84] Canadian farm scientists and enthusiasts like Howard Fraleigh, Cora Hind, and Seager Wheeler wrote to Bolley and gained relatively full access to the kind of research done across the border by Bolley and his colleagues.[85]

Regional and International Oilseed Knowledge Networks

By 1930 the flax-growing industry had all but collapsed in the US Plains and Canada, and Argentinian farmers dominated the global supply chains. Perhaps no one felt its demise as deeply as Henry Bolley. Although he had gone on to do other work in the 1920s, he maintained an interest in flax and at the USDA's suggestion re-examined the problem of wilt by going to a flax-growing country that seemed to have evaded the disease. As South American supplies expanded, linseed oil crushers made it their business to know the crop conditions in Argentina. Industry doubted the accuracy of government crop reports, and it more or less disregarded meteorological forecasts. USDA officials also invested in Argentinian data collection, but industry had limited confidence in those results.[86] The FDC was a business association whose official agenda was to promote the production of flax seed in the Great Plains through government-led propaganda campaigns. Unofficially, its agenda was to fund state scientists in return for environmental knowledge. Ultimately the value of ADM's environmental knowledge and regional networks had its limits. By the end of the First World War, they were convinced that building on the Eastern Seaboard and taking advantage of Atlantic trade was a necessary counterpart to their western operations.

In 1930 Bolley embarked on a reconnaissance journey to Argentina and Uruguay, funded by the USDA and the FDC. He was to study flax seed varieties and production in what was by then the undisputed leader in flax seed production and exports – Argentina. He made extensive preparations for travelling and observing specimens, relying on the advice of Allen Allensworth, a Chicago seed buyer and self-proclaimed "flaxseed specialist" who had offices or correspondents in all of the world's flax centres, including Winnipeg, Minnesota, New York, Rosario, Buenos Aires, Riga, and Bombay. Allensworth had recently spent time in Argentina and had advised Bolley on whom to contact (for example, a buyer at the Grain Exchange in Winnipeg who was a key source of information on the Argentinian interior), how to travel, and where to find flax farmers, distributors, and scientists.[87] Bolley's journey, his industrial sponsors, and his consultants in the North American West illustrate that even with the ebb and flow of grassland flax production, the region remained an exporter of scientific expertise and industrial knowledge. Linseed oil crushers had long been familiar with global markets, but by 1930 actors in many links of the commodity web had an intimate knowledge of the world of flax seed. This scientific and industrial reconnaissance of South American flax represented the movement of knowledge through the global commodity web.

The crushers asked Bolley to pass off as a general researcher, but one of his real assignments was to determine why Argentinian flax crops yielded between 33 and 50 per cent more per acre than North American flax crops and contained a higher oil content per seed.[88] Bolley's trip took him through Chile to the northern provinces of Argentina and into neighbouring districts in Uruguay. What he found was an efficient system of flax cultivation and distribution. Growers faced many of the same diseases and other biological limitations to cultivation as they did in North America,[89] but production thrived in the same areas for decades with relatively short rotations, using what to Bolley was antiquated bag and wagon distribution, and without a system of private land ownership. The Argentinian system operated in a way practically opposite to the American way and added little clarity to the mystery of the disappearance of flax in the Great Plains. Bolley could only point rather weakly at cheap labour and a low standard of living as an explanation.[90]

By the time Henry Bolley journeyed to Argentina, he was quite convinced that flax's predisposition to new land was partly what caused it to decline so rapidly in the Great Plains. By contrast, in Argentina, farmers in sixteen contiguous counties in Santa Fe and Entre Rios Provinces devoted around half of their cropland to flax, and 80 per cent of cropland around Victoria, Entre Rios, was in flax in 1913.[91] Here it was grown widely and consistently. Argentines farmed flax seed (never fibre) almost constantly over large acreages. Low population density, combined with low land value and the tenure system, discouraged intensive farming and highly processed commodities such as dairy products for local markets. The high ratio of land to farmers and the rich soils and rotation system catered to economies of scale and ensured relatively good land productivity. Argentina possessed a matrix of rural development similar to Canada's, including railroads, an influx of agricultural settlers, and plenty of arable land for expansion, but the production of flax, wheat, and corn remained at the seaboard. There was also little expansion in these maritime areas in this period.[92] According to Jeremy Adelman, the main differences for Argentina were a lack of capital investment and the persistence of tenant farming. We know that Argentinian officials expected that a North American model of land distribution would bring prosperity to the Pampas, but their hopes did not materialize. However, Bolley's journey and report to the USDA suggest that US onlookers were equally as convinced that the uncultivated land southwest of the province of Santa Fe was suitable for agricultural development and that flax would be a spearhead in that country.[93]

In Canada and the United States it made sense to replace the cattle ranches with extensive grain cultivation and the old grain lands with intensive farming (dairy, livestock, fruit, etc.), but that displacement process occurred less in Argentina, and moreover, much cereal expansion took place in Santa Fe, a province generally avoided by ranchers.[94] As a result, grain producers remained at the seaboard where they had easy access to export markets and also benefited from efficient labour and handling systems.[95] Why flax producers remained in the north on the reddish soils and did not move west to the dark soils or south to the grey remains a mystery.[96]

Presumably the system of very intensive flax cultivation, distribution, and close rotation with wheat, corn, and alfalfa was successful enough to endure through periods of overproduction, low prices, and any disease or threats presented by the northern Argentinian environment. Argentinian farmers learned to rotate flax adequately with other cereals and alfalfa to mitigate the effect of disease and soil exhaustion. Bolley returned from South America deeply impressed by the short transit from flax fields to river transportation, by the long growing seasons and other crop conditions, and by the generosity of the people "whose every product must meet quite directly or come into competition with our restless production."[97] However, in September 1931 Bolley submitted a fifty-page report to the linseed oil crushers, and this document argued that Argentinian farmers could happily live with less, and US farmers were "quite right in demanding such proper tariff as will permit them to do a better grade of farming on their high grade lands."[98]

Forty-four years after his first work in flax Bolley sketched a chronology of his flax research for a reporter from the *Montreal Star*. He approached the task carefully, including much detail and taking eight days to write the letter.[99] He concluded with his own involvement with flax at the stages of crossbreeding and creating resistant varieties, a breakthrough that he claimed made it possible for flax farming to return to the Red River Valley, from which it had previously disappeared because of flax wilt. Bolley then shifted to the reasons flax had failed to make a significant comeback in the region. Essentially, the Northwest flax farmer "continu[ed] to fail even though he [was] provided with a disease resistant variety of seed" because of the vagaries of climate and the markets. The real problems, according to Bolley, were poor farming and careless seed cleaning and selection. Many farmers considered flax a "substitute crop," he rankled, and only used it when other crops failed. "Its own hardiness makes it more or less a favourite of those who practice shiftless methods of agriculture."[100]

It is significant that Bolley overlooked the entire work of the Flax Development Committee and attributed two decades of declining production during intensive scientific and industrial research to shiftless farming. His proud history of defeating flax wilt ended with a polemic about the enemies of flax that science could not eliminate – cheap labour in India, trade policy favouring Argentinian exports, and imported substitutes for linseed oil in the United States. He justified his years of work in flax by connecting them to another accomplishment, the establishment of a North Dakotan pure seed law, which he said was primarily to secure "a supply of resistant seed" and prevent mixing with ordinary seed. He seemed paranoid about the persistent belief in flax's soil-exhaustive qualities. He reaffirmed that the cry of flax farmers should rather have been "Conserve the pure seed varieties" and, curiously, restore "the natural water supplies in our remaining forest and agricultural areas."[101] Bolley's view of his flax work was much different in his later years. His language changed in the 1930s from the rhetoric of "saving the flax crop" to convincing farmers that flax cultivation made sense if they would just "give the crop a square deal."[102]

Over his career, Bolley found that by treating the seed the spores of the fungus could be destroyed, and he grew flax continuously in the diseased Plot 30 at the state experiment station in order to select and breed more wilt-resistant strains. He enlisted the help of flax industrialists and farmers alike in an effort to "save the flax crop," but the general seed production levels declined in the first two decades of their work. The acre value of flax compared closely with that of other crops, and the cost of Bolley's treatments and special seed varieties all counted against it. The US crop revived in the 1920s, but this was temporary and at least partly due to protective tariffs. The peculiar mobility that relocated the North American flax crop every year until at least 1916 was curtailed, partly by the boundaries of arable land and partly by a new political economy. The scientific triumph over infected soil had limited effect; ultimately the inputs in the flax-paint web favoured new land and demanded a consistent price for its outputs.

David Danbom has argued that Bolley's work in flax was responsible for the rapid increase in production in North Dakota.[103] However, this movement of flax into the northern dry belt was in motion before Bolley picked up a flax seed, and in fact when Bolley began his largest efforts to save flax with the FDC, the crop was experiencing a rapid decline. The FDC's development was about much more than saving the Northwest flax crop with the help of farm scientists and educational propaganda. It represented the importance of trade associations in integrated industries, gave smaller crushers a chance to learn from and about the large operators at a small price, and reflected the importance of backward integration in the linseed oil, paint, and varnish enterprises. To scientists like Bolley, saving flax was really about the professionalization of farm science; to the crushers, it was ultimately another route to the information they desired.

Flax seed cultivation was predisposed to new land primarily – and perhaps only – in North America. Even as prairie settlement continued into the Peace Country of northern Alberta, farmers favoured flax on new land. The Peace River District went from growing 6 per cent of Alberta's flax in 1941 to 26 per cent in 1956. Production was concentrated along the river and near the town of Grand Prairie, and unlike the ubiquitous cereals, the flax crop was grown primarily by a handful of risk-tolerant farmers who watched eastern markets carefully and sought to capitalize on some of the last new farmland in Canada. However, this was not because the plant had some special connection to virgin topsoil. Flax grew best on good land and with good farming, as Henry Bolley observed in Argentina. It was popular on new land in North America because the Northern Great Plains (especially the transborder region of Montana, North Dakota, Alberta, and Saskatchewan, surrounding the Palliser Triangle) produced excellent flax for linseed oil, it had a settlement and distribution system suitable for flax seed, and the new land was free of diseases unique to the flax genus. When disease on "old" flax land raised the costs of treatment beyond the spread between the acre value of flax and that of other crops, it was abandoned. Unlike in South America, northern farmers had quickly established infrastructure and markets for more lucrative and intensive farm commodities. Therefore flax usually only appeared on the newest, cleanest, and, by the 1920s, the most marginal land.

CONCLUSION

In the postwar period the Great Lakes region's fibre production came to a halt, and the much larger linseed oil industry slowed as synthetic substitutes were developed for paints. Flax seed was still produced for linseed oil, but the market for this oil was minimal and other crops like soybean and canola replaced flax's dominance in the oilseed sector. Beginning in the 1990s, however, flax became both Canada's first genetically modified crop and a commodity that no one in the fibre and oil webs could have predicted. Due to its rich content of essential fatty acids, flax seed has recently become an important part of some human diets as a component of the "nutraceutical" industry. A 2011 promotional video from the University of North Dakota calls flax simple and unassuming, "yet it is perhaps one of the most important crops of our time."[1] This will likely prove to be a hasty conclusion, but it does reflect some of the new life entering the flax commodity web through its increased consumption as a health food in the early twenty-first century.

Beginning around 1997, dozens of new food and beverage products containing flax seed and an "omega-3" label started to appear in North American grocery stores. By 2003 that number had risen to 59 per year, and by 2005 it was 178. Similarly, new bread and cereal products began to appear at around half a dozen per year in the late 1990s and up to 59 per year in 2005.[2] Flax appeared in both functional foods and nutraceuticals. A functional food is one that resembles conventional food products but that has demonstrated physiological benefits. This could include omega-3-enriched eggs or bread products that include ground flax seed additives. A nutraceutical, on the other hand, is a medicinal product that is isolated or purified from foods, but it is not usually associated with foods. This includes medicines with a range of flax seed and oil additives.[3]

Health and nutrition specialists continue to debate the exact health benefits of consuming flax seed and related nutraceutical products, but the seed is rich in three important ingredients. It has soluble and insoluble fibres, omega-3 fats, and antioxidant-rich lignans. These are available through direct consumption, although the seeds must be ground in order for humans to fully digest their contents. A high fibre diet is important in regulating both human cholesterol and digestive systems. Lignans are a source of antioxidants that apparently regulate hormones, and recent research suggests that they may also help to reduce the risk of breast and prostate cancer.

Flax is particularly unique in the vegetable kingdom for its high levels of omega-3 fatty acids. Health and agri-food agencies had not advanced the official recommended daily intakes of long-chain omega-3 fatty acids as of 2015, but the scientific literature on the human health benefits of these foods is extensive and it generally recommends higher intakes. Fatty acids are an important source of ATP (Adenosine triphosphate) in human and animal nutrition. The acids that the body needs and can only obtain from food are called *essential* fatty acids. The most valuable essential fatty acids for human consumption are eicosapentaenoic acid (EPA) and docosahexaenoic acid (DHA), but these are only found in fish and fish oil – that is, in a threatened and declining food supply. Alpha-linolenic acid (ALA) is a slightly shorter-chain omega-3 essential fatty acid, and it is derived from many terrestrial plant oils. Of all the commonly produced North American vegetable oils, flax produces the highest yield of alpha-linolenic acid, by weight. About 55 per cent of the oil (23 per cent of the seed) is alpha-linolenic acid, whereas canola oil contains 9.6 per cent and soybean oil contains 7.8 per cent. In 2013 flax was the fourth-largest potential source of ALA from vegetable oils, ranking close to the third-largest source, palm oil. The largest two sources, soybean and canola, were many orders of magnitude larger.[4]

Flax is increasingly being fed to hens in order to produce omega-3-enriched eggs. Consumer product research shows that the best feeding ratio of flax to standard feeds is between 5 and 10 per cent of the hen's total feed.[5] Similarly, dairy cattle that eat a balanced supplement of flax seed meal produce milk that is higher in omega-3 fats and lower in harmful saturated fats. The efficiency of these conversions relative to direct flax seed consumption is still in question. Flax has even been used to beef up the nutritional value of meats, although this is currently in the research stage, as well. Researchers add milled flax seed directly to meats to increase the "product functionality" of that food, again, within the parameters of consumer preference.[6] After a long period of dormancy, flax is back.

Flax Americana explores what allowed a specialty crop like flax to appear in concentrated regions and what caused it to vanish for long periods of time – perhaps to resurface. This is not a history of a commodity that changed the world; rather, this book uses flax's commodity web as a lens through which to examine the world that changed a commodity. In the century after 1850, that world was North America, and like the oilseed sector that followed flax's footsteps of regional and place-based expertise, the flax belt became increasingly "telecoupled" with the South American Pampas. In the Great Lakes region, flax fibre production grew rapidly in mid-nineteenth-century mill complexes, followed by a much larger increase in flax seed. The fibre was manufactured into special textiles like sails, but most North American flax was too coarse for linen. The main problem was the labour demands caused by the need to pull flax by hand and to do it with precise timing, before the seed matured and the fibres coarsened. Farmers began to specialize in maximum seed production, and eventually the fibre was discarded in major flax seed areas such as the West. Flax mill outputs in Waterloo County suggest that the industry focused on semi-processed exports and fed into growth in local manufacturing and markets for consumer and intermediate goods. This in turn encouraged flax cultivation among farmers who for the most part had never grown significant amounts of flax before. Millers like the Perine brothers used this system of flax factorship, as well as the labour of escaped slaves, First Nations families, and other marginalized people, to harvest the crop in the Grand, Kettle, and Credit River valleys. The small but relatively stable system that developed as a result persisted into the 1940s.

By the end of the First World War, flax supplied fibre and oilseed industries in Canada and the northern United States in ways that were all but absent seven or eight decades earlier. Small-town landscapes in southwestern Ontario, Ohio, and Upstate New York were conspicuously marked by these flax fibre mills and warehouses on the edges of town. On the Prairies, Canadian flax seed production rivaled that of the United States, and farmers sowed flax on first breaking on the very edges of arable land. In the growing cities, families lived, worked, and traveled in a world coloured and covered by linseed oil in relatively new commodities, such as ready-mixed paint and linoleum. Once it became an intensive prairie oilseed crop, flax was another tool in the hands of sodbusters like Joe Bellas and Charles Noble, encroaching on the fragile land of the semi-arid grasslands. If wheat made the Northern Great Plains the breadbasket of North

America, then flax seed made Saskatchewan, Montana, and North Dakota the continental paint bucket. Some officials even promoted flax as a living instrument of plains pioneering, a crop that prepared and broke down the soil, even if it seemed to lead quickly to soil exhaustion for flax production itself. Flax seed production in the semi-arid Great Plains and western Prairies shows farmers' surprising involvement in commodities outside of wheat and other foodstuffs and their conversant manipulation of grassland ecosystems from the first days of settlement in these areas.

This research required a deceptively simple methodology: following things around. The best way to see the nuance and meaning of a commodity web is to conduct a thorough examination by region and ecosystem, not by its products. *Flax Americana* set out to tell a story about flax in a particular time and place, but in order to carry the narrative forward, it had to follow the commodity and the people who moved it to the next set of locations. The process often requires a transnational approach. The processors were central to understanding flax's role in both the international and the local economies of the Great Lakes and the Northern Great Plains regions. In the East, most processors and many farmers were businesses in transit. Some had previous experience in flax growing or milling, and many did not. The most successful Canadian millers operated on both dimensions of the socio-ecological system – that is, they understood both people and place. Fibre processors followed and filled in the gaps left by the retreating lumber industry in small-town Ontario mill complexes. Millers used the factorship system and a seasonal labour supply that experts missed. Most important, we see that farmers, workers, and millers adapted. Mennonite families who have previously grown little or no flax produced large amounts after the Perines arrived, both for the market and for new forms of self-provisioning. Farmers in Ontario adjusted their rotations to include flax for fibre, and then in both Ontario and Manitoba they adjusted their harvest practices to accommodate the new market for seed. First Nations families worked in the flax harvest because they were denied other opportunities, but some also used the opportunity itself to extend older practices of summer migration. Millers like the Livingstons emulated the Perines' practice of using small-town mill complexes and local partners to extend the factorship system across southwestern Ontario and into Michigan. These intensely local relationships continued for nearly a century, but the expanding urban consumption of linseed oil and paint required businesses like the Livingstons and ADM to reach out to a new environment in the West.

As in most regions, these ecosystems ignored political borders. In the Canadian case studies in this book, practically all the major figures came from or expanded into similar businesses in the United States. Flax was grown and milled in a transnational part of the Great Lakes region, but then Mennonites grew it in a planned arrangement on a planned settlement in southern Manitoba. South of the border, farmers began to grow flax in a more traditional open market, and like the Mennonites, they found that it thrived on the newly ploughed land of the Midwest and Northern Great Plains. The story followed them there. As linseed oil became the most valuable product from the plant, we find an oilseed empire being built in the West; it retained important capital connections to the East but no longer to the rural mill complexes. The products were now found mainly in urban manufacturing centres. Still another shift occurred when farmers in Argentina began producing large amounts of flax seed during the First World War. Flax was transnational because it was a small but integrated industry, the source of raw material moved frequently, and the border was porous. Knowledge and capital crossed the border easily, farmers and flax fibre millers moved or expanded across it regularly, and US linseed oil crushers set up mills in Winnipeg, Buenos Aires, and other cities.

North American linseed oil production was based in Minneapolis, New York, and a few lakeport cities, but it drew its raw material from an extensive agricultural area in the temperate grasslands of the Northern Great Plains, Canadian prairies, and Argentinian Pampas. Industrialists encouraged the Northern Great Plains flax crop because they had invested in the region. Crushers such as ADM had a strong connection to place and became authorities on western crop and climate knowledge. As South American supplies expanded, linseed oil crushers made it their business to learn about the crop conditions in Argentina. In the 1910s and 1920s Argentina experienced a land-use transition and industrial commodity export boom that mirrors what scientists call the present-day soybean commodity frontier in several important respects. Most important, it was an oilseed supply chain, formerly centred in North America and by then growing in South America. It also involved animal feed, technology transfer, and a strong core of North American buyers, investors, and developers. Businesses like ADM later turned their attention toward soybeans, corn, and other commodities available in the Northern Great Plains, but there is reason to believe that they are again investing heavily in South American assets and once again contributing to the process of creating a telecoupled commodity frontier.

Flax was a specialty crop that appeared and vanished from some regions so quickly that it escaped record. But, what often remained was a network of specialists that serviced the flax fibre and seed industries internationally. Flax regions in Ohio hosted businesses such as the French Company, which supplied oil-processing machinery to linseed oil and cottonseed oil crushers long after flax seed production had disappeared from the Miami River valley.[7] Specialists in southwestern Ontario had a reputation as fibre machinery suppliers for the Oregon flax industry even when Ontario's flax fibre production was in decline in the 1920s.[8] The examination of Henry Bolley's network demonstrates that flax seed buyers, scientists, and mill managers, based mostly in the West, became recognized experts and consultants for the international linseed oil industry in the twentieth century.

Fibre and oil were in many ways the trademark commodities of the First and Second Industrial Revolutions, but to most people *Linum usitatisimum* was simply the nameless intermediate good whose fibres and oils made up so much of their manufactured worlds. The early nineteenth century was marked by a nearly insatiable appetite for fibre, and the general transition from small-scale cottage industry to factory textile production in the United Kingdom and the northeastern United States was one of the early hallmarks of industrialization. Then, in the late nineteenth century, the rapid expansion of urban manufacturing and the important role of chemicals and other advanced goods marked a second phase in the on-going socio-ecological transition.

The only actors who did not seem to understand the two interconnected flax webs were the promoters. They promoted fibre cultivation in the 1860s when it made sense to do so, but thereafter they failed to recognize how much value had shifted to the seed. Promoters often mentioned how much fibre was going to waste in the Prairies, but to them that was almost a separate crop, and it certainly was not suitable for fine linen. Two of the reasons they got it so wrong were geopolitical and scientific. First, in Canada, flax promoters were often politicians or government employees, and as such their international collaborations were mainly in the United Kingdom. Political ideologies and allegiances exacerbated the pattern of fibre promotion that emerged every few decades in Canada. Second, Canadian promoters made the Linnaean mistake of assuming that flax fibre and flax seed were two different commodities, not two parts of the same web. Even those scientists who understood the plant itself had surprisingly limited knowledge of the larger commodity web. In most cases the commodity knowledge was asymmetrical. For example, federally funded scientists worked to improve the varieties available to farmers, gathering information from traditional flax-producing regions like Russia, Belgium, and Ireland. In 1930 Henry Bolley traveled to Argentina to survey the flax industry there and to conduct field experiments. His trip was funded by the Flax Development Committee, one of several committees funded by the major paint and linseed oil trade associations. Two others were the Save the Surface Campaign committee and the Clean-up, Paint-up Campaign committee. Associations disclosed only part of their motives for funding state research, and scientists told farmers (especially non-US farmers) as little as possible as they traveled around the world to gather valuable trade and biological information. Bolley's interactions with linseed oil businesses and trade associations illustrates the role of information asymmetry between actors in these early telecoupled commodity frontiers.

The historical significance of the flax commodity web's emerging health food branch has yet to be realized. If the health benefits of the direct consumption of flax seed and its conversion to milk and eggs are proven, then certain elements of the commodity web are likely to change again. They may even create new rifts in the commodity's structure. Such new findings might also suggest that some historical groups, like those consuming northern Europe's dairy products, might have had prolonged exposure to valuable nutrients that North Americans ignored. The very fact that changing consumer tastes have today produced – again – an increased interest in flax seed should demonstrate the importance of considering the environmental history of commodities in telecoupled systems. Commodity webs are another way to model the flows of material, energy, and information between coupled systems, and when historians follow these things around, they are able to see the complex and interrelated roles of environmental and temporal forces.

For now, the new popularity of flax in functional foods has boosted the crop's importance in the Canadian prairies and the very northern US Plains. Moreover, the growth of associations such as the Flax Council of Canada and AmeriFlax has been distinctly regional, with headquarters in Winnipeg and Bismark, respectively. The Flax Institute of the United States continues to hold annual meetings in the northern Plains or Prairies, and their most recent work focuses on disseminating information about flax in comestible goods.⁹ This geography of cultivation and association reaffirms the role of regional patterns in the flax commodity web as well as the continued importance of environmental knowledge. If the twenty-first century equivalent of the millers in this section of the web follow historical trends, then the health food sector may rekindle the important connection between grassland producers and middle-class urban consumers.

NOTES

ABBREVIATIONS

AO	Archives of Ontario, Toronto, ON
CF	*Canada Farmer*
CIHM	CIHM/ICMH Microfiche Series = CIHM/ICMH Collection de microfiches
DCB	*Dictionary of Canadian Biography*
DHC	Doon Heritage Crossroads – Research Library and Archives, Kitchener, ON
EMA	Esplanade Museum and Archives, Medicine Hat, AB
GMA	Glenbow Museum and Archive, Calgary, AB
HLBP	H.L. Bolley Papers, University Archives, North Dakota State University, Fargo
KPL	Kitchener Public Library, Kitchener, ON
LAC	Library and Archives Canada, Ottawa, ON
MAO	Mennonite Archives of Ontario, Conrad Grebel College, University of Waterloo, ON
MHS	Minnesota Historical Society, Minneapolis, MN
OAC	Ontario Agricultural College, Guelph, ON
SA	Saskatchewan Archives, Regina, SK
TDHS	Tavistock and District Historical Society
TWA	Township of Wilmot – Sir Adam Beck Archives, Castle Kilbride, Baden, ON
UGA	Archival and Special Collections, University of Guelph, Guelph, ON
UWA	University of Waterloo Archives, Waterloo, ON
WCMA	Wellington County Museum and Archives
WHF	Woolwich Heritage Foundation (part of the Waterloo Historical Society)
WHS	Waterloo Historical Society

INTRODUCTION

1 Eliso Kvavadze, Ofer Bar-Yosef, Anna Belfer-Cohen, Elisabetta Boaretto, Nino Jakeli, Zinovi Matskevich, and Tengiz Meshveliani, "30,000-Year-Old Wild Flax Fibers," *Science* 325, no. 5946 (September 2009): 1359; Marion Vaisey-Genser and Diane H. Morris, "Introduction: History of the Cultivation and Uses of Flaxseed," in Alister D. Muir and Neil D. Westcott, eds, *Flax: The Genus Linum* (London: Routledge 2003), 1–2; Füsun Ertuğ, "Linseed Oil and Oil Mills in Central Turkey Flax/Linum and Eruca, Important Oil Plants of Anatolia," *Anatolian Studies* 50 (2000): 171–85.

2 Joshua MacFadyen, "Spinning Flax in Mills, Households, and the Canadian State, 1850–1870," in James Murton, Dean Bavington, and Carly Dokis, eds, *Bringing Subsistence Out of the Shadows: Nature and Economy in Historical and Contemporary Perspectives* (Montreal & Kingston: McGill-Queen's University Press 2016); Henry Youle Hind, "On the Cultivation and Manufacture of Flax and Hemp in Canada," *British American Magazine* (August 1863): 369–80.

3 "Start Pulling Flax on Government's Farm," *Toronto World*, 19 July 1918, 5; "Flax Festival Today at Willowdale Farm," *Toronto World*, 23 July 1918, 3.

4 The North Dakota Agricultural College Extension Service, "FLAX: You can help America win by producing more," *NDAC Special Circular* A-13 (March 1942): 1.

5 Emily Eaton, *Growing Resistance: Canadian Farmers and the Politics of Genetically Modified Wheat* (Winnipeg: University of Manitoba Press 2013), 88; Donna Haraway, "The Promises of Monsters: A Regenerative Politics for Inappropriate/d Others," in L. Grossberg, C. Nelson, and P. Treichler, eds, *Cultural Studies* (New York: Routledge 1992); Michael Pollan, *The Botany of Desire: A Plant's-Eye View of the World* (New York: Random House 2001).

6 Anatoly Marchenkov, Tatiana Rozhmina, Igor Uschapovsky, and Alister D. Muir, "Cultivation of Flax," in Alister D. Muir and Neil D. Westcott, eds, *Flax: The Genus Linum* (London: Routledge 2003), 86–7.

7 William J. Smyth, "Flax Cultivation in Ireland: The Development and Demise of a Regional Staple," in William J. Smyth and Kevin Whelan, eds, *Common Ground: Essays on the Historical Geography of Ireland: Presented to T. Jones Hughes* (Cork, Ireland: Cork University Press 1988), 241.

8 Canada, *Census of Canada, 1911*, Agriculture, Vol. 4 (Ottawa: J. de L. Taché 1914), Table IV, 222–309; B.R. Mitchell, *International Historical Statistics: The Americas and Australasia* (London: Macmillan 1983), Series D3 and D4, 269–75.

9 US Department of the Interior, *Twelfth Census, 1900*, Vol. 5, Manufactures (Washington, DC: United States Census Office 1902), 593.

10 F.H. Leacy, *Historical Statistics of Canada* (Ottawa: Statistics Canada 1983), R759.

11 "Flax," in *The Plough, the Loom, and the Anvil*, 1854, 583; Barbara Hahn, "Paradox of Precision: Bright Tobacco as Technology Transfer," *Agricultural History* (Spring 2007), 232.

12 *Manual of Flax Culture, Comprising Full Information on the Cultivation, Management, and Marketing of the Crop, Together with a Complete Glossary and Index*

(New York: Orange Judd Company 1883), 37; Marchenkov et al., "Cultivation of Flax," 86–7.

13 Jonathan Hamill, "The Irish Flax Famine and the Second World War," in Brenda Collins, Phillip Ollerenshaw, and Trevor Parkhill, eds, *Industry, Trade and People in Ireland, 1650–1950: Essays in Honour of W.H. Crawford* (Belfast: Ulster Historical Foundation 2005), 229.

14 Mavis Atton, *Flax Culture: From Flower to Fabric* (Owen Sound: Ginger Press 1988), 14; Kenneth W. Keller, "From the Rhineland to the Virginia Frontier: Flax Production as a Commercial Enterprise," *Virginia Magazine of History & Biography* 98, no. 3 (1990): 487–511, 496. The United States census differentiated between water- and dew-retted hemp and indicated that the latter process accounted for almost all of hemp production. United States, Department of the Interior, *Seventh Census, 1850: Embracing a Statistical View of Each of the States and Territories, Arranged by Counties, Towns, etc....* (Washington, DC: Robert Armstrong 1853), Table 55, lxxxiii.

15 Alexander Kirkwood, *Flax and Hemp* (Toronto: W.C. Chewett 1864), 28; *Manual of Flax Culture*, 35–6. The toxicity of flax retting to fish has been understood since the fifteenth century; some manuals made the selection of safe water a priority in the retting process. Michael R. Best, ed., *The English Housewife* (Montreal & Kingston: McGill-Queen's University Press 1994), 155; Chevalier Claussen, "The Flax Movement: Its National Importance and Advantages with Directions for the Preparation of Flax-Cotton" (1851), 37; James A. McCracken, *A Review of the Status and Possibilities of Flax Production and Manipulation in Canada* (Ottawa: Government Printing Bureau 1916).

16 Sterling Evans, *Bound in Twine: The History and Ecology of the Henequen-Wheat Complex for Mexico and the American and Canadian Plains, 1880–1950* (College Station: Texas A&M University Press 2007); Sven Beckert, *Empire of Cotton: A Global History* (New York: Vintage 2015), 246–51.

17 N.I. Gasparri, T. Kuemmerle, P. Meyfroidt, P. le Polain, Y. de Waroux, and H. Kreft, "The Emerging Soybean Production Frontier in Southern Africa: Conservation Challenges and the Role of South-South Telecouplings," *Conservation Letters* 9, no. 1 (2016): 21.

18 J. Liu, V. Hull, M. Batistella, R. DeFries, T. Dietz, F. Fu, T.W. Hertel, R.C. Izaurralde, E.F. Lambin, S. Li, L.A. Martinelli, W.J. McConnell, E.F. Moran, R. Naylor, Z. Ouyang, K.R. Polenske, A. Reenberg, G. de Miranda Rocha, C.S. Simmons, P.H. Verburg, P.M. Vitousek, F. Zhang, and C. Zhu, "Framing Sustainability in a Telecoupled World," *Ecology and Society* 18, no. 2 (2013): 26.

19 George Colpitts, "Knowing Nature in the Business Records of the Hudson's Bay Company, 1670–1840," *Business History*, 2017, http://dx.doi.org/10.1080/00076791.2017.1304914; Sanjay Subrahmanyam, "Connected Histories: Notes towards a Reconfiguration of Early Modern Eurasia," *Modern Asian Studies* 31, no. 3 (1997): 735–62.

20 William Cronon, *Nature's Metropolis: Chicago and the Great West* (Norton 1991), 259.

21 The theory was proposed by Immanuel Wallerstein and Terence K. Hopkins and introduced in Gary Gereffi and Miguel Korzeniwicz, eds, *Commodity Chains and Global Capitalism* (Westport: Praeger Publishers 1994).

22 Matthew Evenden, "Aluminum, Commodity Chains, and the Environmental History of the Second World War," *Environmental History* 16, no. 1 (January 2011): 71.

23 Philip Scranton, *Proprietary Capitalism: The Textile Manufacture at Philadelphia, 1800–1885* (Cambridge: Cambridge University Press 1983), 5–7; John Brewer and Roy Porter, eds, *Consumption and the World of Goods* (London: Routledge 1993).

24 Y. Eyüp Özveren, "Shipbuilding, 1590–1790," *Review (Fernand Braudel Center)* (2000): 15–86; Philip Ollerenshaw, "Stagnation, War and Depression: The UK Linen Industry, 1900–1930," in Brenda Collins and Philip Ollerenshaw, eds, *The European Linen Industry in Historical Perspective* (Oxford: Oxford University Press 2003), 285–307, Figure 13.1.

25 Some recent commodity historians have used the language of networks or "webs" of production. John Soluri, *Banana Cultures: Agriculture, Consumption, and Environmental Change in Honduras and the United States* (Austin: University of Texas Press 2005), 218; Michael Roche, "The Commodity Chain at the Periphery: The Spar Trade of Northern New Zealand in the Early 19th Century," in Richard Le Heron and Christina Stringer, eds, *Agri-Food Commodity Chains and Globalising Networks* (Aldershot, UK: Ashgate Publishing 2012): 210.

26 Tina Loo, "Missed Connections: Why Canadian Environmental History Could Use More of the World, and Vice Versa," *Canadian Historical Review* 95, no. 4 (2014): 621; Albert G. Way, "'A Cosmopolitan Weed of the World': Following Bermudagrass," *Agricultural History* 88, no. 3 (2014): 354–67.

27 See, for example, Andrew Rimas and Evan Fraser, *Beef: The Untold Story of How Milk, Meat, and Muscle Shaped the World* (New York: HarperCollins 2008); Mark Kurlansky, *Cod: A Biography of the Fish That Changed the World* (Toronto: Alfred A. Knopf 1997); Bruce Robbins, "Commodity Histories," *PMLA* 120, no. 2 (2005): 455.

28 Jared Diamond, *Guns, Germs, and Steel: The Fates of Human Societies* (New York: W.W. Norton & Company 1997); Michael Pollan, *The Botany of Desire: A Plant's-Eye View of the World* (New York: Random House 2001).

29 Arjun Appadurai, "Introduction: Commodities and the Politics of Value," in Appadurai, ed., *The Social Life of Things: Commodities in Cultural Perspective* (Cambridge: Cambridge University Press 1988), 29. See also Igor Kopytoff, "The Cultural Biography of Things: Commoditization as Process," in Appadurai, ed., *The Social Life of Things*.

30 Graeme Wynn, *Canada and Arctic North America: An Environmental History* (Santa-Barbara: ABC-Clio 2007), xv.

31 Thomas M. Truxes, *Irish-American Trade, 1660–1783* (Cambridge: Cambridge University Press 2004), 201.

32 Joshua MacFadyen, "Long-Range Forecasts: Linseed Oil and the Hemispheric Movement of Market and Climate Data, 1890–1939," *Business History* (2017): 1–24.

33 Liu et al., "Framing Sustainability in a Telecoupled World," 26.
34 Jim Clifford, *West Ham and the River Lea: A Social and Environmental History of London's Industrialized Marshland, 1839–1914* (Vancouver: UBC Press 2017), 26–33.
35 Canada, *Census of Canada, 1911*, Agriculture, Vol. 4 (Ottawa: J. de L. Taché 1914), Table 4, 222–309; B.R. Mitchell, *International Historical Statistics: The Americas and Australasia* (London: Macmillan 1983), Series D3 and D4, 269–75.
36 Allan Kulikoff, "The Transition to Capitalism in Rural America," *William and Mary Quarterly* 46, no. 1 (1989): 120–44; Albert Schrauwers, "The Gentlemanly Order & the Politics of Production in the Transition to Capitalism in the Home District, Upper Canada," *Labour/Le Travail* 65 (2010): 9–45.
37 Appadurai, "Introduction," in *The Social Life of Things*, 20–1.
38 Margaret Conrad, Kadriye Ercikan, Gerald Friesen, Jocelyn Létourneau, Delphin Muise, David Northrup, and Peter Seixas, *Canadians and Their Pasts* (Toronto: University of Toronto Press 2013).
39 Carole Shammas, "How Self-Sufficient Was Early America?" *Journal of Interdisciplinary History* 13, no. 2 (Autumn 1982): 247–72; Adrienne D. Hood, "The Material World of Cloth: Production and Use in Eighteenth-Century Rural Pennsylvania," *The William and Mary Quarterly* 53, no. 1 (January 1996): 50, 52; Laurel Thatcher Ulrich, *The Age of Homespun: Objects and Stories in the Creation of an American Myth* (New York: Alfred A. Knopf 2001), 4.
40 Michel Boisvert, "La Production textile au Bas-Canada: L'Exemple Laurentien," *Cahiers de géographie du Québec* 40, no. 111 (1996): 421–37; Fernand Ouellet, *Economic and Social History of Quebec, 1760–1850: Structures and Conjonctures* (Toronto: Gage Publishing 1980), 265; Louise Dechêne, *Habitants and Merchants in Seventeenth-Century Montreal* (Montreal & Kingston: McGill-Queen's University Press 1992), 78–80; Alice Lunn, "Economic Development in New France, 1713–1760," PhD thesis, McGill University (1942), 82; David-Thiery Ruddel, "The Domestic Textile Industry in the Region and City of Quebec, 1792–1835," *Material Culture Review/Revue de la culture matérielle* 17 (1983).
41 W.H.K. Turner, "Flax Cultivation in Scotland: An Historical Geography," *Transactions of the Institute of British Geographers* 55 (March 1972): 135.
42 Richard Walker, *The Conquest of Bread: 150 Years of Agribusiness in California* (New York: New Press 2004); William Cronon, *Nature's Metropolis: Chicago and the Great West* (New York: W.W. Norton 1991), 259; Douglas A. Harper, *Working Knowledge: Skill and Community in a Small Shop* (Chicago: University of Chicago Press 1987), 17–25.
43 An excellent treatment of this process in agriculture is in Wynn, *Canada and Arctic North America*, 200–3.
44 MacFadyen, "Spinning Flax in Mills."
45 Stéphane Castonguay and Matthew Evenden, eds, *Urban Rivers: Remaking Rivers, Cities, and Space in Europe and North America* (Pittsburgh: University of Pittsburgh Press 2012), 4.
46 Robert B. Kristofferson, *Craft Capitalism: Craftworkers and Early Industrialization in Hamilton, Ontario, 1840–1872* (Toronto: University of Toronto Press 2007), 51, 77–9.

47 John L. Riley, *The Once and Future Great Lakes Country: An Ecological History* (Montreal & Kingston: McGill-Queen's University Press 2013), 348.
48 Merle Massie, *Forest Prairie Edge: Place History in Saskatchewan* (Winnipeg: University of Manitoba Press 2014).
49 Recent HGIS research in Canada is available in Jennifer Bonnell and Marcel Fortin, eds, *Historical GIS Research in Canada* (Calgary: University of Calgary Press 2014). For similar approaches in US agriculture, see Geoff Cunfer, *On the Great Plains: Agriculture and Environment* (College Station: Texas A&M University Press 2005); and Brian Donahue, *The Great Meadow: Farmers and the Land in Colonial Concord* (New Haven: Yale University Press 2004). For the incorporation of milling technology and geography, see Anne Kelly Knowles, *Mastering Iron: The Struggle to Modernize an American Industry, 1800–1868* (Chicago: University of Chicago Press 2013).
50 Patricia Bowley, "A Century of Soybeans: Scientific Research and Mixed Farming in Agricultural Southern Ontario, 1881–1983," PhD dissertation, University of Guelph (Guelph 2013); Jason Bennett, "Blossoms and Borders: Cultivating Apples and a Modern Countryside in the Pacific Northwest, 1890–2001," PhD dissertation, University of Victoria (Victoria 2008); Ross D. Fair, "A Most Favourable Soil and Climate: Hemp Cultivation in Upper Canada, 1800–1813," *Ontario History* 96, no. 1 (2004): 41–61. For some recent international examples of commodity histories, see Juan Infante-Amate, "The Ecology and History of the Mediterranean Olive Grove: The Spanish Great Expansion, 1750–2000," *Rural History* 23, no. 2 (2012): 161–84; Steven Topik, Carlos Marichal, and Zephyr L. Frank, eds, *From Silver to Cocaine: Latin American Commodity Chains and the Building of The World Economy, 1500–2000* (Durham: Duke University Press 2006).
51 Whitney Eastman, *The History of the Linseed Oil Industry in the United States* (Minneapolis: T.S. Denison 1968).
52 Kathryn Hansuld Lamb, "When Flax Was King," *Waterloo Historical Society* 83 (1995): 75–97; Keller, "From the Rhineland to the Virginia Frontier," 487–511; Ulrich, *The Age of Homespun*, 282–305.
53 David-Thiery Ruddel, "The Domestic Textile Industry in the Region and City of Quebec, 1792–1835," *Material Culture Review/Revue de la culture matérielle* 17 (1983); Sophie-Laurence Lamontagne and Fernand Harvey, *La production textile domestique au Québec, 1827–1941: Une approche quantitative et régionale* (Ottawa: Musée National des Sciences et de la Technologie 1998); Boisvert, "La production textile au Bas-Canada," 435.
54 See, for example, Boisvert, "La production textile au Bas-Canada," 421–37; Ben Forster and Kris Inwood, "The Diversity of Industrial Experience: Cabinet and Furniture Manufacture in Late Nineteenth-Century Ontario," *Enterprise & Society* 4, no. 2 (2003): 326–71; Philip Scranton, *Figured Tapestry: Production, Markets, and Power in Philadelphia Textiles, 1885–1941* (Cambridge: Cambridge University Press 1989); William N.T. Wiley, "The Blacksmith in Upper Canada, 1784–1850: A Study of Technology, Culture and Power," *Canadian Papers in Rural History* 7

(1990): 17–214; J.H. Galloway, *The Sugar Cane Industry: An Historical Geography from Its Origins to 1914* (Cambridge: Cambridge University Press 1989); Peter A. Coclanis, "The Poetics of American Agriculture: The United States Rice Industry in International Perspective," *Agricultural History* 69, no. 2 (1995): 140–62; John Brewer and Roy Porter, eds, *Consumption and the World of Goods* (London: Routledge 1993); Douglas McCalla, *Consumers in the Bush: Shopping in Rural Upper Canada* (Montreal & Kingston: McGill-Queen's University Press 2015).

55 Gordon Darroch and Lee Soltow, *Property and Inequality in Victorian Ontario: Structural Patterns and Cultural Communities in the 1871 Census* (Toronto: University of Toronto Press 1994), 6–12; Peter A. Russell, *How Agriculture Made Canada: Farming in the Nineteenth Century* (Montreal & Kingston: McGill-Queen's University Press 2012), 167.

CHAPTER ONE

1 "Upper Canada" and "Lower Canada" will be used in reference to the areas that became the provinces of Ontario and Quebec, respectively, in 1867.
2 The Perine-Young flax mill photograph is undated, but it was likely taken in 1865, its first full season and likely busiest year ever before the end of the Civil War and the temporary mill shutdown in 1869–70. St Thomas archivists confirm that the photo style dates this image to the 1860s.
3 S. Edwards Todd, "First Prize Essay," in *Manual of Flax Culture*.
4 H. Brosius, "St. Thomas" (Madison: J.J. Stoner 1875); J.C. Young, "City of St. Thomas" (J.C. Young 1886).
5 Robert Leslie Jones, *History of Agriculture in Ontario, 1613–1880* (Toronto: University of Toronto Press 1946), 186 and Chap 7.
6 Experts recommended that pullers always start at the lowest point of a field and work upward. *Manual of Flax Culture*, 38.
7 Ibid.
8 J. Tracey Power, "Brother against Brother: Alexander and James Campbell's Civil War," *South Carolina Historical Magazine* 95, no. 2 (April 1994).
9 Carl Campbell, "Early Post-Emancipation Jamaica: The Historiography of Plantation Culture, 1834–1865," in Kathleen E.A. Monteith and Glen Richards, eds, *Jamaica in Slavery and Freedom: History, Heritage and Culture* (Mona, Jamaica: University of the West Indies Press 2002); Kathleen Butler, *The Economics of Emancipation: Jamaica and Barbados* (Chapel Hill: UNC Press 1995).
10 Canada, *Census of Canada 1861*, LAC, Manuscript Schedule 1, St Thomas, Elgin County, Canada West, Roll C-1019, 16.
11 Maria Jane's marriage certificate indicates that her mother's name was Maria, the 1891 census manuscripts reveal that her mother was born in Jamaica, and the *1827 Jamaica Almanac* shows a Mary Pinckney (the only Pinckney plantation owner on the island) who owns a plantation and seven slaves in Cornwall County, Westmoreland, Jamaica. AO, *Canada, Registrations of Marriages, 1869–1928*, MS 932, Reel 13 (1874): 274, No. 001696; Canada, *Census of Canada, 1891*, LAC,

Manuscript Schedule 1, Port Dover, Norfolk South, Ontario; Roll T-6356, 18; National Library of Jamaica, "Return of Givings-In for the March Quarter, 1826, County of Cornwall, Parish of Westmoreland," *1827 Jamaica Almanac*, accessed on Jamaica Family Search Geneaology Research Library, http://www.jamaicanfamilysearch.com/Members/a/a1827al10.htm (accessed December 2014).

12 Maria Jane's marriage to Charles Battersby in St Thomas in 1874 and her family's entry in the 1891 census at Port Dover suggest a longer-term presence in Canada and confirms the birthplaces of her parents. Ibid.

13 *Kent, Lambton, and Essex Counties Directory and Gazetteer, 1866–1867*, 172; Canada, *Census of Canada, 1871*, LAC, Manuscript Schedule 1, St Davids Ward, Toronto East, Ontario; Roll C-9972, p. 54.

14 LAC, Campbell to Walker, December 1862, BAG minutes, 11 April 1865.

15 General examinations of Central Canadian manufacturing are now available in Michael Bliss, *Northern Enterprise: Five Centuries of Canadian Business* (Toronto: McClelland & Stewart 1990), 225–52; Ian M. Drummond, *Progress without Planning: The Economic History of Ontario from Confederation to the Second World War* (Toronto: University of Toronto Press 1987), 105–15; Robert B. Kristofferson, *Craft Capitalism: Craftworkers and Early Industrialization in Hamilton, Ontario, 1840–1872* (Toronto: University of Toronto Press 2007); Gerald J.J. Tulchinsky, *The River Barons: Montreal Businessmen and the Growth of Industry and Transportation, 1837–53* (Toronto: University of Toronto Press 1977), 203–5, 213–31; A.B. McCullough, *Primary Textile Industry in Canada: History and Heritage* (Ottawa: Canadian Parks Service National Historic Parks 1992).

16 K.W. Taylor, "Tariffs," in W. Stewart Wallace, ed., *The Encyclopedia of Canada*, Vol. 6 (Toronto: University Associates of Canada 1948), 102–8.

17 Drummond, *Progress without Planning*, 6; Douglas McCalla, *Planting the Province: The Economic History of Upper Canada, 1784–1870* (Toronto: University of Toronto Press 1993), 113–15.

18 Even a proper definition of flax and flax manufacturing has been elusive. One definition of a flax mill simply reads, "A mill that grinds flax." Jane Turner, Janine Grant, and Barbara Sibley, compilers, "Research Report 5: Glossary of Industrial Language," in Elizabeth Bloomfield, ed., *Canadian Industry in 1871* (Guelph, ON: Department of Geography, University of Guelph 1989), 22. Flax seed was ground in linseed oil mills, but in the nineteenth century a "flax mill" referred to a scutching mill whose primary purpose was to process flax fibre.

19 Evans, *Bound in Twine*; David Sutherland, "The Stanyan Ropeworks of Halifax, NS: Glimpses of a Pre-Industrial Manufactory," *Labour/Le Travail* 6 (Autumn 1980): 149–58. The Dartmouth Ropeworks supplied cordage to the shipping industry and fisheries and by 1888 claimed to control over a third of the Canadian market for binder-twine. James D. Frost, "Dartmouth Ropeworks, 1869–1958," in James E. Candow, ed., *Industry and Society in Nova Scotia: An Illustrated History* (Halifax: Fernwood 2001), 117–20.

20 Ralph H. Estey, *Essays on the Early History of Plant Pathology and Mycology in Canada* (Montreal & Kingston: McGill-Queen's University Press 1994), 110;

McCullough, *Primary Textile Industry in Canada*, 54; Kathryn Hansuld Lamb, "When Flax Was King," *Waterloo Historical Society* 83 (1995): 75–97.

21 Jonathan Hamill, "The Irish Flax Famine and the Second World War," in Brenda Collins, Phillip Ollerenshaw, and Trevor Parkhill, eds, *Industry, Trade and People in Ireland, 1650–1950: Essays in Honour of W.H. Crawford* (Belfast: Ulster Historical Foundation 2005), 228.

22 Robert C.H. Sweeny, with David Bradley and Robert Hong, "Movement, Options and Costs: Indexes as Historical Evidence, a Newfoundland Example," *Acadiensis* 22, no. 1 (Autumn 1992): 111–21; Beatrice Craig, Judith Rygiel, and Elizabeth Turcotte, "The Homespun Paradox: Market-Oriented Production of Cloth in Eastern Canada in the Nineteenth Century," *Agricultural History* 76, no. 1 (Winter 2002): 28–57; Douglas McCalla, *Consumers in the Bush: Shopping in Rural Upper Canada* (Montreal and Kingston: McGill-Queen's University Press 2015), 18–20; Elizabeth Mancke, "At the Counter of the General Store: Women and the Economy in Eighteenth-Century Horton, Nova Scotia," in *Intimate Relations: Family and Community in Planter Nova Scotia, 1759–1800* (Fredericton, NB: Acadiensis Press 1995); Alan Greer, *Peasant, Lord, and Merchant: Rural Society in Three Quebec Parishes, 1740–1840* (Toronto: University of Toronto Press 1985); Rosemary E. Ommer, "The Truck System in Gaspé, 1822–77," *Acadiensis* 19, no. 1 (1989): 91–114.

23 The American connections in Montreal's commercial sector are well documented in Tulchinsky, *The River Barons*, 187–9, 237–8; Drummond, *Progress without Planning*, 132–3.

24 Russell D. Smith, "The Early Years of the Great Western Railway, 1833–1857," *Ontario History* 60 (1968): 217–18. McCalla, "Railways and the Development of Canada West, 1850–1870," in Allan Greer and Ian Walter Radforth, eds, *Colonial Leviathan: State Formation in Mid-Nineteenth-Century Canada* (Toronto: University of Toronto Press 1992): 192–229.

25 Lewis Publishing Company, *The Bay of San Francisco: The Metropolis of the Pacific Coast and Its Suburban Cities: A History*, Vol. 1 (Chicago: Lewis Publishing Company 1892), 670; "The Flax Interest: Complimentary Dinner to Messrs. Perine and Young," *Globe*, 18 November 1864, 2.

26 DHC, "Daybook," Perine Brothers, 2006.023.024.1, 86.

27 "M.B. Perine," *Canadian Journal of Fabrics* 12, no. 3 (1895): 74. See also "Moses Billings Perine," FindAGrave.com, Find A Grave Memorial #153713934, https://www.findagrave.com/cgi-bin/fg.cgi?page=gr&GRid=153713934 (accessed August 2017).

28 John Perine (1878–1948) is buried at Revolutionary Cemetery, Salem, Washington County, New York. See "John Perine," FindAGrave.com, Find A Grave Memorial #45546004, https://www.findagrave.com/cgi-bin/fg.cgi?page=gr&GRid=45546004 (accessed August 2017).

29 DHC, "Daybook," Perine Brothers, 2006.023.024.1, 72–4.

30 Cuyler Reynolds, *Hudson-Mohawk Genealogical and Family Memoirs*, Vol. 4 (New York: Lewis Historical Publishing Company 1911) 1429; Congressional Series of United States Public Documents, Vol. 53, "Report of the Comptroller of the

Currency" (Washington, DC: Government Printing Office 1877), 326. By 1866 the R.G. Dun credit ledgers show that the Perine brothers had a credit rating of 2, or "good." R.G. Dun, *Mercantile Agency Reference Book for the British Provinces, 1866* (Montreal: M. Longmoore 1866), 71.
31 *Manual of Flax Culture*, 37, 40.
32 Archives of Ontario, *Registrations of Deaths, 1869–1938*, MS 935, Reels 1-615, Reel 11, 125. See also "Hannah Perine," FindAGrave.com, Find A Grave Memorial #177054854, https://www.findagrave.com/cgi-bin/fg.cgi?page=gr&GRid=177054854 (accessed August 2017).
33 DHC, "Daybook," Perine Brothers, 2006.023.024.2, 318, 381.
34 Lewis Publishing Company, *The Bay of San Francisco*, 670–1.
35 MAO, Entry, 25 January 1877, "Diary of Elias Eby (1810–1878)," Elias Eby Diary, 1872–78, Historical MSS 17.8, 47.
36 This did not include eight rope and cordage factories recorded elsewhere in the directory. *Mitchell & Co.'s Canada Classified Directory for 1865–66* (Toronto: Mitchell 1866), 252, 381.
37 "Flax Interest: Complimentary Dinner to Messrs. Perine and Young," *Globe*, 18 November 1864, 2.
38 See, for example, a summary of a letter written to W.D. Perine on 9 April 1866. "Alexander Young, Notebook, 1866," Alexander Young Records, LAC, Ottawa, R7316-21-7-E.
39 "Central School, 1857 to 1871," *First Annual Report of the Waterloo Historical Society* (1913); "Death of Mr. Alex Young," *Galt Reporter*, 2 June 1882, 1–2.
40 "The Flax Mills," *Canadian Home Journal*, 15 September 1864.
41 John A. Donaldson, "Flax Works at Norval; Harvesting Flax," *CF*, 1 August 1864, 211; Henry Beaumont Small, *The Products and Manufactures of the New Dominion* (Toronto: George E. Desbarats 1868), 31–2; A.H. Richardson, B.B. Blake, et al., *Credit Valley Conservation Report, 1956* (Toronto: Department of Planning and Development 1956), 116.
42 James Sutherland, *County of Waterloo Gazetteer and General Business Directory for 1864* (Toronto: Lovell and Gibson 1864), 162, 179.
43 Henry McEvoy, *Province of Ontario Gazetteer and Directory* (Toronto: Robertson & Cook 1869), 328.
44 "Charles Hendry & Co., Merchants, Millers, and Lumber Merchants, Canistoga" [*sic*], in William Henry Smith, *Business Directory of Canada West, 1851* (Toronto: T. Maclear 1851), 85.
45 Tulchinsky, *The River Barons*, 231.
46 "Flax-Growing in Canada," *Canada Farmer*, February 1864, 51.
47 Benjamin Walker, "The Lancashire Relief Fund – Flax vs. Cotton," *Canadian Agriculturist, and Journal of the Board of Agriculture of Upper Canada*, January 1863, 8, reprinted from the *Canadian Home Journal* (St Thomas), 24 November 1862.
48 George Maclean Rose, ed., *A Cyclopaedia of Canadian Biography: Being Chiefly Men of the Time. A Collection of Persons Distinguished in Professional and Political Life; Leaders in the Commerce and Industry of Canada, and Successful Pioneers*, Vol. 1 (Toronto: Rose Publishing Company 1886), 132.

49 H. Belden, *Illustrated Historical Atlas of the County of Huron, Ontario* (Toronto: H. Belden Company 1879), xviii.
50 George Farwell, *Hay Township: A Study of Industry, Past and Present* (Zurich: Opportunities for Youth Technological Survey 1972), 16, 26.
51 *Province of Ontario Gazetteer and Directory 1910–1911* (Ingersoll: Union 1910); Charles E. Goad, *Insurance Plan of Zurich, Ontario, Canada* (Toronto: C.E. Goad 1896 [1913]).
52 Charles E. Goad, *Insurance Plan of Wellesley, Ontario, Canada* (Toronto: C.E. Goad 1894); Charles E. Goad, *Insurance Plan of Crediton, Ontario, Canada* (Toronto: C.E. Goad 1896); Charles E. Goad, *Insurance Plan of Dashwood, Ontario, Canada* (Toronto: C.E. Goad 1896).
53 W.W. Smith, *Gazetteer and Directory of the County of Grey, 1865–66* (Toronto: Globe Steam Press 1865).
54 In a letter dated March 1863, John A. Donaldson replied to request by a member of the Legislative Council for information on "the present and future prospects of flax cultivation in Canada." John A. Donaldson to Hon. G. Alexander, *Journals of the Legislative Council of the Province of Canada* 26, Appendix 18 (Quebec: R. Stanton 1863).
55 A spinning frame from A. Sesson & Co. cost $422.25. A drawing frame cost $200.00. A flax scutcher cost $305.85. Two Ceiras from A. Sesson & Co. cost $332.00. A comber bellows from Holley cost $43.70, and a brake from Colone & Co. cost $450 plus $10 for packaging. DHC, "Daybook," Perine Brothers, 2006.023.024.2, 414.
56 See, for example, 8 August 1863, when Frederick Lake was credited $145.75 for stonework and for laying 50,000 bricks for the boarding house. Ibid., 418.
57 "Flax Brake," *Canada Farmer*, 15 July 1864, 201.
58 DHC, "Daybook," Perine Brothers, 2006.023.024.2, 164; "Flax-Growing in Canada," *Canada Farmer*, 15 February 1864, 51.
59 DHC, "Daybook," Perine Brothers, 2006.023.024.2, 204.
60 Sidney T. Fisher, *The Merchant-Millers of the Humber Valley: A Study of the Early Economy of Canada* (Toronto: NC Press 1985), xvii.
61 Florence Feldman-Wood, "A Long Linen Thread: Smith & Dove in Andover," *Newsletter* of Andover Historical Society 23, no. 4 (Winter 1999): 1; "Hemp and Manilla Cordage," United States, Department of Interior, *Eighth Census, 1860*, Vol. 3, Manufactures (Washington, DC: Government Printing Office 1865), cxii.
62 J.A. Blyth, "The Development of the Paper Industry in Old Ontario, 1824–1867," *Ontario History* 62 (1970): 124; A.H. Richardson, B.B. Blake, et al., *Credit Valley Conservation Report, 1956* (Toronto: Department of Planning and Development 1956), 128. Flax tow is one of the strongest materials for paper, but beginning in 1866 it was replaced by the more abundant wood pulp in Upper Canadian paper mills. Nina L. Edwards, "The Establishment of Papermaking in Upper Canada," *Ontario History* 39 (1947): 69; Raymond A. Young, "Processing of Agro-Based Resources into Pulp and Paper," in Roger M. Rowell and Raymond A. Young, *Paper and Composites from Agro-Based Resources* (Boca Raton: CRC Press 1997), 230–1.

63 Joseph Nathan Kane, Steven Anzovin, and Janet Podell, *Famous First Facts*, 6th edn (New York: H.W. Wilson 2006), 94.
64 Robert F. Trent, "17th Century Upholstery in Massachusetts," in Edward S. Cooke, *Upholstery in America & Europe: From the Seventeenth Century to World War I* (New York: Norton 1987), 44.
65 Ben Forster and Kris Inwood, "The Diversity of the Industrial Experience: Cabinet and Furniture Manufacture in Late Nineteenth-Century Ontario," *Enterprise & Society* 4 (June 2003): 326–71; Ian Radforth, "Confronting Distance: Managing Jacques and Hay's New Lowell Operations, 1853–1873," *Canadian Papers in Business History* 1 (1989): 75–100.
66 William Cronon, *Nature's Metropolis: Chicago and the Great West* (New York: W.W. Norton 1991), 120.
67 "The Cheapest Bags in the Dominion of Canada," *Trade Review and Intercolonial Journal of Commerce* 3, no. 44 (1867): 697; W. Imlach, "Flax Culture vs. Wheat," letter to the editor, *Canada Farmer*, 1 May 1865, 131.
68 James Buik, "Flax Thread," *Canada Farmer*, 15 March 1864, 67. Buik also argued that "the sooner the young lasses of Canada begin to rattle away at the *twa-handed wheel*, the better both for back and bed, and bags to hold their father's wheat in, for the bags they buy now are nothing but trash"; Buik, "The Twa-handed Wheel," *Canada Farmer*, 15 February 1865, 62 (emphasis in original).
69 Quoted in Beatrice Craig, *Backwoods Consumers and Homespun Capitalists: The Rise of a Market Culture in Eastern Canada* (Toronto: University of Toronto Press 2009), 191; *Saint John Evening Globe*, 25 September 1866.
70 McCullough, *Primary Textile Industry in Canada*, 51.
71 John A. Donaldson, *Practical Hints on the Cultivation and Treatment of the Flax Plant* (Toronto: Globe Steam Press 1865), 13. Henry Beaumont Small wrote that Barber Brothers were "using large quantities of yarn from the linen mills of Messrs. Perrine [*sic*] for warp and filling woolen weft, thus producing a much stronger good than can be manufactured with cotton and woollen." Henry Beaumont Small, *The Products and Manufactures of the New Dominion* (Toronto: George E. Desbarats 1868), 31.
72 *Canada Farmer*, 16 November 1868, 347.
73 "Flax Culture," *Canada Farmer*, 15 January 1869, 3.
74 Harry Livingston, "James Livingston," *Waterloo Historical Society* 9 (1921): 190.
75 A.H. Richardson, B.B. Blake, et al., *Credit Valley Conservation Report, 1956* (Toronto: Department of Planning and Development 1956), 116.
76 R.G. Dun, *Mercantile Agency Reference Book for the British Provinces, 1866* (Montreal: M. Longmoore 1866), 227.
77 "The Streetsville Flax Works," *CF*, 2 July 1866, 1.
78 "The Streetsville Flax Works," *Journal of the Board of Arts and Manufactures for Upper Canada* (August 1866): 198–9.
79 "City News – Flax Mills Closed," *Globe*, 7 November 1867, 2; "Destructive Fire at Streetsville: Gooderham & Worts' Linen Mills Burned Down," *Globe*, 27 January 1868, 2.

80 "The Streetsville Flax Works," *CF*, 2 July 1866, 1; "The Ice in the River Credit," *Acton Free Press*, 9 March 1876.
81 The R.G. Dun credit ledgers show that the mill was worth between $5,000 and $10,000 and had a credit rating of 2, or "good." Dun, *Mercantile Agency Reference Book for the British Provinces, 1866*, 399; "Death of Mr. Alex Young," *Galt Reporter*, 2 June 1882, 1–2; "Close of the St Thomas Flax Mills," *Canada Farmer*, 16 August 1868; "Flax Pulling Heavy Job but Boys Liked It," *St. Thomas Times-Journal*, 1 August 1952.
82 Rose, *A Cyclopaedia of Canadian Biography*, Vol. 1, 127.
83 *1869 Province of Ontario Gazetteer and Directory*, 328; Canada, *Census of Canada 1871*, LAC, Manuscripts, Industrial Schedule, Grey County.
84 Evans, *Bound in Twine*, 29–30.
85 Canada, *Canada, Census, 1860–61*, LAC, Personal Schedule for Waterloo County, Canada West, Reels C-1077–C-1080.
86 Thomas Dublin, "Women's Work and the Family Economy: Textiles and Palm Leaf Hatmaking in New England, 1830–1850," *Tocqueville Review* 5, no. 2 (1983): 297–316.
87 DHC, "Daybook," Perine Brothers, 2006.023.024.2, 72–4, 208.
88 This was possibly George Ellinton (or Ellantun), who appeared in the 1871 census as a forty-year-old farmer in Erin Township, Wellington Centre, ON. Canada, *Census of Canada 1871*, LAC, Manuscripts, Schedule 1, Erin, Wellington Centre, Ontario; Roll C-9947, 60; DHC, "Ledger," Perine Brothers, 2006.023.024.6, 164.
89 William T. Ruddock, *Linen Threads and Broom Twines: An Irish and American Album and Directory of the People of the Dunbarton Mill, Greenwich, New York, 1879–1952* (Bowie, MD: Heritage Books 1997).
90 Keller, "From the Rhineland to the Virginia Frontier," 487–511; Charlotte Erickson, *Invisible Immigrants: The Adaptation of English and Scottish Immigrants in Nineteenth-Century America* (Coral Gables, FL: University of Miami Press 1973), 344, 368.
91 Canada, *Census of Canada, 1870–71*, Vol. 3 (Ottawa: I.B. Taylor 1875), Table LI, 431; Canada, *Census of Canada, 1880–81*, Vol. 3 (Ottawa: Maclean, Roger 1883), Table LI, 475.
92 Canada, *Fifth Census of Canada, 1911*, LAC, Manuscripts, Schedule 1, Sub-district 7, Waterloo South, Ontario, 18–25.
93 These data do not include the three urban districts in Waterloo (Bloomfield, *Waterloo Township*, 410–11). In 1866 the R.G. Dun credit ledgers show that the Doon and Canestoga [*sic*] mills were worth between $25,000 and $50,000, and the brothers had a credit rating of 2, or "good." No mill was listed for Baden. R.G. Dun, *Mercantile Agency Reference Book for the British Provinces, 1866* (Montreal: M. Longmoore 1866), 71.
94 Henry Lye, "The Agriculture of Canada," *Markdale Standard*, 17 July 1890, 6.
95 Moses's attending doctor recorded the eldest Perine's cause of death as apoplexy, noting that he had been ailing for two years. Edward's cause of death was given as acute trauma, eventually leading to heart failure. Edward was survived by his

wife, their two children, and his mother, Sarah Ann Baily. Sarah died two years later, in 1913. AO, "Registrations of Deaths, 1869–1938," MS 935, Reels 1-615, Reel 90, 223, and Reel 170, 232; AO, "Schedule B – Marriages," MS 932, Reel 70, No. 001427.

96 Lamb, "When Flax Was King," 88–9; Jean Haalboom, "Rope-Making Company Had a Long History," *Kitchener-Waterloo Record*, 19 April 2007, A.10; Greg Mercer, "Historic Rope Maker Shuts after 151 Years," *Kitchener-Waterloo Record*, 2 April 2007, A.1.

97 *Industry '67: Centennial Perspective* (Toronto: Canadian Manufacturers' Association 1967).

98 Samson, *A Spirit of Industry and Improvement*, 263, 267.

99 Philip Scranton, *Proprietary Capitalism: The Textile Manufacture at Philadelphia, 1800–1885* (Cambridge: Cambridge University Press 1983), 5–7.

CHAPTER TWO

1 "A Street Named Nathan," *St. Thomas Times-Journal*, 30 January 1959; "Death of Nathan Ryan: Born in Slavery in the State of Maryland," *St. Thomas Evening Journal*, 18 December 1903.

2 Robert A. Dodgshon, *Society in Time and Space: A Geographical Perspective on Change* (Cambridge, UK: Cambridge University Press 1998), 17.

3 Kenneth Michael Sylvester, *The Limits of Rural Capitalism: Family, Culture, and Markets in Montcalm, Manitoba, 1870–1940* (Toronto: University of Toronto Press 2001), 4; Peter A. Russell, *How Agriculture Made Canada: Farming in the Nineteenth Century* (Montreal & Kingston: McGill-Queen's University Press 2012), 32–3; Colin A.M. Duncan, *The Centrality of Agriculture: Between Humankind and the Rest of Nature* (Montreal & Kingston: McGill-Queen's University Press 1996), 45; Deborah Fitzgerald, *Every Farm a Factory: The Industrial Ideal in American Agriculture* (New Haven, NH: Yale University Press 2003).

4 Keller, "From the Rhineland to the Virginia Frontier," 487–511.

5 Laurel Thatcher Ulrich, *The Age of Homespun: Objects and Stories in the Creation of an American Myth* (New York: Alfred A. Knopf 2001); Michel Boisvert, "La production textile au Bas-Canada: L'exemple Laurentien," *Cahiers de géographie du Québec* 40, no. III (1996): 421–37; Beatrice Craig, Judith Rygiel, and Elizabeth Turcotte, "The Homespun Paradox: Market-Oriented Production of Cloth in Eastern Canada in the Nineteenth Century," *Agricultural History* 76, no. 1 (Winter 2002): 28–57.

6 Beth Light and Alison L. Prentice, *Pioneer and Gentlewomen of British North America, 1713–1867* (Toronto: New Hogtown Press 1980); *Documents in Canadian Women's History*, Vol. 1 (1980), 121–2.

7 Upper Canada and Lower Canada are the regions that in 1867 became Ontario and Quebec, respectively, although after 1840 they were formally known as Canada West and Canada East. J. David Wood, *Making Ontario: Agricultural Colonization and Landscape Re-Creation before the Railroad* (Montreal & Kingston: McGill-Queen's University Press 2000), 88–90, 100.

8 Kris E. Inwood and Phyllis Wagg, "The Survival of Handloom Weaving in Rural Canada circa 1870," *Journal of Economic History* 53, no. 2 (1993): 346–58.
9 Béatrice Craig, *Backwoods Consumers and Homespun Capitalists: The Rise of a Market Culture in Eastern Canada* (Toronto: University of Toronto Press 2009), 187–9; Judith Rygiel, "'Thread in Her Hands – Cash in Her Pockets': Women and Domestic Textile Production in 19th Century New Brunswick," *Acadiensis* 30, no. 2 (Spring 2001), 56–70; Cyril Simard, "Le lin au Madawaska au 19e siècle: Les 'Brayons' n'étaient pas les seuls a ce 'mauvais coton,'" *Revue de la Société historique du Madawaska* 15 (October–December 1987): 9–25.
10 Craig, *Backwoods Consumers*, 187–9.
11 Based on 1861 produce prices in McCalla, *Planting the Province: The Economic History of Upper Canada, 1784–1870*, 340, Table C.2. The price of 11 cents per pound of flax is based on the Perine accounts of that year. DHC, "Daybook," Perine Brothers, 2006.023.024.2, 378.
12 Elliot & Co., *Elliot & Co., no. 3 Front St. East, Toronto, Exhibit in Class 79 as Samples, and Not for Competition, the Following Linseed and Its Products* [1880] 1982 (CIHM #27283), 1; William Saunders, "Flax," Central Experimental Farm, *Bulletin* 25 (May 1896): 8. The difference is not clearly explained by linseed oil mill locations because although mills were crushing flax seed in Waterloo County in 1861, they were in Bridgeport and Chicopee, in the east. Ken E. Seiling, "Early Building in the Bridgeport, Ontario Area" (BA, Department of History, University College, Waterloo Lutheran University 1969), 13.
13 Canada, *Census of Canada, 1860–61*, LAC, Agricultural Schedule for Waterloo County, Canada West, Reels C-1077–C-1080. The amount of raw material used is as follows: Doon, 2,000 tons flax; Baden, 370 tons flax; Conestogo, 350 acres flax (using average flax yields in that township in 1860, 275 pounds per acre, I estimate this acreage would produce 48 tons of flax). This tonnage used by the three Perine mills equals 4.8 million pounds of flax, or thirteen times the amount recorded in Waterloo County.
14 Canada, *Census of Canada, 1870–71*, LAC, Industrial Schedules.
15 M.C. Urquhart et al., *Gross National Product, Canada, 1870–1926: The Derivation of Estimates* (Montreal & Kingston: McGill-Queen's University Press 1993), 70.
16 Benjamin Walker, "Elgin Flax Association," *Canadian Agriculturist and Journal of the Board of Agriculture of Upper Canada*, 16 March 1862, 167.
17 Alexander Kirkwood, *Flax and Hemp* (Toronto: W.C. Chewett 1864), 28.
18 E.F. Deman, *The Flax Industry: Its Importance and Progress. Also Its Cultivation and Management and Instructions in the Various Belgian Methods of Growing and Preparing It for Market, with Extracts from the Annual Report of the Royal Irish Flax Society, and a Word on Chevalier Claussen's Invention of Cottonizing Flax* (London: J. Ridgway 1852), 98.
19 *Manual of Flax Culture, Comprising Full Information on the Cultivation, Management, and Marketing of the Crop, Together with a Complete Glossary and Index* (New York: Orange Judd Company 1883), 52.
20 Deman, *The Flax Industry*, 111–12.
21 Ibid., 155–7 (emphasis in original).

22 See Frank E. Wright, "Plants of Great Commercial Value" (1912), Library of Congress, Copyright Office, *Catalog of Copyright Entries: Part 4, Works of Art ...* (Washington, DC: Government Printing Office 1913), 300.
23 Craig Heron, in Paul Craven, ed., *Labouring Lives: Work and Workers in Nineteenth-Century Ontario* (Toronto: University of Toronto Press 1995), Table 1, 491.
24 It is possible that these were farmers who scutched flax manually for a small clientele. *Census of Prince Edward Island 1871.*
25 Courville et al., eds, *Atlas historique du Québec,* 56; MacFadyen, "Spinning Flax in Mills."
26 Christopher Andreae, *Lines of Country: An Atlas of Railway and Waterway History in Canada* (Erin, ON: Boston Mills Press 1997).
27 Charles Richard Dodge, "Profits in Flax," *Technical World* 16 (1911): 567.
28 "Flax Wanted," *Berlin Chronicle,* 12 March 1856.
29 Mennonite Archives of Ontario (MAO), Conrad Grebel College, Entry, 24 November 1856, Diary of Christian B. Snyder, 1856 (Historical MSS 17.12 – photocopy).
30 DHC, "Daybook," Perine Brothers, 2006.023.024.2, 192.
31 The price of seed in both the Hilborn and Furtney accounts was 12 shillings or $1.50 per bushel. Ibid., 143.
32 Sowing estimates (0.8, 1, and 1.3 bushels per acre) are from the *Canada Farmer* (15 April 1864, 19; 15 June 1864, 163; and 1 February 1864, 18) and mean that Hilborn had enough seed to sow 35 to 55 acres of flax.
33 DHC, "Daybook," Perine Brothers, 2006.023.024.1, 183, 395.
34 Canada, *Census of Canada, 1861,* LAC, Manuscripts, Schedule 1, Waterloo South, Roll C1078–1079, 1; George R. Tremaine and G.M. Tremaine, *Tremaine's Map of the County of Waterloo, Canada West. Compiled & Drawn from Actual & Original Surveys by the Publishers* (Toronto: George R. Tremaine and G.M. Tremaine 1861), Waterloo South. The 1851 census shows that David and Jacob were the oldest of ten siblings, including three sets of twins, on the farm of their Tunker parents, Jesse and Maria Hilborn. Canada, *Census of Canada, 1851,* LAC, Manuscript, Schedule 1.
35 Ibid., 2 December 1856, 155.
36 See, for example, 8 August 1864, when A. Knox is credited $40 for leased land and John Smith is paid $6.75 for cutting Knox's field and $1.00 for spreading flax. LAC, "Alexander Young, Notebook, 1864," Alexander Young Records, R7316-21-7-E.
37 DHC, "Daybook," Perine Brothers, 2006.023.024.2, 278, 295.
38 "M.B. Perine," *Canadian Journal of Fabrics* 12, no. 3 (1895): 74.
39 W. Imlach, "Flax Culture vs. Wheat," Letter to the editor, *Canada Farmer,* 1 May 1865, 131.
40 DHC, "Daybook," Perine brothers, 2006.023.024.1, 72–4, 176; DHC, "Daybook," Perine brothers, 2006.023.024.2, 72–4, 232, 237.
41 William Leslie, "Flax in and about Meadowvale," *Canada Farmer,* 1 July 1865, 195; "More About Meadowvale Conservation Area," Credit Valley Conservation website, http://www.creditvalleyca.ca/enjoy-the-outdoors/conservation-areas/meadowvale-conservation-area/more-about-meadowvale-conservation-area (accessed December 2014).

42 Canada, *Census of Canada 1871*, LAC, Industrial Schedule, Oxford West, Oxford County, Kris E. Inwood dataset.
43 *Canada Farmer*, 1 April 1864, 83.
44 William H. Smith, *Canada: Past, Present and Future; Being a Historical, Geographical, Geological and Statistical Account of Canada West* (Toronto: Thomas MacLear 1851), 245–6.
45 "Flax: The Fall River Company," *The Plough, the Loom, and the Anvil* 7, no. 1 (July 1854): 21–2; "Linen," in United States, Department of the Interior, *Eighth Census, 1860*, Vol. 3, Manufactures (Washington, DC: Government Printing Office 1865), cx.
46 Nian-Sheng Huang, "Financing Poor Relief in Colonial Boston," *Massachusetts Historical Review* 8 (2006): 72–103.
47 Deman, *The Flax Industry*, 155–7 (emphasis in original).
48 Jan de Vries, "The Industrial Revolution and the Industrious Revolution," *Journal of Economic History* 54, no. 2 (June 1994): 249–70; Daniel Vickers, *Farmers and Fishermen: Two Centuries of Work in Essex County, Massachusetts, 1630–1850* (Chapel Hill: UNC Press Books 1994).
49 James A. Byrne, "Flax-Pulling," *Farmer's Advocate*, 15 July 1909, 1140; David Vaught, *Cultivating California: Growers, Specialty Crops, and Labor, 1875–1920* (Baltimore, MD: Johns Hopkins University Press 1999), 213n24.
50 "In Canada the limited amount of flax grown for fiber is usually pulled by Indians." Walter S. Barker, "Flax: The Fiber and Seed. A Study in Agricultural Contrasts," *Quarterly Journal of Economics* 31, no. 3 (May 1917): 506.
51 F.W. Stock, "The Economics of Flax Production in Ontario," *OAC Review* 33, no. 3 (November 1920): 113–14. Austin L. McCredie argued that "pulling is left a week too long in Ontario, to produce a quality suitable for the British market." Austin L. McCredie, "Flax for Fibre," thesis, Department of Chemistry, Ontario Agricultural College, 6 February 1906, 18; TDHA, "Hanke to Byron and Homer Hanke, 27 October, 1933," Archival Collection.
52 "Local News: Bosworth," *Drayton Advocate*, 22 August 1907, 3; "Correspondence: Parker (received too late for last week's issue)," *Drayton Advocate*, 29 August 1907, 2.
53 Ottis Wright, "An Aching Back and a Big Night," in Sheila McMurrich Koop and Alvin Koop, *Older Voices among Us* (Erin, ON: Boston Mills Press 1981), 52.
54 WCMA (Wellington County Museum and Archives), "The Flax Industry: History of Maryborough Township, 1851–1976," anonymous (n.d.); Bonnie Elliott, "The Lost Industry: History of Maryborough" (no date).
55 Terry Crowley, "Rural Labour," in Craven, ed., *Labouring Lives*, 87n133.
56 "Harvesting Flax," *Canada Farmer*, 1 August 1864, 211; *Canada Farmer*, 15 April 1864, 99; *Canada Farmer*, 15 June 1864, 163; *Canada Farmer*, 15 July 1864, 195; "A Flax Puller Invented," *Canada Farmer*, 1 April 1864, 82; John M. Wilson, *Lecture on Flax: Its Treatment, Agricultural and Technical* (New York: C.M. Saxton 1853), 14; three failed attempts are recorded in *Manual of Flax Culture*, 38.
57 It is difficult to know whether the most common rate of pay was per day or per acre.

58 "Experience in Flax Growing," *Canada Farmer*, 2 January 1865, 2; *Canada Farmer*, 15 February 1865, 57; "'E.M's' Experience in Raising Flax," *Canada Farmer*, 1 February 1865, 35; *Canada Farmer*, 15 April 1865, 123.
59 *Canada Farmer*, 1 August 1864.
60 LAC, *Canada, Census, 1860–61*, Agricultural Schedule for Waterloo County, Canada West, Reels C-1077–C-1080.
61 DHC, "Daybook," Perine Brothers, 2006.023.024.2, 143.
62 Miriam H. Snyder and Joseph M. Snyder, *Hannes Schneider and his wife Catharine Haus Schneider; their descendants and times, 1534–1939* (Kitchener: Miriam H. Snyder 1937), 160G.
63 *Manual of Flax Culture*, 35; J.H. Grisdale, "Flax for Grain," *Drumheller Mail*, 13 April 1916, 9.
64 Keller, "From the Rhineland to the Virginia Frontier," 499; Ulrich, *The Age of Homespun*; Adrienne D. Hood, *The Weaver's Craft: Cloth, Commerce, and Industry in Early Pennsylvania* (Philadelphia: University of Pennsylvania Press 2003).
65 Byrne, "Flax-Pulling," 1140.
66 Ibid.
67 Donaldson, *Practical Hints* (1865), 10.
68 "Thamesford," *Woodstock Sentinel-Review*, 30 August 1888, 2.
69 WCMA, Jean Boyle, "Flax Mill," in *Alma Women's Institute Tweedsmuir History*, A1980.21, 73, 121.
70 However, Mary might have been paid room and board by her employers. Alan A. Brookes and Catharine A. Wilson, "'Working Away' from the Farm: The Young Women of North Huron, 1910–1930," *Ontario History* 77, no. 4 (December 1985): 281–300.
71 Cecilia Danysk, *Hired Hands: Labour and the Development of Prairie Agriculture, 1880–1930* (Toronto: McClelland & Stewart 1995), 83, 85. See also Edward Dunsworth, "Green Gold, Red Threats: Organization and Resistance in Depression-Era Ontario Tobacco," *Labour/Le Travail* 79 (2017): 105–42, 114.
72 "Alma Flax Mill," *Drayton Advocate*, 2 May 1907, 3.
73 TWA, James McColl to James Livingston, 19 July 1907, 1.
74 Ibid., 22 August 1907, 1.
75 Ibid., 5 September 1907, 2.
76 TWA, International Flax Twine Co. to W.D. Stoner, 8 August 1907, 2.
77 TWA, James McColl to James Livingston, 7 August 1907, 2.
78 George V. Haythorne, *Labor in Canadian Agriculture* (Cambridge, MA: Harvard University Press 1960), 26, based on Canada, *Census of Canada, 1901*, Vol. 2 (Ottawa: S.E. Dawson 1904), Table 8.
79 Alicja Muszynski, *Cheap Wage Labour: Race and Gender in the Fisheries of British Columbia* (Montreal & Kingston: McGill-Queen's University Press 1996); Avis Mysyk, "The Role of the State in Manitoba Farm Labour Force Formation," in Jill L. Findeis, Ann M. Vandeman, Janelle M. Larson, and Jack L. Runyan, eds, *The Dynamics of Hired Farm Labour: Constraints and Community Responses* (Wallingford: CABI Publishing 2002): 169–82. Mysyk argues that the Department of Indian Affairs's agricultural policy and mismanagement of reserve land

in Manitoba disadvantaged the Indigenous population and "created what would become a permanent reserve army of labour for Manitoba farms" (173). However, she later states that during the Second World War some First Nations had become such skilled harvesters that they earned even more than the average worker (175).

80 See also Sally M. Weaver, "The Iroquois: The Grand River Reserve in the Late Nineteenth and Early Twentieth Centuries, 1875–1945," in Edward S. Rogers and Donald B. Smith, eds, *Aboriginal Ontario: Historical Perspectives on the First Nations* (Toronto: Ontario Historical Studies Series, Government of Ontario 1994), 224.

81 Peter S. Schmalz, *The Ojibwa of Southern Ontario* (Toronto: University of Toronto Press 1991), 224–5, referring to a statement made by Chief W. Wawanosh in 1900.

82 LAC, W.B. MacLean, "Dominion of Canada Annual Report of the Department of Indian Affairs [DIA] for the Year Ended June 30, 1901," Brantford, 31 July 1901, DIA, Indian Agent Reports, 35; LAC, John Scoffield, "Dominion of Canada Annual Report of the Department of Indian Affairs for the Year Ended June 30, 1901," Komoka, 23 August 1901, DIA, Indian Agent Reports, 9; "Indians on the Thames," *Dutton Advance*, 30 September 1909, 1.

83 "News Items," *Drayton Advocate*, 25 July 1907, 4; Kathryn Hansuld Lamb, "When Flax Was King," *Waterloo Historical Society* 83 (1995): 89.

84 "Floradale," *Waterloo Historical Society* 49 (1961): 48; Lamb, "When Flax Was King," 75–97.

85 Byrne, "Flax-Pulling," 1140.

86 Diane Newell, *Tangled Webs of History: Indians and the Law in Canada's Pacific Coast Fisheries* (Toronto: University of Toronto Press 1993), 76; John Sutton Lutz, *Makuk: A New History of Aboriginal-White Relations* (Vancouver: UBC Press 2008), Chap. 6, 147. In Oregon, flax was variously pulled by Chinese, Japanese, and Hispanic itinerant work gangs and processed by inmates at the Oregon State Penitentiary. Steve M. Wyatt, "Flax and Linen: An Uncertain Oregon Industry," *Oregon Historical Quarterly* 95, no. 2 (1994): 154, 157–9, 170.

87 The Drayton train station photographs came from the same collection, and since they were taken at the same location, they were almost certainly images from the same meeting. The photos were donated by Allen D. Martin to the WHF (Woolwich Heritage Foundation) collection of the Waterloo Historical Society, KPL PH595, "WHF Three men on train platform," n.d.; PH597 "WHF Group on railway platform," n.d.

88 *Stratford Beacon*, 30 June 1886, 1.

89 Johnston and Johnston, *History of Perth County to 1967* (Stratford, 1967), 29.

90 Ibid., 33.

91 Canada, *Fifth Census of Canada, 1911*, Southwold Township, Elgin West, http://data2.collectionscanada.gc.ca/1911/jpg/e001984116.jpg

92 William J. Gladding, ed., *Crossroads in Time: A Pictorial History of Tavistock (1890–1920): The Lemp Studio Collection* (Tavistock: Rotary Club of Tavistock 1998), 61.

93 Thomas Vennum Jr, *Wild Rice and the Ojibway People* (St Paul: Minnesota Historical Society Press 1988), 166. Berry, tobacco, and fruit picking continued in the 1960s for the Anishinaabe, but flax gangs declined along with the larger fibre industry. Schmalz, *The Ojibwa of Southern Ontario*, 250.

94 Paul O'Donnell and Frank D. Coffey, *Portrait: A History of the Arthur Area* (1971); WCMA, photo used by permission of the Wellington County Museum and Archives, A1976.87, ph 9536.

95 "The Cultivation of Flax for Fibre," *Farmer's Advocate*, 2 May 1918, 769.

96 Paige Raibmon, *Authentic Indians: Episodes of Encounter from the Late-Nineteenth Century Northwest Coast* (Durham, NC: Duke University Press 2005).

97 Johnston and Johnston, *History of Perth County to 1967*, 29.

98 *Advertiser* [London], 5 August 1880, 3.

99 Julia Roberts, *In Mixed Company: Taverns and Public Life in Upper Canada* (Vancouver: UBC Press 2009), 108.

100 John Gamble was a mill or harvest gang manager who lived in Stratford with his wife and three children. He hired one of the non-Native flax pullers, who later was a witness in the case, and he arranged to house several Native and non-Native workers together in a "shack" near the mill. "Killed in a Fierce Fight," *Windsor Evening Record*, 24 August 1905, 1; "A Bloody Tragedy," *Windsor Evening Record*, 25 August 1905, 5; "Seneca and Gamble," *Windsor Evening Record*, 26 August 1905, 4; "The Gamble Murder," *Windsor Evening Record*, 8 September 1905, 4; "Poyner Let Go, Corfield Held," *Windsor Evening Record*, 14 September 1905, 1.

101 "Indians on the Thames," 1.

102 *Woodstock Sentinel-Review*, 4 August 1887, 4.

103 Koop and Koop, *Older Voices among Us*, 58.

104 Gladding, ed., *The Lemp Studio Collection*.

105 Magee, *The Belgians in Ontario*, 178; D. Hoerder, *Creating Societies: Immigrant Lives in Canada* (Montreal & Kingston: McGill-Queen's University Press 2000), 116; Lois De Shield and Dionne Brand, *No Burden to Carry: Narratives of Black Working Women in Ontario, 1920s–1950s* (London: Women's Press 1991), 53–4; AO, "Marriage of James William Aylestock and Jemima Lawson," Marriages, Series MS 932 (1909), Reel 144, 607, No. 022034; Canada, *Census of Canada 1911*, LAC, Manuscripts, Schedule 1, Sub-district 33, Peel, Wellington North, Ontario, 7; AO, "Birth of Mabel Adeline Aylestock," Births, Series MS 929 (1909), Reel 29, 727.

106 "Town Local," *Woodstock Sentinel-Review*, 23 July 1888, 1.

107 "Some Ontario Flax Fields," *Farmer's Advocate*, 22 August 1918, 1361.

108 Canada, Parliament, *Sessional Papers of the Dominion of Canada*, 1895, "Report of Department of Indian Affairs," Part 1, 14-1.

109 *Tavistock Gazette*, 23 August 1928; "Fire at Flax Mill," *St Mary's Journal*, 4 August 1921, 8.

110 Roy Paisley, "New Seed and Different Methods of Handling May Revive Canada's Flax Industry Which Supplied Many European Countries during First World War" (n.d.); Clipping in Warwick History Township Committee, *The*

Township of Warwick: A Story through Time (Warwick History Township Committee, 2008), accompanying website http://warwickhistory.ca/picture/crops/897-article-on-flax-part-1-of-2.html (accessed March 2017).
111 P. Whitney Lackenbauer, *Battle Grounds: The Canadian Military and Aboriginal Lands* (Vancouver: UBC Press 2007), 122–6.
112 Magee, *The Belgians in Ontario: A History*, 28; Stephanie Bangarth, "The Long, Wet Summer of 1942: The Ontario Farm Service Force, Small-Town Ontario and the Nisei," *Canadian Ethnic Studies* 37, no. 1 (2005): 40–62; Nick Johnson. "Workers' Weed: Cannabis, Sugar Beets, and Landscapes of Labor in the American West, 1900–1946," *Agricultural History* 91, no. 3 (2017): 320–41.
113 Patti Tamara Lenard and Christine Straehle, eds, *Legislated Inequality: Temporary Labour Migration in Canada* (Montreal & Kingston: McGill-Queen's University Press 2012), 7–8; Jenna L. Hennebry and Kerry Preibisch, "A Model for Managed Migration? Re-examining Best Practices in Canada's Seasonal Agricultural Worker Program," *International Migration* 50, no. 1 (2012).
114 Beatrice Craig, "Women, Children, and the Calculation of Labour Productivity in Europe and North America," *Histoire et mesure* 15, nos. 3–4 (2000): 280, 282.
115 Newell, *Tangled Webs of History*, 76.
116 Crowley, "Rural Labour," 40.
117 Lee A. Craig and Thomas Weiss, "Agricultural Productivity Growth during the Decade of the Civil War," *Journal of Economic History* (1993): 544; R.M. McInnis, "Perspectives on Ontario Agriculture, 1815–1930," *Canadian Papers in Rural History* 8 (1992): 105; Marjorie Griffin Cohen, *Women's Work, Markets and Economic Development in Nineteenth-Century Ontario* (Toronto: University of Toronto Press 1999).
118 Keller, "From the Rhineland to the Virginia Frontier," 499.
119 Craig Heron, "Factory Workers," in Craven, ed., *Labouring Lives*, 491.
120 "Close of the St. Thomas Flax Mills," *Canada Farmer*, 16 August 1868.
121 Nathan's youngest son, also Nathan, worked for the New York Central and occupied the family home until his death in 1956. Funerals, "Nathan Ryan," *St. Thomas Times-Journal*, 2 April 1956, 7; "A Street Named Nathan," *St. Thomas Times-Journal*, 30 January 1959. Nathan Sr's wife and their daughter, also Frances, worked "in St. Thomas homes, where they were much liked and respected." The mother lived to eighty-six. "A Street Named Nathan," Funerals, Frances Ryan, *St. Thomas Times-Journal*, 13 February 1939, 9. The younger Frances married twice, once to a farmer in Malahide and the second time to a Mr Gordon. In 1959 she was living on St George Street across from her family home. She was "afflicted with blindness but still possessing much of her father's cheerfulness." "A Street Named Nathan," Marriage certificate of Frances Ryan and Frank Graves (Yeoman, Malahide Township, Elgin County), Ontario Marriages, Schedule B, County of Elgin, Series MS 932; Reel 113, 182, No. 006185.
122 Canada, *Second Census of Canada, 1881*, Vol. 1, Table III, 286; "Nathan Ryan," in *Farmers and Business Directory for the Counties of Elgin, Middlesex and Oxford, 1883–1884*, 193.

123 Canada, *Census of Canada, 1891*, LAC, Manuscripts, Schedule 1, Yarmouth, Elgin East, ON, Roll T-6333, 56.
124 "St Thomas Flax Mill Burns," *Victoria Daily Colonist* [*British Colonist*], 28 December 1893, 8.
125 Dodgshon, *Society in Time and Space*.
126 "A Street Named Nathan," *St. Thomas Times-Journal*, 30 January 1959.
127 *Stratford Beacon*, 21 March 1990, 4; "Town and Vicinity," *Essex Free Press*, 27 May 1904, 5; Canada, *Fifth Census of Canada, 1901*, LAC, Woolwich, Waterloo County, Reels T-6428 to T-6556; *Chronicle-Telegraph*, 19 January 1911, 1; TWA, James McColl to James Livingston, 23 September 1907, 1; "Flax Mill Worker Loses Leg," *Elmira Signet*, 1948.
128 Chris Yantzi account books, A, B, C, and D, personal collection of Mrs Yantzi, Tavistock, ON.
129 "The Flax Mills," *Canadian Home Journal*, 21 July 1864; Advertisement, *Canadian Home Journal*, 15 April 1870.

CHAPTER THREE

1 Mark Clark and Henry Nielsen, "Crossed Wires and Missing Connections: Valdemar Poulsen, the American Telegraphone Company, and the Failure to Commercialize Magnetic Recording," *Business History Review* 69, no. 1 (1995): 1–41; "Raid Men Who Sold $10,000,000 in Stock," *New York Times*, 21 December 1912, 1.
2 Keller, "From the Rhineland to the Virginia Frontier," 498; Hood, *The Weaver's Craft*, 44; G.C. Bolton, "The Hollow Conqueror: Flax and the Foundation of Australia," *Australian Economic History Review* 8, no. 1 (1968): 3–16; Alfred W. Crosby, *America, Russia, Hemp, and Napoleon: American Trade with Russia and the Baltic, 1783–1812* (Columbus: Ohio State University Press 1965).
3 For parallels in the iron and steel industries, see Kris E. Inwood, "The Iron and Steel Industry," in Ian M. Drummond, *Progress without Planning: The Economic History of Ontario from Confederation to the Second World War* (Toronto: University of Toronto Press 1987), 204.
4 Keller, "From the Rhineland to the Virginia Frontier," 495; Colin Coates, *The Metamorphoses of Landscape and Community in Early Quebec* (Montreal & Kingston: McGill-Queen's University Press 2000), 43–5; Nian-Sheng Huang, "Financing Poor Relief in Colonial Boston," *Massachusetts Historical Review* 8 (2006): 72–103; J.M. Bumsted, "David Lawson," *Dictionary of Canadian Biography*, Vol. 5; Evelyn Simpson, *Stanhope: Sands of Time* (Stanhope, PEI: Stanhope Women's Institute, History Committee 1984), 1–6; G. Melvin Herndon, "A War-Inspired Industry: The Manufacture of Hemp in Virginia during the Revolution," *Virginia Magazine of History and Biography* 74 (1966): 301–11.
5 Bumsted, "David Lawson," DCB, Vol. 5; Simpson, *Stanhope: Sands of Time*, 1–6; Jonas Howe, "Early Attempts to Introduce the Cultivation of Hemp in Eastern British America," paper presented to the New Brunswick Historical Society

(CIHM, Vol. A07250, 1892), 2; Coates, *The Metamorphoses of Landscape*; Norman MacDonald, "Hemp and Imperial Defence," *Canadian Historical Review* 17, no. 4 (1936): 385–98; Ross D. Fair, "A Most Favourable Soil and Climate: Hemp Cultivation in Upper Canada, 1800–1813," *Ontario History* 96, no. 1 (2004): 41–61; Eleanor Darke, *"A Mill Should Be Built Thereon": An Early History of the Todmorden Mills* (Toronto: Dundurn 1995), 42; R. Alan Douglas, *Uppermost Canada: The Western District and the Detroit Frontier, 1800–1850* (Detroit: Wayne State University Press 2001), 14, 18; Bumsted, *The Collected Writings of Lord Selkirk, 1810–1820*, Vol. 2, 17–18, 24, 42n414; Randy Richmond and Tom Villemaire, *Colossal Canadian Failures 2* (Toronto: Dundurn 2006), 123; George Bryce, "The Old Settlers of Red River," *Manitoba Historical and Scientific Society Transactions* 9 (November 1885): 5; "The Settlements on the Red and Assiniboine Rivers," *Journals of the Legislative Council of the Province of Canada*, 21 Vic A. (Toronto: Hunter Rose 1858), Appendix 3.

6 According to R.L. Jones, agricultural leaders in the mid-nineteenth century "believed that flax would make an ideal staple." Jones, *History of Agriculture in Ontario, 1613–1880*, 221; "Linen," in United States, Department of the Interior, *Eighth Census, 1860*, Vol. 3, Manufactures (Washington, DC: Government Printing Office 1865); Bruce Curtis, *The Politics of Population: State Formation, Statistics, and the Census of Canada, 1840–1875* (Toronto: University of Toronto Press 2001), 253; Ste Anne de la Pocatière, "Direction pour la culture du lin et du chanvre," *Gazette des Campagnes* (1863): 1–8.

7 LAC, "Finance Dept., Ottawa. Referring Imperial Correspondence with J.C. Allen on Formation of a Company to Grow Flax in the North West," 15 April 1885, General Correspondence, RG17, Vol. 437, File 47643; LAC, "Sir Chas. Tupper, London England. Mr. O'Leary's Correspondence on Flax Growing in North West Territory," 30 April 1885, General Correspondence, RG17, Vol. 438, File 47795; Charles Arthur Fontaine, "De la culture du lin dans la province de Québec," *Circulaire* 32 (1920), Ministère de l'agriculture de la province de Québec; Service de l'Horticulture, Section des Jardins Scolaires.

8 Steve M. Wyatt, "Flax and Linen: An Uncertain Oregon Industry," *Oregon Historical Quarterly* 95, no. 2 (1994): 172; Fred C. Slater, "Revival in Flax Growing in Ontario," in J. Frederic Thorne, ed., "The Culture and Manufacture of Flax for Fibre and Seed," *University of Oregon Bulletin – New Series* 8, no. 13 (1 August 1916): 51.

9 USDA, "Height of Hemp Doubled: Improved Fiber Results," *Weekly News Letter* (USDA), 7 September 1921, 2; Steve Vogel, *The Pentagon: A History* (New York: Random House 2008), 36.

10 Suzanne Elizabeth Zeller, *Inventing Canada: Early Victorian Science and the Idea of a Transcontinental Nation* (Toronto: University of Toronto Press 1987), 197, 201, 215, 319n57.

11 Sven Beckert, "From Tuskegee to Togo: The Problem of Freedom in the Empire of Cotton," *Journal of American History* 92, no. 2 (September 2005): 498–526.

12 Allan Cameron, *The Cultivation of Flax, and Preparation of Flax Cotton by the Chevalier Claussen Process* (New York: John A. Gray 1852), 3.
13 Marie-Louise Bauchot, Jacques Daget, and Roland Bauchot, "Ichthyology in France at the Beginning of the 19th Century: The *Histoire Naturelle des Poissons* of Cuvier (1769–1832) and Valenciennes (1794–1865)," in Theodore W. Pietsch and William Dewey Anderson, eds, *Collection Building in Ichthyology and Herpetology* (Lawrence, KS: Allen Press 1997), 27–80, 60; Joakim Garff, *Soren Kierkegaard: A Biography* (Princeton, NJ: Princeton University Press 2013), 51; "Clausen, Pedro Cláudio Dinamarquez (Peter) (1801–1872)," *The International Plant Names Index (2012)*, published on the Internet, http://www.ipni.org/ipni/idAuthorSearch.do?id=12382-1 (accessed March 2017).
14 Chevalier Claussen, "The Flax Movement: Its National Importance and Advantages with Directions for the Preparation of Flax-Cotton" (1851); W. Newton, *London Journal of Arts, Sciences, and Manufactures*, Vol. 40 (1852) (London: R. Folkard 1952), 138; Quoted in "The Inventor of Flax-Cotton a Lunatic," *Scientific American* 7, no. 18 (November 1862): 275.
15 Cameron, *The Cultivation of Flax*, 3, 6.
16 *Morning Chronicle*, 12 May 1852.
17 George E. Waring, *The Elements of Agriculture: A Book for Young Farmers, with Questions Prepared for the Use of Schools* (New York: D. Appleton & Co. 1854), 28; Margaret W. Rossiter, *The Emergence of Agricultural Science: Justus Liebig and the Americans, 1840–1880* (New Haven: Yale University Press 1975), 25–6.
18 Walter S. Barker, "Flax: The Fiber and Seed. A Study in Agricultural Contrasts," *Quarterly Journal of Economics* 31, no. 3 (May 1917): 515–16.
19 For more on cottonization, Schenck, and Claussen, see Alexander Kirkwood, "Report on the System of Cultivation and Preparation of the Flax, as Practised in Belgium and the British Islands," *Journals of the Legislative Assembly of the Province of Canada* 18, Appendix 1.1 (1854), 34–40; "Linen," in United States, Department of the Interior, *Eighth Census, 1860*, Vol. 3, Manufactures (Washington, DC: Government Printing Office 1865), cviii–cix.
20 "The Claussen Flax Process," *Morning Chronicle*, 27 May 1852.
21 William Richardson, "Township of Hamilton Farmer's Club," *Canadian Agriculturalist* 7, no. 4 (April 1855): 97–9.
22 Rod Bantjes, *Improved Earth: Prairie Space as Modern Artefact, 1869–1944* (Toronto: University of Toronto Press 2005); Daniel Samson, *The Spirit of Industry and Improvement: Liberal Government and Rural-Industrial Society, Nova Scotia, 1790–1862* (Montreal & Kingston: McGill-Queen's University Press 2008), 253. For agricultural societies in Upper Canada, see Ross Fair, "Gentlemen, Farmers, and Gentlemen Half-Farmers: The Development of Agricultural Societies in Upper Canada, 1792–1846," PhD thesis, Queen's University, Kingston, 1998. For Prince Edward Island, see Elinor Vass, "The Agriculture Societies of Prince Edward Island," *Island Magazine*, no. 7 (Fall/Winter 1979), 31–7.
23 "Mr. McDougall's Report to the Bureau of Agriculture," *Canadian Agriculturalist* 7, no. 4 (April 1855): 101; Elsbeth Heaman, *The Inglorious Arts of Peace: Exhibitions*

in Canadian Society during the Nineteenth Century (Toronto: University of Toronto Press 1999), 154.
24 Richard A. Jarrell, "Hind, Henry Youle," *Dictionary of Canadian Biography*, Vol. 13 (1994).
25 Alexander Kirkwood, *Flax and Hemp* (Toronto: W.C. Chewett & Co. 1864).
26 Gerald Killan, "Alexander Kirkwood," DCB, Vol. 8; Richard S. Lambert and A. Paul Pross, *Renewing Nature's Wealth: A Centennial History of the Public Management of Lands, Forests & Wildlife in Ontario, 1763–1967* (Toronto: Ontario Department of Lands and Forests 1967), 277–80; Kirkwood, *Flax and Hemp*, 9.
27 Donaldson, *Practical Hints*. See, for example, John A. Donaldson, "Flax Prospects," *Canada Farmer*, 16 May 1864, 130, and Donaldson, "Flax Culture," *The Irish Canadian*, 11 August 1887, 4; John A. Donaldson to Hon. G. Alexander in *Journals of the Legislative Council of the Province of Canada, 1863* (Quebec: R. Stanton 1863), Appendix 18.
28 Quoted in Kirkwood, "Report on the System of Cultivation and Preparation of the Flax," 46.
29 Barker, "Flax: The Fiber and Seed," 518. See more on Schenck's chemical retting in E.F. Deman, *The Flax Industry*, 98–109. Alexander Kirkwood's report on flax to the Province of Canada mentioned that his samples of flax sent to Scotland were processed by the Schenck method and "the later method of Watt." Kirkwood, "Report on the System of Cultivation and Preparation of the Flax," 13.
30 "Flax Progress," *Canada Farmer*, 1 November 1864, 318.
31 Beckert, *Empire of Cotton*, 246–51.
32 "Editorial," *London Daily News*, 29 August 1862, 4.
33 The US census of 1860 claimed that cottonization was still a viable option. "Flax-Wool," *Canada Farmer*, 16 January 1865, 18.
34 "Flax Companies," *Canada Farmer*, 1 February 1865, 46.
35 Ibid., 57; "Stratford Flax Mill," *Canada Farmer*, 15 February 1865, 58.
36 For example, "Coe's Super Phosphate of Lime as a Manure for Flax," *Canada Farmer*, 2 January 1865, 1.
37 Lambert and Pross, *Renewing Nature's Wealth*, 168–9.
38 W.H. Graham, *Greenbank: In the Country of the Past* (Peterborough, ON: Broadview Press 1988), 224.
39 Jones, *History of Agriculture in Ontario, 1613–1880*, 174.
40 LAC, "Circular of Agricultural Queries from the Minister of Agriculture," RG 17, Vol. 2326.
41 Ibid.
42 LAC, Campbell to Thompson, 1 April 1862, Letterbooks A12; Campbell to Belleau, 9 April 1862, Letterbooks A12; Campbell to J.A. Donaldson, 1 September 1863, Letterbooks A12; Jean-Guy Nadeau, "Joseph-Charles Taché," *Dictionary of Canadian Biography*, Vol. 7.
43 LAC, Donaldson, Londonderry, to Vankoughnet, 10 February 1862, RG 17, Vol. 1663; Campbell to Benjamin Walker, St Thomas, 29 December 1862, RG 17, Letterbooks A12.

44 LAC, H.C. Thomson, 29 December 1865, RG 17, Vol. 7; "Order in Council," 6 January 1866, RG 17, Vol. 7, File 12; "Riga Pamphlet," RG 17, Vol. 8, File 586; Georges Leclerc to BAS, 30 January 1866, RG 17, Vol. 8.

45 Norman MacDonald, "Hemp and Imperial Defence," *Canadian Historical Review* 17, no. 4 (1936): 387–8.

46 LAC, E. Simays, Montreal, to A.A. Dorion, 9 August 1862, RG 17, Vol. 2393.

47 LAC, Taché à T. Robitaille MPP, 4 February 1865, RG 17, Letterbooks A12; and Curtis, *The Politics of Population*, 253.

48 Margaret Fairly, ed., *The Selected Writings of William Lyon Mackenzie, 1824–1837* (Oxford: Oxford University Press 1960), 171.

49 LAC, "J.C. Taché to the Commissioner of Crown Lands, Prince Edward Island," 7 November 1865, RG 17, Series A12, "Letter Books," Vol. 1495, Reel T1113.

50 LAC, Perine and Young to Fleming & Co., 7 May 1866, RG 17, Vol. 11, File 840.

51 Clinton Evans, *The War on Weeds in the Prairie West: An Environmental History* (Calgary: University of Calgary Press 2002), 78–9. For contemporary advice on weed treatment and prevention in Upper Canada, see "Thistles," *Canadian Agriculturalist* 12, no. 24 (December 1860): 643–5; "Weeds," *Canadian Agriculturalist* 13, no. 4 (February 1861): 97–8; "Adulteration of Seeds," *Canadian Agriculturalist* 14, no. 2 (January 1862): 36–8; "How Farmers Cultivate Weeds," *Canadian Agriculturalist* 15, no. 4 (April 1863): 127–30.

52 LAC, Clerk ex co, 1 September 1866, RG 17, Vol. 11; Thomas Leeming, 6 September 1866, RG 17, Vol. 12; Buchanan, 13 September 1866, RG 17, Vol. 12.

53 "Flax Seed Growing," *Country Gentleman's Magazine*, Vol. 4 (London: Simpkin Marshall & Company 1870), 32.

54 "Destructive Fire at Streetsville: Gooderham & Worts' Linen Mills Burned Down," *Globe*, 27 January 1868, 2; "Ku-Klux in Crediton," *London Advertiser*, 14 December 1884, 4.

55 LAC, C. Cambie to J. Manning Esq., Great Flax Agent, England, 9 July 1862.

56 "Charles Hendry & Co., merchants, millers, and lumber merchants, Canistoga" [*sic*] William Henry Smith, *Business Directory of Canada West, 1851* (Toronto: T. Maclear 1851), 85.

57 RWA, *Journal of the Proceedings and By-Laws of the Municipal Council of the County of Waterloo 1864*, Third Session (16 June 1864), 19.

58 LAC, J.A. Donaldson, 19 September 1865, RG 17, Vol. 8.

59 Samson, *The Spirit of Industry and Improvement*, 263.

60 "An Act for Appropriating Certain Moneys therein mentioned for the Service of the Year of Our Lord One Thousand Eight Hundred and Sixty-six," *Act of Assembly, P.E.I., Victoria 29, 1866*; see also Elinor Vass, "The Agricultural Societies of Prince Edward Island," *Island Magazine* 7 (Fall–Winter 1979): 33; "Report of the Commissioners appointed by the Government to Report on certain Proprietary Estates offered to the Government for Sale with Memorandum of Annual Rent thereof Estate of Daniel Hodgson, Esq, Township No. 23," *Journal of the PEI House of Assembly for 1870*, Appendix X.

61 Quoted in Whitman and Leeson, *Flax Culture*, 24.

62 Ibid.

63 "The Inventor of Flax-Cotton a Lunatic," *Scientific American* 7, no. 18 (November 1862): 275.

64 Gordon M. Winder, *The American Reaper: Harvesting Networks and Technology, 1830–1910* (Aldershot, UK: Ashgate Publishing 2013), 160; Crowley "Rural Labour," 40; Evans, *Bound in Twine*, 2–3; Olmstead and Rhode, *Creating Abundance*, 4–6.

65 1899 Directory; *Ontario Commercial Year Book and Gazetteer, 1906* (Toronto).

66 William Johnston, *History of Perth County 1825–1902* (Stratford: W.M. O'Beirne 1903), 479; "Market and Financial Notes," *Fibre and Fabric* 37, no. 948 (2 May 1903): 138; "Town and Vicinity," *Essex Free Press*, 15 January 1904, 5; "The Essex Board of Trade," *Essex Free Press*, 22 January 1904, 4; "Town and Vicinity," *Essex Free Press*, 27 May 1904, 5; "Town and Vicinity," *Essex Free Press*, 8 July 1904, 5; "Mill News," *Textile World Record* 27, no. 3 (June 1904): 167.

67 "Tenders for Sale of Flax Crops," *Essex Free Press*, 1 July 1904, 1; "Local News," *Essex Free Press*, 5 August 1904, 1; "Local News," *Essex Free Press*, 19 August 1904, 5; *Ontario Commercial Year Book and Gazetteer, 1906*, 475.

68 Mark Clark and Henry Nielsen, "Crossed Wires and Missing Connections: Valdemar Poulsen, The American Telegraphone Company, and the Failure to Commercialize Magnetic Recording," *Business History Review* 69, no. 1 (1995): 1–41; "Raid Men Who Sold $10,000,000 in Stock," *New York Times*, 21 December 1912, 1; "Held for Using Mails to Defraud," *Times-Dispatch* (Richmond), 21 December 1912, 9; "F.W. Shumaker Gets Bail," *New York Times*, 22 December 1912, 5; "Six Arrested: Six Year Investigation, Say Government Officials, Income of Promoters $1,000,000," *Lewiston Daily Sun*, 21 December 1912, 1; "Stock Boosters get $20,000,000," *Cold Spring Recorder*, 27 December 1912, 6.

69 D. Hamilton Hurd, *History of Essex County, Massachusetts: With biographical sketches of many of its pioneers and prominent men* (Philadelphia: J.W. Lewis 1888), 357; Oxford Linen Mills, "From Flax to the Fabric" (1908), 4.

70 Benjamin Mudge, "Flax stock and method of preparing same," US Patent Office, 30 June 1903, US 732103 A.

71 "New Linen Experiment," *New York Times*, 17 April 1904.

72 Oxford Linen Mills, "From Flax to the Fabric" (1908), 8.

73 Ibid., 5.

74 "Stock Boosters get $20,000,000," *Cold Spring Recorder*, 27 December 1912, 6; "Guide to the Elwyn A. Barron Papers 1877–1921," Special Collections Research Center, University of Chicago Library, http://www.lib.uchicago.edu/e/scrc/findingaids/view.php?eadid=ICU.SPCL.BARRONEA (accessed September 2017).

75 Oxford Linen Mills, "From Flax to the Fabric" (1908), 2.

76 John F. Varty, "On Protein, Prairie Wheat, and Good Bread: Rationalizing Technologies and the Canadian State, 1912–1935," *Canadian Historical Review* 85, no. 4 (2004): 721–54, 740; Oxford Linen Mills, "From Flax to the Fabric" (1908), 7.

77 Oxford Linen Mills, "From Flax to the Fabric" (1908), 3. Similar boosterist language was used to describe Mudge in "Making Linen: A New American Triumph," *Sacred Heart Review*, 23 May 1908, 352.

78 It is not clear whether Barron was ever tried, but others, like promoter and bond-seller Clarence M. Smith, received lighter sentences, such as six months in the local prison on Blackwell Island. "Bond Seller Sentenced," *New York Times*, 19 April 1914, 9.
79 "Eighth Skyrocket Financier Taken," *New York Call*, 22 December 1912, 1; "Bond Seller Sentenced," *New York Times*, 19 April 1914, 9.
80 "Winchell Held for Fraud," *Boston Evening-Transcript*, 21 December 1912, 3; "Winchell Proceeding Dismissed," *Boston Evening-Transcript*, 7 January 1913, 12.
81 "Raid Men Who Sold $10,000,000 in Stock," *New York Times*, 21 December 1912, 1.
82 HLBP, William J. Robinson to H.L. Bolley, 2 April 1910, Box 45, File 6.
83 HLBP, Goad to H.L. Bolley [22 March 1910].
84 On the Sydenham, Alan Mann and Frank Mann, *Settlement on the Sydenham: The Story of Wallaceburg* (Wallaceburg: Mann Historical Files 1984), 6; J.H. Beers, *Commemorative Biographical Record of the County of Kent, Ontario: Containing Biographical Sketches of Prominent and Representative Citizens and Many of the Early Settled Families* (Toronto: J.H. Beers 1904), 568.
85 United States, Department of Commerce and Labor, "Commercial Relations of the United States with Foreign Countries during the Year 1903, Vol. 2, *United States Congressional Serial Set, Issue 4730* (Washington, DC: Government Printing Office 1904), 91.
86 "Town and Vicinity," *Essex Free Press*, 18 November 1904, 5; "New Jury Sitting," *Essex Free Press*, 2 December 1904, 1.
87 *Province of Ontario Gazetteer and Directory 1910–1911* (Ingersoll, ON: Union 1910).
88 McCredie, "Flax for Fibre," 22–3.
89 Ibid., 70–2.
90 Austin L. McCredie, "Facts about Flax," *Canadian Courier*, March 1919.
91 James A. McCracken, *A Review of the Status and Possibilities of Flax Production and Manipulation in Canada* (Ottawa: Government Printing Bureau 1916).
92 J.H. Grisdale, "Report of the Director," in *Annual Report of the Experimental Farms, Sessional Papers of the Dominion of Canada, 1917*, 7 George V, A. (1917), 10.
93 "The Cultivation of Flax for Fibre," *Farmer's Advocate*, 2 May 1918, 769.
94 "Seek Crowd to Pull Ripe Flax," *Windsor Evening Record*, 2 August 1918, 1; David Roberts, *In the Shadow of Detroit: Gordon M. McGregor, Ford of Canada, and Motoropolis* (Detroit: Wayne State University Press 2006), 185.
95 "Flax Festival Today at Willowdale Farm," *Toronto World*, 23 July 1918, 3.
96 City of Toronto Archives, Fonds 1244, f1244_it4511.
97 Mike Filey, *From Horse Power to Horsepower: Toronto: 1890–1930* (Toronto: Dundurn Press 1993), 107.
98 By 1919 the Ontario Motion Picture Bureau and Pathéscope featured flax in an early film titled *Why Not Use a Tractor?* It included scenes of a tractor pulling a mechanical flax harvester, which the bureau promised would help with "taking the 'ache' out of flax pulling." LAC film collection and YouTube channel, https://www.youtube.com/watch?v=tEK3iiQduIY (accessed August 2017); "Facts about Flax."

99 McCracken, *Review of the Status and Possibilities*, 16–17.
100 "Indians, Not English," *Windsor Evening Record*, 14 December 1907, 6.
101 HLBP James A. McCracken to H.L. Bolley, 6 January 1915.
102 See also "James A. McCracken," FindAGrave.com, Find A Grave Memorial# 51242899, https://www.findagrave.com/cgi-bin/fg.cgi?page=gr&GRid=51242899 (accessed August 2017).
103 J.H. Grisdale, "Report of the Director," and "Annual Report of the Experimental Farms," in *Sessional Papers of the Dominion of Canada 1917*, 7 George V, A. 10.
104 Ibid., 9.
105 Catherine Carstairs, *Jailed for Possession: Illegal Drug Use, Regulation, and Power in Canada, 1920–1961* (Toronto: University of Toronto Press 2006).
106 Canada, *Fifty Years of Progress on Dominion Experimental Farms, 1886–1936* (Ottawa: J.O. Pateneude 1939), 72.
107 T.H. Anstey, *One Hundred Harvests: Research Branch, Agriculture Canada, 1886–1986* (Ottawa: Canadian Government Publishing Centre 1986), 257.
108 Canada, *Sixth Census of Canada, 1921*, LAC, Manuscripts, Caradoc (Township), Middlesex West, ON; Reel (folder number) 72, 18; *Illustrated historical atlas of the county of Middlesex, Ont.* (Toronto: H.R. Page & Co. 1878).
109 Roos UWO MA; Helen Roos, "'My parents, they became poor': The Socio-Economic Effects of the Expropriation and Relocation of Stoney Point Reserve 42, 1942," *Past Imperfect* 7 (1998): 155–75; P. Whitney Lackenbauer, *Battle Grounds: The Canadian Military and Aboriginal Lands* (Vancouver: UBC Press, 2007), 122–3.

CHAPTER FOUR

1 Evans, *Bound in Twine*; Steven Topik, Carlos Marichal, and Zephyr L. Frank, eds, *From Silver to Cocaine: Latin American Commodity Chains and the Building of the World Economy, 1500–2000* (Durham, NC: Duke University Press 2006); John Tully, "A Victorian Ecological Disaster: Imperialism, the Telegraph, and Gutta-Percha," *Journal of World History* 20, no. 4 (2009): 559–79.
2 Eric Wolf, *Europe and the People without History* (Berkeley: University of California Press 1982), 225, 335, 352; Frances Moore Lappé, *Diet for a Small Planet* (New York: Ballantine Books 1978), 27.
3 Arjun Appadurai, ed., *The Social Life of Things: Commodities in Cultural Perspective* (Cambridge: Cambridge University Press 1988), 21, 38.
4 Vaclav Smil, *Made in the USA: The Rise and Retreat of American Manufacturing* (Cambridge, MA: MIT Press 2013).
5 William Duane Ennis, *Linseed Oil and Other Seed Oils: An Industrial Manual* (New York: D. Van Nostrand 1909), 262.
6 Alfred D. Chandler, *The Visible Hand: The Managerial Revolution in American Business* (Cambridge, MA: Belknap Press 1977), 230–1; Michael Bliss, *Northern Enterprise: Five Centuries of Canadian Business* (Toronto: McClelland & Stewart 1990), 354.
7 Elliot & Co., *Elliot & Co., no. 3, Front St. East, Toronto, Exhibit in Class 79 as Samples, and Not for Competition, the Following Linseed and Its Products* [1880]

1982 (CIHM #27283), 1; Ann Marie Low, *Dust Bowl Diary* (Lincoln: University of Nebraska Press 1984), 29. The oil could also be fed to livestock when heavy feeding schedules caused constipation. Hugh G. Van Pelt, *How to Feed the Dairy Cow: Breeding and Feeding Dairy Cattle* (Waterloo, ON: Fred L. Kimball Co., 1919), 72–6; Dianne Elizabeth Dodd and Deborah Gorham, eds, *Caring and Curing: Historical Perspectives on Women and Healing in Canada* (Ottawa: University of Ottawa Press 1994), 76.

8 Ennis, *Linseed Oil and Other Seed Oils*, 263; David Scott, interview with Bruce Church, "Recalling Past Days of Zurich Mills," *Lakeshore Advance* 15, no. 32 (12 August 1998): 2.

9 Marion Vaisey-Genser and Diane H. Morris, "Introduction: History of the Cultivation and Uses of Flaxseed," in Alister D. Muir and Neil D. Westcott, eds, *Flax: The Genus Linum* (London: Routledge 2003), 16.

10 Whitney Eastman, *The History of the Linseed Oil Industry in the United States* (Minneapolis: T.S. Denison 1968), 123.

11 Vaisey-Genser and Morris, "Introduction," 16; Charles Lock Eastlake, *Methods and Materials of Painting of the Great Schools and Masters*, Vols 1 and 2 (New York: Courier Dover Publications 2001), 259–60.

12 Füsun Ertuğ, "Linseed Oil and Oil Mills in Central Turkey Flax/Linum and Eruca, Important Oil Plants of Anatolia," *Anatolian Studies* 50 (2000): 171–85; Keller, "From the Rhineland to the Virginia Frontier," 492. Lyman Clare & Company used leftovers from their linseed oil presses as fuel for "the gas required for the establishment." Canada Railway Advertising Company, *Montreal Business Sketches with a Description of the City of Montreal* (Montreal: Canada Railway Advertising Company 1864), 15.

13 Abbott Lowell Cummings and Richard M. Candee, "Colonial and Federal America: Accounts of Early Painting Practices," in Roger W. Moss, ed., *Paint in America: The Colors of Historic Buildings* (Washington, DC, and New York: Preservation Press and John Wiley & Sons 1994), 20.

14 Carter Litchfield, *The Bethlehem Oil Mill, 1745–1934: Oilseed Mill, Hemp Mill, Tanbark Mill, Groat Mill, Snuff Mill, Waterworks: German Technology in Early Pennsylvania* (Kemblesville: Olearius Editions 1984), 30, 48–51, 76–82, 85.

15 Williams Haynes, *American Chemical Industry* (New York: Van Nostrand 1945), 60–1, 196–7; Daniel F. Tiemann, "The Paint, Oil, and Varnish Trade," in Chauncey M. Depew, ed., *1795–1895: One Hundred Years of American Commerce ... a History of American Commerce by One Hundred Americans* (New York: Greenwood Press 1968 [1895]), 621–2; Eastman, *History of the Linseed Oil Industry*, 143–5.

16 Keller, "From the Rhineland to the Virginia Frontier," 492.

17 Tiemann, "The Paint, Oil, and Varnish Trade," 621; Theodore Zuk Penn, "Decorative and Protective Finishes, 1750–1850: Materials, Process, and Craft," *Bulletin of the Association for Preservation Technology* 16, no. 1 (1984): 2–46, 8; white lead figures are William P. Thompson's estimates, "The Lead Industry," in Depew, ed., *1795–1895: One Hundred Years of American Commerce*, 439.

18 Haynes, *American Chemical Industry*, 190.

19 Tiemann, "The Paint, Oil, and Varnish Trade," 621.
20 Cummings and Candee, "Colonial and Federal America," 23.
21 Tiemann, "The Paint, Oil, and Varnish Trade," 621.
22 "Painting," from the *Southern Planter*, in *The British American Cultivator* 2, no. 2 (February 1846): 46.
23 "Visit to the Indian Preacher Peter Jones, and His English Lady," *The Bee* 1, no. 32 (30 December 1835): 250.
24 Peter S. Schmalz, *The Ojibwa of Southern Ontario* (Toronto: University of Toronto Press 1991), facing p. 226.
25 Cummings and Candee, "Colonial and Federal America," 41.
26 St Marys Museum, Carter & Isaac Collection, photo no. 12.
27 Ibid., 38–9.
28 Haynes, *American Chemical Industry*, 196.
29 "Painting," from the *Southern Planter*, 46.
30 Douglas McCalla, *Consumers in the Bush: Shopping in Rural Upper Canada* (Montreal & Kingston: McGill-Queen's University Press 2015), 109–11.
31 D.C. Mackay, "Valentine, William," *DCB*, Vol. 7.
32 Carriage painting was about 15 per cent of the Rea and Johnston business (1767–1803) at first, but in the last decade it lost its share. Cummings and Candee, "Colonial and Federal America," Table 1, 31.
33 Hezekiah Reynolds, *Directions for House and Ship Painting: A Facsimile Reprint of the 1812 Edition with New Introduction by Richard M. Candee* (Worcester, MA: American Antiquarian Society 1978), 22; Susan Buggey, "Merrick, John," *DCB*, Vol. 7. The importance of regularly repainting wooden ships is evident in Stephen J. Hornsby, *Surveyors of Empire: Samuel Holland, J.F.W. Des Barres, and the Making of the Atlantic Neptune* (Montreal & Kingston: McGill-Queen's University Press 2011).
34 United States, Department of the Interior, *Seventh Census, 1850: Embracing a Statistical View of Each of the States and Territories, Arranged by Counties, Towns, etc.* ... (Washington: Robert Armstrong 1853), Table 10, 119.
35 On putty and glass work, see "Making Putty," *The Commercial* 13, no. 13 (10 December 1894): 302; "A Two-Story Farm-House," *Canada Farmer*, 15 April 1865, 117. Ideas about health and painted homes are found in "Painted Residences," *Canadian Magazine of Science and the Industrial Arts, Patent Office Record* 17, no. 3 (March 1889): 92–3.
36 Zuk Penn, "Decorative and Protective Finishes, 1750–1850," 8.
37 Ian C. Bristow, "House Painting in Britain: Sources for American Paints, 1615 to 1830," in Roger W. Moss, ed., *Paint in America: The Colors of Historic Buildings* (Washington, DC, and New York: Preservation Press and John Wiley & Sons 1994), 38.
38 "Painting," from the *Southern Planter*, 46.
39 Cummings and Candee, "Colonial and Federal America," 118.
40 Thomas Green Fessenden, *The New England Farmer and Horticultural Journal* 11, no. 48 (June 1833): 382; quoted in Cummings and Candee, "Colonial and Federal America," 38–9.

41 These seemingly obvious instructions showed readers how to paint the bottom edges of surfaces by holding the brush "with the handle inclining downward." "Painting," from the *Southern Planter*, 46. This rather elementary pedagogical approach continued for at least seventy years, at which point farmers were still reading articles such as "Paint: What to Buy and How to Use It," *Canadian Countryman* 3 (31 January 1914): 5.
42 "A Two-Story Farm-House," 117.
43 "Painting and Oiling Furniture and Floors," *Canada Farmer*, 15 September 1870, 353–4.
44 "Practical Painting: Certain Properties of Oil Paints," *Scientific Canadian Mechanics Magazine and Patent Office Record* 7, no. 7 (July 1879): 334; "Painting," from the *Southern Planter*, 46. Spencer Kellogg Linseed Oil Company claimed raw oil would dry in as little as three to five days. Alexander Schwarcman, *Laboratory Letters* (Buffalo: Spencer Kellogg and Sons 1921), 10.
45 Walter Needham, *A Book of Country Things* (Brattleboro, VT: Stephen Greene Press 1965), 124.
46 H.W. Beecher, *All Around the House, or, How to Make Homes Happy* (Toronto: J. Ross Robertson 1881), 155.
47 Cummings and Candee, "Colonial and Federal America," 13–14.
48 "Painting and Oiling Furniture and Floors," 353–4. Floor paint was very rarely used in the United States before the American Revolution. Cummings and Candee, "Colonial and Federal America," 27.
49 The ingredients were "50 pounds of best white lead, 10 quarts linseed oil; ½ lb., driers; 50 lbs. finely sifted clean white sand; [and] 2 lbs. raw umber." "Paint and Sand," *Canadian Agriculturalist* 2, no. 9 (September 1850): 210.
50 Susan M. Burke et al., *This Old Haus: A Place in Time* (Kitchener: Friends of the Joseph Schneider Haus 2008), 48–9.
51 Only a few articles on paint were found in Edwinna Von Baeyer's survey of these journals in Ontario, and almost all of these were written after 1900. Edwinna Von Baeyer, *Ontario Rural Society 1867–1930: A Thematic Index of Selected Ontario Agricultural Periodicals* (Ottawa: E. Von Baeyer 1985), 37.
52 See, for instance, "Home Beautification," *OAC Review* 31, no. 8 (1919): 355–7.
53 Julie Harris and Jennifer Mueller, "Making Science Beautiful: The Central Experimental Farm, 1886–1939," *Ontario History* 89, no. 2 (June 1997): 103–23, 109.
54 "Johnson's Pure Paints," *Farmer's Advocate*, 5 September 1890, 266.
55 According to the journal's recipe, 100 pounds of white lead paste, 5 gallons of oil, 1 pint of drier, and 1 gallon of turpentine would have produced about 7.5 gallons of paint at a cost of $11 or $12. The cost per gallon would have been at least 40 cents higher and a lot more troublesome than purchasing a gallon of "prepared paint." H.M. Tandy, "Paint Insurance for the Farmer," *Farmer's Advocate*, 5 March 1908, 396–7.
56 Jacob Mellis, "A Plea for Ready Mixed Paints," *Farm and Dairy*, 21 May 1914, 607.
57 McCalla, *Consumers in the Bush*, 110–11.
58 Haynes, *American Chemical Industry*, 196.

59 MAO, "Entries, 29–30 September 1856," Diary of Christian B. Snyder, 1856, Historical MSS 17.12.
60 "Provincial Exhibition," *Trade Review and Intercolonial Journal of Commerce* 4, no. 40 (2 October 1868): 632–3; "Advertisement," *Canadian Pharmaceutical Journal* 2, no. 11 (March 1869): 30.
61 "Chemical Manufactures and Preparations at the Provincial Exhibition," *Canadian Pharmaceutical Journal* 3, no. 30 (October 1870): 152.
62 Moss admits that these early mixed paints remained difficult to use and unpredictable in quality. Roger W. Moss, "Nineteenth Century Paints: A Documentary Approach," in Roger W. Moss, ed., *Paint in America: The Colors of Historic Buildings* (Washington, DC, and New York: Preservation Press and John Wiley & Sons 1994), 56–7.
63 Joseph Nathan Kane, Steven Anzovin, and Janet Podell, *Famous First Facts*, 6th edn (New York: H.W. Wilson 2006), 151.
64 Tandy, "Paint Insurance for the Farmer," 396–7.
65 "Johnson's Pure Paints," 266.
66 Haynes, *American Chemical Industry*, 204; Moss, "Nineteenth Century Paints," 56–7.
67 Canada, *Census of Canada, 1870–71*, LAC, Industrial Schedules.
68 Christian Warren, *Brush with Death: A Social History of Lead Poisoning* (Baltimore, MD: Johns Hopkins University Press 2000), 56.
69 Pittsburgh Plate Glass Company, *Glass, Paints, Varnishes and Brushes: Their History, Manufacture and Use* (Chicago: Lakeside Press 1923), 60–4.
70 Douglas McCalla, *Planting the Province: The Economic History of Upper Canada, 1784–1870* (Toronto: University of Toronto Press 1993), 280.
71 Carriage painters were selected by finding those who identified with both paint establishments and carriage works, as well as those who did specified work in a shop and mentioned "carriage" or "voiture" in their trade description.
72 "Coach-Painting," *Scientific Canadian Mechanics Magazine and Patent Office Record* 7, no. 7 (July 1879): 208.
73 These were the four mentioned above as well as Quebec City and Longueuil, Quebec, and one near St Andrews, New Brunswick. The other tenth of the nation's paint shop output was remarkably even, the result of a distribution of rural districts in Quebec and of a few being distributed around the Baie de Chaleurs and the Lower Saint John River.
74 McCalla, *Planting the Province*, 280. For carriage makers and painting shops in Brantford, see David G. Burley, *A Particular Condition in Life: Self-Employment and Social Mobility in Mid-Victorian Brantford, Ontario* (Montreal & Kingston: McGill-Queen's University Press 1994), 31–3, 39–40.
75 Canada, *Census of Canada, 1890–91*, Vol. 4 (Ottawa: S.E. Dawson 1897), Table 1, 234.
76 In 1871 one painter in Toronto Gore Township claimed that his main service was "country painting," setting himself apart from the painters who serviced the cities and towns surrounding him.

77 Axel Diederichsen and Ken Richards, "Cultivated Flax and the Genus *Linum* L.: Taxonomy and Germplasm Conservation," in Alister D. Muir and Neil D. Westcott, eds, *Flax: The Genus Linum* (London: Routledge 2003), 30.

78 Litchfield, *The Bethlehem Oil Mill, 1745–1934*, 50–1.

79 Ennis, *Linseed Oil and Other Seed Oils*, 110; Edward H. Knight, *Knight's American Mechanical Dictionary: Being a Description of Tools, Instruments, Machines, Processes, and Engineering* (Missouri: Early American Industries Assoc.,1881), 1554. Improved processing technology increased the yield of oil to about 20 pounds per bushel of seed. Eastman, *History of the Linseed Oil Industry*, 277.

80 Elliot & Co, *Linseed and Its Products*, 2; A. Gordon Spencer, "Commercial Linseed Oils," *Canadian Engineer* 26, no. 1 (1 January 1914): 114; Zuk Penn, "Decorative and Protective Finishes," 2–46, 20–1.

81 Some painters preferred "cold pressed" oil, and the industry marketed both forms of oil. George B. Heckel, *The Paint Industry; Reminiscences and Comments* (St Louis: American Paint Journal Company 1931), 7; Spencer, "Commercial Linseed Oils," 114.

82 Diederichsen and Richards, "Cultivated Flax and the Genus *Linum*," 30.

83 Elliot & Co, *Linseed and Its Products*, 2.

84 Cottonseed oil was apparently far less variable by region, although manufacturers in the southwest branded their product as a "butter" oil. Lynette Boney Wren, *Cinderella of the New South: A History of the Cottonseed Industry, 1855–1955* (Knoxville: University of Tennessee Press 1995), 72–8.

85 Spencer, "Commercial Linseed Oils," 114; James R. Scobie, *Revolution on the Pampas: A Social History of Argentine Wheat, 1860–1910* (Austin: Published for the Institute of Latin American Studies by the University of Texas Press 1964), 76–8.

86 Canada, "An Act to amend the General Inspection Act so as to provide a grade for Flax Seed," *Statutes of Canada*, 63 Vic., Chap. 38. The best-quality seeds for oil were grown in northern latitudes. I.F. Laucks, *Commercial Oils, Vegetable and Animal; With Special Reference to Oriental Oils* (New York: John Wiley & Sons 1919), 31.

87 Spencer, "Commercial Linseed Oils," 115. Linseed oil could also be charred with high heat. Heating to temperatures over 400°F was necessary for certain (especially varnish) oils, but caused problems like porosity and "breaks" or hazy substances in the oil. Schwarcman, *Laboratory Letters*, 16, 23.

88 Spencer, "Commercial Linseed Oils," 115; Zuk Penn, "Decorative & Protective Finishes, 1750–1850," 2–46, 21.

89 Canada, *Census of Canada, 1870–71*, LAC, Industrial Schedules, Kris E. Inwood dataset. The Elliot linseed oil company was a branch of Lyman and Clare in Montreal. "Advertisement," *Canadian Pharmaceutical Journal* 2, no. 11 (March 1869): 30; R.G. Dun, *Mercantile Agency Reference book for the British provinces, 1866* (Montreal: M. Longmoore 1866), 450.

90 The average mill's output was $11,600, so if 70 per cent of the output came from the oil ($8,120 per mill) and oil sold for 75 cents per gallon, the average mill could have produced 10,826 gallons of raw oil per year. "Wholesale Prices, Current – Oct 1869," *Canadian Pharmaceutical Journal* 2, no. 18 (October 1869): 154; United

States, Department of the Interior, *Seventh Census, 1850: Embracing a Statistical View of Each of the States and Territories, Arranged by Counties, Towns, etc. ...* (Washington: Robert Armstrong 1853), lxxii.

91 Elizabeth Bloomfield, *Waterloo Township through Two Centuries* (Kitchener: Waterloo Historical Society 1995), 190.

92 John N. Jackson, *The Welland Canals and Their Communities: Engineering, Industrial, and Urban Transformation* (Toronto: University of Toronto Press 1997), 94.

93 Canada, *Census of the Canadas, 1851–52*, Vol. 2 (Quebec: Lovell and Lamoureux 1855), Table 7, 170–396; Canada, *Census of the Canadas, 1860–61*, Vol. 2 (Quebec: S.B. Foote 1863), Table 13, 226–83; Canada, *Census of Canada, 1870–71*, Vol. 3 (Ottawa: I.B. Taylor 1875), Table LV, 458–63; *Census of Canada, 1880–81*, Vol. 3 (Ottawa: Maclean, Roger & Company 1883), Table LVI, 512–15.

94 Kathryn Lamb, "Bridgeport's First 50 Years, 1829–1879," *Waterloo Historical Society* 67 (1979): 46.

95 William Henry Smith, *Business Directory of Canada West, 1851* (Toronto: T. Maclear 1851), 84.

96 Canada, *Census of Canada, 1851*, LAC, Manuscripts, quoted in Ken E. Seiling, "Early Building in the Bridgeport, Ontario Area" (BA, Department of History, University College, Waterloo Lutheran University), 13.

97 Sutherland, *County of Waterloo Gazetteer and General Business Directory*, 111–13.

98 Frederick H. Armstrong, "Lyman, Benjamin," *Dictionary of Canadian Biography*, Vol. 10; *Montreal in 1856: a sketch prepared for the celebration of the opening of the Grand Trunk Railway of Canada* (Montreal: Lovell 1856); Gerald J.J. Tulchinsky, *The River Barons: Montreal Businessmen and the Growth of Industry and Transportation, 1837–53* (Toronto: University of Toronto Press 1977), 223–4.

99 The judges regretted that the oil cake entries were "so few in number," but they assured readers that Lyman & Clare's "oil cake is very excellent, carefully prepared, free from grit, &c., and worthy of the premium awarded"; "List of Prizes Awarded at the Provincial Exhibition at Kingston. September 22 to 25, 1863," *Canadian Agriculturist, and Journal of the Board of Agriculture of Upper Canada* 15, no. 11 (November 1863): 437–8. The company also dominated this class in 1868. "Provincial Exhibition," *The Trade Review and Intercolonial Journal of Commerce* 4, no. 40 (2 October 1868): 633.

100 Canada Railway Advertising Company, *Montreal Business Sketches with a Description of the City of Montreal* (Montreal: Canada Railway Advertising Company 1864), 17.

101 Two of the presses were horizontal and two were vertical, although both types had the same output. There were also two 4,000-gallon storage tanks outside the factory. These details suggest that the output was doubled at some point since construction in the early 1850s. Ibid., 16–17; Henry Beaumont Small, *The Products and Manufactures of the New Dominion* (Ottawa: G.E. Desbarats 1868), 32.

102 Canada Railway Advertising Company, *Montreal Business Sketches*, 17.

103 "Cultivation and Manufacture of Flax: The Perine's Works," *Canada Farmer*, 15 January 1864, 2; John A. Donaldson, Quebec, to the Honourable George Brown,

13 June 1864, *Journals of the Legislative Assembly of the Province of Canada*, 1864 (2nd session of the 8th Provincial Parliament), Appendix 11 (5–7), 6.

104 *The Presbyterian* 25, no. 10 (October 1872): 249; *Trade Review and Intercolonial Journal of Commerce* 3, no. 12 (5 April 1867): i.

105 They received a 1-A1 rating in the Dun credit ledgers. Dun, *Mercantile Agency Reference book*, 253.

106 Canada, *Census of Canada, 1870–71*, LAC, Industrial Schedules, Kris E. Inwood dataset. Their awards at the Provincial Exhibition also included "ground plaster for manure." "List of Prizes," *Canadian Agriculturist*, 437.

107 "Advertisement," *Canadian Pharmaceutical Journal* 2, no. 11 (March 1869): 30.

108 "Many Important Drug Companies Started in Little Back Shops," *Canadian Pharmaceutical Journal* 75 (June 1942): 82.

109 LAC, J.A. Donaldson, Toronto, to Taché, 2 January 1865, RG 17, Vol. 3; LAC, J.A. Donaldson, Toronto, to Taché, 7 May, 1866, RG 17, Vol. 10; "Flax Seed and Oil Cake," *Canada Farmer*, 15 December, 1864; LAC, "J. C. Taché to the Commissioner of Crown Lands, Prince Edward Island," 7 November 1865, RG 17, Series A12 "Letter Books," Vol. 1495, Reel T113.

110 Castle Kilbride later became a National Historic Site. The local press joked that it was a little plain on the outside, but more important they considered the rural location "a great mistake" and believed Livingston should have relocated to the city. "A Fine House" *Berlin Daily News*, 14 August 1878; "County of Waterloo Gazetteer and Directory for 1877–8," in Ross Cumming ed., *Illustrated Atlas of the County of Waterloo, 1881* (Owen Sound: Richardson, Bond & Wright 1972); Geoffrey Hayes, *Waterloo County: An Illustrated History* (Kitchener: Waterloo Historical Society 1997).

111 Canada, *Census of Canada, 1890–91*, Vol. 4 (Ottawa: S.E. Dawson, 1897), Table 1, 234, 298.

112 TWA, Brian Rawding, "Through the Keyhole: Historical Analysis of the Livingston's Life Within Castle Kilbride," unpublished paper, 38.

113 Although it had been higher in 1866, Lyman's business was now estimated by Dun and Bradstreet to be worth $250,000 to $300,000. Armstrong, "Lyman, Benjamin"; *Livingston v. Livingston* (1912), 4 D.L.R. (old series), 346; "Sale of Flax Mills, Farms, Farming Stock, Etc., to Close an Estate," *Globe*, 6 March 1897. It seems that offering the estate for sale was more a way to determine the value of the mills than to dispose of them. In any case it had the effect of shutting the mills down. "The manager here [Linwood] has been instructed to give out no seed," it reported, and "the farmers in this neighbourhood will feel the stoppage of the mill very much," *Daily Mail and Empire*, 15 April 1897, 6.

114 Michael D. Longo, "Castle Kilbride," *Waterloo Historical Society* 73 (1985): 42, 47.

115 Shirk & Snider, the new owners, are listed as flour and saw millers in 1878, "County of Waterloo Gazetteer and Directory for 1877–8," 38.

116 It is not clear if this reference is to rented land or simply to flax purchased from a crop of about 3,000 acres. Ibid., 10; William Rothwell Plewman, *Adam Beck and the Ontario Hydro* (Toronto: Ryerson Press 1947), 17.

117 Longo, "Castle Kilbride," 44.

118 "County of Waterloo Gazetteer and Directory for 1877–8," 38.
119 By 1897 their Ontario properties included flax mills and farms in ten towns. "Sale of Flax Mills, Farms, Farming Stock, Etc., To Close an Estate."
120 The US census notes that local flax farmers earned a high value per acre from their flax fibre in 1889, United States, Department of the Interior, *Eleventh Census, 1890*, Vol. 5, Agriculture (Washington, DC: Government Printing Office 1895), 93.

CHAPTER FIVE

1 John H. Warkentin, *The Mennonite Settlements of Southern Manitoba* (Steinbach: Hanover Steinbach Historical Society 2000), 89–90; *Manitoba Weekly Free Press*, 21 December 1878. A *Winnipeg Times* article about the operation suggests that Church was a flax partner of Gooderham & Worts and was planning to build a mill in Niverville, Manitoba, to make use of the Mennonite flax fibre. *Winnipeg Times*, 3 May 1880, 1; *Winnipeg Times*, 20 July 1880, 4. Charles Hendry will be familiar from the earlier chapters on flax fibre in Waterloo County.
2 TWA, "Livingston Flax Seed Products, Known the World Over," Clippings File, *Hardware and Metal*, 4 August 1917; Jesse Edgar Middleton and Fred Landon, *The Province of Ontario: A History, 1615–1927* (Toronto: Dominion Publishing Co. 1927), 297.
3 Little is known about James's largest customers, but eventually Dominion Linseed Oil sold to paint companies such as CIL and Alcraft. "Notes from the Attic: Dominion Linseed Oil Company," *Baden Outlook* 6, no. 3 (15 October 2005): 12–13. The Livingstons measured linseed oil at 9 pounds per gallon, according to their letterhead and receipts.
4 The Livingstons' outputs in 1891 were valued at a little more than the products of all painters and paint manufacturers in rural counties west of Toronto. The combined products and services in these trades would have only cost the local population about 25 cents per person that year. Canada, *Census of Canada, 1890–91*, Vol. 4 (Ottawa: S.E. Dawson 1897), Table 1, 234.
5 Middleton and Landon, *The Province of Ontario*, 294.
6 Longo, "Castle Kilbride," 47.
7 "County of Waterloo Gazetteer and Directory for 1877–8," 10.
8 Longo, "Castle Kilbride," 47.
9 Ibid., 45.
10 The Perines apparently had a preferential agreement with Lyman & Company, shipping all surplus seed to Montreal in 1863 and likely causing what Elias Eby called the "many adversities in our business with the Perrines [sic]," MAO, Entry, 25 January 1877, "Diary of Elias Eby (1810–1878)," Elias Eby Diary, 1872–1878, Historical MSS 17.8, 47.
11 William Saunders, "Flax," *Central Experimental Farm Bulletin* 25 (May 1896): 8; Austin L. McCredie, "Flax for Fibre," thesis, Department of Chemistry, Ontario Agricultural College, 6 February 1906, 21.
12 TWA, James Livingston, "Letter Book," 1897.

13 Ibid.
14 McCredie, "Flax for Fibre," 21.
15 *Linseed Mill*, Carl E. Johnson, 1934, Minnesota Historical Society, Location No. AV1981.20.56, Negative No. 36584.
16 J.P. Van Niekerk, *The Development of the Principles of Insurance Law in the Netherlands from 1500 to 1800*, Vol. 1 (Hilversum: Uitgeverij Verloren 1998), 423–4, 637.
17 FPL, Ralph May, "The W.P. Orr Linseed Oil Mill, Piqua," Clippings File No. 751, n.d; Tom Millhouse, "Piqua Once Was a Major Linseed Oil Producer," *Piqua Daily Call*, 24 January 1986.
18 UWA, "Toronto Elevators Ltd., Baden Linseed Oil Mills," Canadian Underwriters Association, Special Risk Department, Toronto, Ontario, File No. 494, G3464.B23G475.B3X 1894 UML [microform].
19 EMA, "Alberta Linseed Oil Mills fire (October 1914)," Fred Forster family fonds, Accession No. 0206.0015.
20 "Notes from the Attic," 12–13.
21 SA, D.C. Macdonald, Canadian Pacific Railway, to A.F. Mantle, 19 July 1913; SA, C.E. Highley, International Linseed Company, Moose Jaw, to A.F. Mantle, 30 July 1913, "Flax: Seed Processing, 1913–1942," File 1, Field Crops Branch, Ag 3, No. 40.
22 Heckel, *The Paint Industry*, 195.
23 The Canada Paint Company was called "proprietors of the businesses of Fergusson, Alexander & Co., the William Johnson Company, and A.G. Peuchen Company," *Lovell's Montreal Directory for 1894–95*, Vol. 37069 (Montreal: Lovell 1983 [1894]), 516.
24 Robert Lewis, *Manufacturing Montreal: The Making of an Industrial Landscape, 1850–1930* (Baltimore, MD: Johns Hopkins University Press 2000), 202, 206–8.
25 Lorenzo Prince, *Montreal Old and New* (Montreal: International Press 1915), 370; Lewis, *Manufacturing Montreal*, 232–6.
26 Lewis, *Manufacturing Montreal*, 206.
27 Ibid., 208; Prince, *Montreal Old and New*, 370.
28 "Montreal Paints, Oils and Glass Market," *The Commercial* 19, no. 40 (8 June, 1901): 955; *Lovell's Directory for Montreal* (Montreal: Lovell 1903), 708.
29 *Lovell's Directory for Montreal* (Montreal: Lovell 1907), 791, 919.
30 TWA, James Livingston, Letter File, H.P. Livingston to James Livingston, 17 January 1908.
31 TWA, James Livingston, Letter File, H.P. Livingston to James Livingston, 14 January 1908.
32 "Official report of the debates of the House of Commons of the Dominion of Canada: first session, sixth Parliament ... comprising the period from the thirteenth day of April to the twenty-seventh day of May, 1887" (Ottawa: MacLean, Roger and Company 1887), 547–8.
33 *Census of Manitoba 1886* (Ottawa: Printed by Maclean, Roger and Company 1887), 164. "Body & Noakes, of the Winnipeg Linseed Oil Works" was selling oil and cake made from Manitoba flax seed in 1892. *The Commercial* 10, no. 3 (5 October 1891): 70; *The Commercial* 11, no. 6 (24 October 1892): 162.

34 "Body & Noakes, of the Winnipeg Linseed Oil Works"; *Lovell's Directory for Montreal* (Montreal: Lovell 1910), 836.
35 C.W. Parker, ed., *Who's Who in Western Canada: A Biographical Dictionary of Notable Living Men and Women of Western Canada*, Vol. 1 (Vancouver: Canadian Press Association 1911).
36 EMA, "Canada Needs More Flax," Poster; "Canada Needs More Flax," *Lethbridge Herald*, 16 April 1941, 11; J.H. 'Hop' Yuill, "Alberta Linseed Oil Co., *Trade and Commerce* 51, no. 9 (September 1957): 8; Paul Leonard Voisey, *Vulcan: The Making of a Prairie Community* (Toronto: University of Toronto Press 1987), 87–8; David C. Jones, *Empire of Dust: Settling and Abandoning the Prairie Dry Belt* (Edmonton: University of Alberta Press 1987), 124–5.
37 "White Lead Prices," *The Commercial* 12, no. 36 (21 May 1894): 862.
38 Michael Bliss, *A Living Profit: Studies in the Social History of Canadian Business, 1883–1911* (Toronto: McClelland & Stewart 1974); Mark Cox, "Associationalism Canadian Style: Flour Millers, Self-Regulation and the State, 1920–1935," *Canadian Historical Association Journal* 1 (1990): 119–43.
39 The paint was "not free," the journal exclaimed, "but we understand it has been offered for delivery in the near future at 9 to 10c less." "The Demoralized Paint Trade," *The Commercial* 12, no. 36 (21 May 1894): 862.
40 This was apparently a very close group. See Tiemann, "Paint, Oil and Varnish Industry," 624.
41 Only four years were with the government in power, the Liberals, and those came only after he had expressed interest in retirement, so it is not clear how much impact he could have had on legislation affecting the flax industry. "Mr. James Livingston, M.P.," *Globe*, 5 February 1895, 2; TWA, H.P. Livingston to James Livingston, 20 December 1907.
42 HLBP, George A. Archer to H.L. Bolley, 6 March 1906, Box 39, File 9.
43 Marion E. Cross, *From Land, Sea, and Test Tube: The Story of Archer-Daniels-Midland Company* (Minneapolis, MN: ADM 1954), 13.
44 United States, Department of the Interior, *Thirteenth Census, 1910*, Vol. 8, *Manufactures: General Report and Analysis* (Washington: Government Printing Office 1913), 52.
45 Ennis, *Linseed Oil and Other Seed Oils*, 213.
46 The average price of a bushel from 1902 to 1909 was $1.05, so freight on seed could have represented from 2.5 (at the Great Lakes) to 9.5 per cent (on the seaboard) of raw material cost. The earliest annual record of flax seed prices I have found for the United States is 1902. Bureau of the Census Library, *Statistical Abstracts of the United States 1902, Twenty-fifth Number* 25 (Washington, DC: Government Printing Office 1903), 399.
47 One gallon of linseed oil typically weighed between 7.7 and 7.8 pounds, depending on the amount of sediment and the degree of refining. I.F. Laucks, Table "Weight per Gallon of Oils," *Commercial Oils, Vegetable and Animal; With Special Reference to Oriental Oils* (New York: John Wiley & Sons 1919), 133; Ennis, *Linseed Oil and Other Seed Oils*; 7.7 according to Alvah Horton Sabin, *House Painting:*

Glazing, Paper Hanging, and Whitewashing, a Book for the Householder, 3rd edn (New York: John Wiley & Sons 1924), 6.
48 Ennis, *Linseed Oil and Other Seed Oils*, 119, 271.
49 Ibid., 270.
50 The value of outputs in the Buffalo linseed oil industry was only surpassed by the city's meat-packing, copper-refining, and foundry and machine shop industries. United States Census Bureau, *Twelfth Census, 1900*, Vol. 5: *Manufactures: United States by Industries* (Washington, DC: United States Census Office 1902), 593.
51 Eastman, *History of the Linseed Oil Industry*, 31; Victor S. Clark, *History of Manufactures in the United States*, Vol. 2 (New York: P. Smith 1949 [1929]), 523–4; Alfred D. Chandler, *The Visible Hand: The Managerial Revolution in American Business* (Cambridge, MA: Belknap Press 1977), 327.
52 Unidentified crusher, member of National Linseed Oil Company, in Charles Baker, *Monopolies and the People* (New York: G.P. Putnam's Sons 1889), 9–14.
53 Ibid., 12.
54 "Spontaneous Combustion," *Journal of Commerce, Finance and Insurance Review* 1, no. 1 (20 August 1875): 13–14; "Spontaneous Combustion," *Scientific Canadian Mechanics' Magazine and Patent Office Record* 9, no. 11 (November 1881): 342.
55 As the monopoly was so complete, some trust members desired a price hike, but because "trust-makers are not entirely selfish," the crusher promised, most members wanted fair prices. Baker, *Monopolies*, 14.
56 Chandler, *Visible Hand*, 327.
57 Eastman, *History of the Linseed Oil Industry*, 32.
58 D. Jerome Tweton, "The Business of Agriculture," in Clifford E. Clark Jr, ed., *Minnesota in a Century of Change: The State and Its People Since 1900* (St Paul: Minnesota Historical Society Press 1989), 279–80; Eastman, *History of the Linseed Oil Industry*, 40–1.
59 National's 45 mills and 280 hydraulic presses were merged with an additional 9 crushers for a total of 180 presses.
60 John Moody, *The Truth about the Trusts: A Description and Analysis of the American Trust Movement* (New York: Greenwood Press 1904), 54–5; Cross, *From Land, Sea, and Test Tube*, 13.
61 Moody claimed that American Linseed controlled 85 per cent of the industry (*The Truth about the Trusts*, 55), and Eastman downplayed that to two-thirds in his typical defence of small independent operations like his own (William O. Goodrich Company), *History of the Linseed Oil Industry*, 32; Chandler, *Visible Hand*, 327. John D. Rockefeller Jr gained a controlling interest in American Linseed in 1909, when his father granted him a $16-million gift, one of the first real transfers of wealth from the senior Rockefeller to his son. Ron Chernow, *Titan: The Life of John D. Rockefeller Sr* (New York: Vintage Books 1998), 511.
62 Moody, *The Truth about the Trusts*, 56; Clark, *History of Manufactures*, 370.
63 Eastman, *History of the Linseed Oil Industry*, 43.
64 Ennis, *Linseed Oil and Other Seed Oils*, 8.
65 Western crushers were the only ones to grind their cake, as it incurred extra expense and was only offset by the fact that it shipped more cheaply and could

be mixed with some of the seed screenings to increase oil content. Crushing the cake just before feeding also maintained its "rich, nutty flavour." Hugh G. Van Pelt argued that Americans should also prefer the cake for the same reasons and because it cost $5.00 less per ton than ground meal. *How to Feed the Dairy Cow: Breeding and Feeding Dairy Cattle* (Waterloo, IA: Fred L. Kimball Co. 1919), 75.

66 Ennis, *Linseed Oil and Other Seed Oils*, 274; Elliot & Company claimed that "thousands of English farm leases contain a stipulation that so many tons of Oil-cake shall be fed to animals" and that "if the Oil-cake be not fed, it shall be spread on the land as manure," and they valued the manure at $18.50 per ton in England, $3.25 more than the second place manure from Peas, and three time the value of manure from Indian corn. Elliot & Co, *Elliot & Co., no. 3 Front St. East, Toronto, Exhibit in Class 79 as Samples, and Not for Competition, the Following Linseed and Its Products*, [1880] 1982 (CIHM #27283), 3; "Flax Seed and Oil Cake," *Canada Farmer*, 15 December 1864.

67 McCredie, "Flax for Fibre"; Eastman, *History of the Linseed Oil Industry*, 128, 21. In 1909 all but 20 per cent of the oil cake produced in the United States was exported. Ennis, *Linseed Oil and Other Seed Oils*, 115.

68 These estimations are from 1906 prices according to an interview with James Livingston by Austin L. McCredie. "Linseed oil sells now around 45 cents per gallon of nine lbs. The oil-cake sells, at factory retail, at $24.00 per ton; and the linseed meal (oil not extracted) at $60.00 per ton," "Flax for Fibre," 21.

69 William Cronon, *Nature's Metropolis: Chicago and the Great West* (New York: W.W. Norton 1991), 226.

70 Elliot & Co, *Linseed and Its Products*, 2.

71 Western Historical Company, *The History of Rock County Wisconsin, Its Early Settlement, Growth, Development, Resources, etc.* (Western Historical Company 1879), 157.

72 Van Pelt, *How to Feed the Dairy Cow*, 76.

73 "Editorial Notes," *The Commercial* 14, no. 24 (24 February 1896): 509.

74 John Lowe, Deputy Minister of Agriculture, "Report on Canadian Flax Industries," in Canada, *Sessional Papers No. 8*, 58 Victoria, 1895, 152–3. The Lower Canada Agricultural Society had been interested in an early form of confinement stock fattening involving heavy doses of linseed cake compounds. "Cornwall Agricultural Association," *Agricultural Journal and Transactions of the Lower Canada Agricultural Society* 1, no. 3 (March 1848): 76. See a similar method from Norfolk described in "On the Use of Linseed," *British American Cultivator* 2, no. 6 (June 1846): 172–3.

75 Ian MacLachlan, *Kill and Chill: Restructuring Canada's Beef Commodity Chain* (Toronto: University of Toronto Press 2001), 69–70.

76 Linseed meal was second only to cottonseed meal in the twentieth century in terms of highest protein content in feed. C.H. Eckles and G.F. Warren, *Dairy Farming* (New York: Macmillan 1916), 109.

77 This is an extremely high estimate of returns and was likely meant to indicate total gain from all feeds combined. The meal's primary use in his estimation was for "horned cattle" (beef) being fattened in stalls. These animals would thrive on

a diet of two to five pounds of meal per day in combination with hay, oats, and feed barley. Quoted in Elliot & Co, *Linseed and Its Products*, 2–3.
78 Ian MacLachlan figured that 10 Kg of solid feed can add up to 1.7 Kg or 3.75 lbs body weight per day, and that is a modern, high-energy ration. *Kill and Chill*, 67.
79 In Canada, the average cattle population between 1918 and 1923 was 9.8 million, Dominion Bureau of Statistics, *The Canada Yearbook, 1924* (Ottawa: F.A. Acland 1925), 225–30.
80 Ennis, *Linseed Oil and Other Seed Oils*, 113; John B. Gordon claimed that drawbacks were part of the linseed oil trust's conspiracy. HLBP, Bureau of Raw Materials, speech, 29 December 1925, Box 45, File 3, 5. See also drawback on cake that lessened the duty on imported Canadian seed: SA, A.F. Mantle to Thomas Thompson, 3 December 1910, "Flax, 1906–1945," File 1, Field Crops Branch, Ag 3. No. 38, 2–3.
81 Bolley was one of the first scientists to tout the importance of domestic cake consumption to the linseed-crushing business, although some crushers had been advertising the process to farmers since the 1890s: HLBP, H.L. Bolley to Archer-Daniels Linseed Oil, 13 April 1913, Box 39, File 1; HML, Pioneer Linseed Oil Company, "Tests for Linseed Oil – Linseed Oil Meal and Oil Cake," 1910, Id. 2270388, Litchfield Collection on the History of Fatty Materials (Call Number Pam 2010.200), Published Collections Department; HML, National Linseed Oil Company, "Ration Tables for Feeding Old Process Ground Linseed Cake," 1896, Id. 2270383, Litchfield Collection on the History of Fatty Materials (Call Number Pam 2010.200), Published Collections Department.
82 The average value of all exported Canadian linseed cake from 1921 to 1924 was almost $900,000 each year, and in some years, such as 1923, over three-quarters of the value of cake came from exports to countries other than the United States and the United Kingdom. *Canada Yearbook, 1924*, 464.
83 *U.S. v. American Linseed Oil Co.*, 262 US 371 (1923); Eastman, *History of the Linseed Oil Industry*, 38–9.
84 "Gold Dust and Best Foods," *Time*, 18 June 1928.
85 Pittsburgh Plate Glass Company, *Glass, Paints, Varnishes and Brushes*, 31, 36–7, 39, 41, 45, 53.
86 Heckel, *The Paint Industry*, 479–81; Tristan Abbott, "Bomb Media, 1953–1964," *Postmodern Culture* 18, no 3 (May 2008); *The House in the Middle*, United States Civil Defense Administration's National Clean-up, Paint-up, Fix-up Bureau, 1954.
87 Margaret Hobbs and Ruth Roach Pierson, "'A Kitchen That Wastes No Steps …': Gender, Class and the Home Improvement Plan, 1936–40," *Histoire sociale/ Social History* 21 (May 1988): 8–38; Jill Wade, *Houses for All: The Struggle for Social Housing in Vancouver, 1919–50* (Vancouver: UBC Press 1994), 71–2; Richard Harris, *Building a Market: The Rise of the Home Improvement Industry, 1914–1960* (Chicago: University of Chicago Press 2012), 143, 202; Mariana Valverde, *The Age of Light, Soap, and Water: Moral Reform in English Canada, 1885–1925* (Toronto: University of Toronto Press 2008), 42–3.

88 These slogans first appeared in 1914, according to Herbert W. Rice of Gutta Percha Paint Co., quoted in Heckel, *The Paint Industry*, 160.
89 Pittsburgh Plate Glass Company, *Glass, Paints, Varnishes and Brushes*, 57.
90 Ibid., 31.
91 Von Baeyer, *Ontario Rural Society 1867–1930*, 37, K01.
92 Pittsburgh Plate Glass Company, *Glass, Paints, Varnishes and Brushes*, 37.
93 Jacob Mellis, "A Plea for Ready Mixed Paints," *Farm and Dairy*, 21 May 1914, 607.
94 H.M. Tandy, "Paint Insurance for the Farmer," *Farmer's Advocate*, 5 March 1908, 396–7.
95 N.C. Campbell, "Paint, Its Place on the Farm," *Farm and Dairy*, 6 May 1909, 6.
96 Tiemann, "The Paint, Oil, and Varnish Trade," 623.
97 Pittsburgh Plate Glass Company, *Glass, Paints, Varnishes and Brushes*, 122.
98 From an advertisement by Delmar Lumber and Builder's Supply Inc., in Delmar, NY, "Pittsburgh Paint Advertisement," *Enterprise* [Altamont], 10 June 1938, 8.
99 Pittsburgh Plate Glass Company, *Glass, Paints, Varnishes and Brushes*, 57, 122.
100 Alan A. MacEachern, *Natural Selections: National Parks in Atlantic Canada, 1935–1970* (Montreal & Kingston: McGill-Queen's University Press 2001), 73.
101 Mary Rubio and Elizabeth Waterston, *The Selected Journals of L.M. Montgomery*, Vol. 3 (Toronto: Oxford University Press 1998), 193 (entry for 9 July 1924); Edward MacDonald, "A Landscape ... with Figures: Tourism and Environment on Prince Edward Island," *Acadiensis* 40, no. 1 (2011): 76.
102 The three grades in between general and common were simply numbered. "Wholesale Prices, Current – Oct 1869," *Canadian Pharmaceutical Journal* 2, no. 18 (October 1869): 154.
103 "Advertisement," *Canadian Pharmaceutical Journal* 2, no. 11 (March 1869): 30.
104 "Johnson's Pure Paints," *Farmer's Advocate*, 5 September 1890, 266.
105 *The Commercial* 10, no. 10 (23 November 1891): 227; *The Commercial* 12, no. 10 (20 November 1893): 227.
106 Sabin, *House Painting*, 6.
107 "Inodorous Turpentine," *Agricultural Journal and Transactions of the Lower Canada Agricultural Society* 2, no. 10 (October 1849): 300.
108 Victor Biart, "On the Quantitative Analysis of White Lead Ground in Linseed Oil," *Canadian Pharmaceutical Journal* 5, no. 4 (November 1871): 115, 116; Theodore Zuk Penn, "Decorative and Protective Finishes, 1750–1850: Materials, Process, and Craft," *Bulletin of the Association for Preservation Technology* 16, no. 1 (1984): 10.
109 Ennis, *Linseed Oil and Other Seed Oils*, 232.
110 Ibid., 264.
111 Ernst R. Root, "Paint for Bee Hives," *Canadian Bee Journal* 8, no. 3 (1 May 1892): 38.
112 Zuk Penn, "Decorative and Protective Finishes, 1750–1850," 10; Biart, "Quantitative Analysis of White Lead," 116.
113 Mitchell Okun, *Fair Play in the Marketplace: The First Battle for Pure Food and Drugs* (Dekalb: Northern Illinois University Press 1986), 23–4.

114 Christian Warren, *Brush with Death: A Social History of Lead Poisoning* (Baltimore, MD: Johns Hopkins University Press 2000); James Harvey Young, *Pure Food: Securing the Federal Food and Drugs Act of 1906* (Princeton, NJ: Princeton University Press 1989); Christopher Sellers and Joseph Melling, eds, *Dangerous Trade: Histories of Industrial Hazard across a Globalizing World* (Philadelphia: Temple University Press 2011).

115 Charles Whiting Baker, *Monopolies and the People* (New York: Putnam 1889), 11–12.

116 Michael Bliss, *A Living Profit: Studies in the Social History of Canadian Business, 1883–1911* (Toronto: McClelland & Stewart 1974), 45.

117 Quoted in "White Lead Prices," *The Commercial* 12, no. 36 (21 May 1894): 862.

118 "Montreal Paints, Oils and Glass Market," *The Commercial* 19, no. 40 (8 June 1901): 955.

119 Ennis, *Linseed Oil and Other Seed Oils*, 234–5.

120 Okun, *Fair Play in the Marketplace*, 100.

121 Warren, *Brush with Death*, 54.

122 Heckel, *The Paint Industry*, 321, 323.

123 Lorine Swainston Goodwin, *The Pure Food, Drink, and Drug Crusaders, 1879–1914* (Jefferson: McFarland 1999), 70–1.

124 Young, *Pure Food*, 181–2.

125 Bill G. Reid, *Five for the Land and Its People* (Fargo: North Dakota Institute for Regional Studies, North Dakota State University 1989), 154.

126 Warren, *Brush with Death*, 54.

127 See H.L. Bolley's reference to the concern at the early FDC meetings over linseed oil substitutes coming into the United States: HLBP, H.L. Bolley, "Outline," 20 January 1936, 3.

128 HLBP, Archer-Daniels Linseed to H.L. Bolley, 7 June 1910, Box 39, File 9.

129 A station publication entitled "Some Ready Mixed Paints," *Bulletin*, No. 86, was published in 1909 claiming certain paints were one third water or were otherwise mislabelled. Heckel, *The Paint Industry*, 365. Christian Warren has likened these documents to "a sort of *The Jungle* for the paint industry." Warren, *Brush with Death*, 54.

130 Warren, *Brush with Death*, 54–7.

131 Ibid., 46.

132 TWA, Sherry Veitch, "Livingston Biography," unpublished paper, 2; TWA, Brian Rawding, "Through the Keyhole: Historical Analysis of the Livingstons' Life within Castle Kilbride," unpublished paper, 41.

133 Veitch, "Livingston Biography," 2; Michael D. Longo, "Castle Kilbride," *Waterloo Historical Society* 73 (1985): 42, 46.

134 Ennis, *Linseed Oil and Other Seed Oils*, 263.

135 Eastman, *History of the Linseed Oil Industry*, 112; Ennis, *Linseed Oil and Other Seed Oils*, 181.

136 Lynnette Boney Wren, *Cinderella of the New South: A History of the Cottonseed Industry, 1855–1955* (Knoxville: University of Tennessee Press 1995), 9, 62; Ennis, *Linseed Oil and Other Seed Oils*, 263.

137 "Gold Dust and Best Foods," *Time*, 18 June 1928; Clifford Edward Clark, ed., *Minnesota in a Century of Change: The State and Its People since 1900* (St Paul: Minnesota Historical Society Press 1989), 280–1.
138 Heckel, "Century of Progress in the Paint Industry," 7.
139 Alexander Schwarcman, *Laboratory Letters* (Buffalo: Spencer Kellogg and Sons 1921), 11.
140 Ibid., 10.
141 Ibid., 7.
142 Ennis, *Linseed Oil and Other Seed Oils*, 235–7.
143 Wren argued that the same trend occurred in cottonseed oil mills in the 1920s. *Cinderella of the New South*, 64.
144 Schwarcman, *Laboratory Letters*, 13, 15.
145 See Image 3, "Cincinnati, Ohio, Tank Station and Warehouse," in Schwarcman, *Laboratory Letters*, 14.
146 Peter Sharp, *Flax, Tow, and Jute Spinning: A Handbook Containing Information on the various Branches of these Trades: With Rules, Calculations, and Tables* (Dundee: J.P. Mathew & Co. 1896), 231.
147 TWA, "Livingston Flax Seed Products, Known the World Over," Clippings File, *From Hardware and Metal*, 4 August 1917, 95.
148 Hagley Museum, Carter Litchfield Collection, "Midland Linseed Oil Co.," Acc. 2413, Box 34.
149 The W.P. Orr Linseed Oil Company produced "Old Process" oil when it was bought by American Linseed Oil in 1899. Tom Millhouse, "Piqua Once Was a Major Linseed Oil Producer," *Piqua Daily Call*, 24 January 1986.
150 Schwarcman, *Laboratory Letters*, 9.
151 EMA, "Alberta Linseed Oil Co. Limited," Photographs, Nos 17–21.
152 "British Association," *The Globe*, 31 July 1897, 14; Frank T. Shutt, "Chemistry and Canadian Agriculture," *Science*, August 1907, 265–76.
153 Pittsburgh Plate Glass Company, *Glass, Paints, Varnishes and Brushes*, 2.
154 Ibid., 5.
155 Ibid., 6.
156 Ibid., 12.
157 Heckel, *The Paint Industry*.
158 Philip Scranton and Walter Licht, *Work Sights: Industrial Philadelphia, 1890–1950* (Philadelphia: Temple University Press 1986), 279.
159 Ennis, *Linseed Oil and Other Seed Oils*.
160 Schwarcman, *Laboratory Letters*.
161 Eastman, *History of the Linseed Oil Industry*, 118, 120.
162 Ibid., 117.
163 Ibid., 120.
164 Nineteenth-century paint manufacturers seemed well aware of the dangers of working with lead. Lyman Clare & Company emphasized their effort to keep the colour department separate in their preparation rooms, although the "castor oil filtering-room" was on the same floor and carried oil to lower levels for preparing medicines. In the factory, the paint- and drug-grinding machines were

enclosed in "tightly-fitting wooden cases, in order to prevent the escape of the impalpable powder, which if allowed to be expelled would prevent the workmen from continuing their labors." Canada Railway Advertising Company, *Montreal Business Sketches with a Description of the City of Montreal* (Montreal: Canada Railway Advertising Company 1864), 14, 16. In 1895 Tiemann claimed that mill improvements "obviate much of the danger to the workmen arising from the poisonous nature of the substances used, notably white lead." Daniel F. Tiemann, "The Paint, Oil, and Varnish Trade," in Chauncey M. Depew, ed., *1795–1895: One Hundred Years of American Commerce ... a History of American Commerce by One Hundred Americans* (New York: Greenwood Press 1968 [1895]), 624.
165 Quoted in Warren, *Brush with Death*, 55.
166 "Save the Surface and You Save All – Paint and Varnish," Pittsburgh Plate Glass Company, *Glass, Paints, Varnishes and Brushes*, 122.

CHAPTER SIX

1 Lynette Boney Wren, *Cinderella of the New South: A History of the Cottonseed Industry, 1855–1955* (Knoxville: University of Tennessee Press 1995), 10–21.
2 See especially Kenneth Norrie, "Dry Farming and the Economics of Risk Bearing: The Canadian Prairies, 1870–1930," *Agricultural History* 51, no. 1 (1977): 134–48; Tony Ward, "Farming Technology and Crop Area on Early Prairie Farms," *Prairie Forum* 20, no. 1 (1995): 19–36; Paul Voisey, *Vulcan: The Making of a Prairie Community* (Toronto: University of Toronto Press 1987); David Jones, *Empire of Dust: Settling and Abandoning the Prairie Dry Belt* (Edmonton: University of Alberta Press 1987); Peter A. Russell, "The Far-from-Dry Debates: Dry Farming on the Canadian Prairies and the American Great Plains," *Agricultural History* 81, no. 4 (Fall 2007): 493–521; Alan L. Olmstead and Paul Webb Rhode, *Creating Abundance: Biological Innovation and American Agricultural Development* (New York: Cambridge University Press 2008), 10.
3 Steve M. Wyatt, "Flax and Linen: An Uncertain Oregon Industry," *Oregon Historical Quarterly* 95, no. 2 (1994): 154.
4 Canada, *Census of Canada, 1890–91*, Vol. 4 (Ottawa: S.E. Dawson 1897); William Saunders, "Flax," Central Experimental Farm, *Bulletin* 25 (May 1896): 8.
5 This did not have the effect of increasing raw material or reducing labour for the Great Lakes fibre industry; scutch mills still rented land and coordinated work gangs into the early twentieth century. In some places the fibre and seed industries fed each other, and the Livingstons took advantage of materials from both.
6 Samuel J. Steiner, *Vicarious Pioneer: The Life of Jacob Y. Shantz* (Winnipeg: Hyperion Press 1988), 224.
7 MAO, Entries, 31 August 1874, "Diary of Elias Eby (1810–1878)," Elias Eby Diary, 1872–1878, Historical MSS 17.8, 28; ibid., 11 November 1874, 30; ibid., 28 May 1875, 34. See also Royden K. Loewen, *From the Inside Out: The Rural Worlds of Mennonite Diarists, 1863 to 1929* (Winnipeg: University of Manitoba Press 1999), 133.

8 Canada, *Census of Canada, 1870–71*, LAC, Agricultural Schedule for Waterloo County, Reels C-9943–C-9945.
9 Warkentin, *The Mennonite Settlements of Southern Manitoba*, 89.
10 *Winnipeg Times*, 20 July 1880, 4.
11 Archives of Ontario, Registrations of Marriages, 1869–1928, MS 932, Reel 16, Schedule B – Marriage, 296, No. 009749.
12 *Winnipeg Times*, 6 October 1883, 8.
13 "Manitoban Flax," *Minnedosa Tribune*, 30 November 1893, 1.
14 TWA, "Estate of J & J Livingston, Baden Ont., Statement of Affairs 31st December 1898," 19.
15 "County of Waterloo Gazetteer and Directory for 1877–8," in Ross Cumming, ed., *Illustrated Atlas of the County of Waterloo, 1881* (Owen Sound, ON: Richardson, Bond & Wright 1972).
16 "Flax," *West Lynn Times*, in *Winnipeg Times*, 23 March 1881, 1.
17 Manitoba, Department of Agriculture, *Report for the Year 1883* (Winnipeg: Queen's Printer 1884), 84–5.
18 Ibid., 85; Manitoba, Department of Agriculture, *Report for the Year 1882* (Winnipeg: Queen's Printer 1883), 40.
19 Canada, Department of Agriculture, *Colonial and Indian Exhibition of 1886: A Revelation of Canada's Progress and Resources* (Ottawa: n.p. 1887), 42–3.
20 "Emerson," *Manitoba Free Press*, 12 May 1877, 1; "Flax Culture," *Manitoba Free Press*, 19 May 1877, 2; "Emerson," *Manitoba Free Press*, 1 September 1877, 5.
21 Warkentin, *The Mennonite Settlements of Southern Manitoba*.
22 The Manitoba Department of Agriculture estimated 5,124 bushels for 1878, but Warkentin indicates that 4,000 bushels left Emerson in one week alone that December. Ibid., 90.
23 "The Flax Crop," *Daily Mail and Empire*, 19 August 1895, 8.
24 "News Items," *Minnedosa Tribune*, 26 November 1886, 2; *Winnipeg City Directory, 1883* (Winnipeg: Steen & Boyce 1883), 260.
25 Ann Marie Low, *Dust Bowl Diary* (Lincoln: University of Nebraska Press 1984), 29; Ruby Delores (Stanek) Arsenault, "The Frank Staneks," in Verla Nevay, *Our Czech Heritage* (Rosetown, SK: V. Nevay 1980), 53. Flax seed was occasionally used as makeshift watertight plugs because when the seeds were bagged and soaked in water, they expanded. Hugh Shaw's gusher in Oil Springs, Ontario, was drilled in 1862 using flax seed with this bagging technique to form a shaft plug. Dianne Newell, *Technology on the Frontier: Mining in Old Ontario* (Vancouver: UBC Press 1986), 123.
26 Royden K. Loewen, "Ethnic Farmers and the 'Outside' World: Mennonites in Manitoba and Nebraska, 1874–1900," *Journal of the Canadian Historical Association* 1 (1990): 195–213.
27 John D. Rempel and William Harms, *Atlas of Original Mennonite Villages: Homesteaders and Some Burial Plots of the Mennonite West Reserve, Manitoba* (Altona, MB: J. Rempel and W. Harms 1990), 47.
28 Ibid., 50.

29 Ibid., 51.
30 Royden K. Loewen, *Family, Church, and Market: A Mennonite Community in the Old and the New Worlds, 1850–1930* (Toronto: University of Toronto Press 1993), 76–9; Warkentin, *The Mennonite Settlements of Southern Manitoba*, 73.
31 William Harms and John Dyck, *1880 Village Census of the Mennonite West Reserve, Manitoba, Canada* (Winnipeg: Manitoba Mennonite Historical Society 1998), 96–7, 124–5.
32 W.L. Morton, *Manitoba: A History* (Toronto: University of Toronto Press 1961), 161; Emerich K. Francis, *In Search of Utopia: The Mennonites in Manitoba* (Altona, MB: D.W. Friesen 1955), 116.
33 The Manitoba, Department of Agriculture, *Bulletin* also lists the activity of the CPR, Western Division. Flax was one of the largest cargoes carried in 1882 at 60,343 bushels, or 1,739 tons out of a total of 359,402 tons carried for public use. Wheat amounted to 525,210 bushels or 13,848 tons; oats were 423,446 bushels or 6,958 tons; rye, corn, barley, beans, and even potatoes were minimal amounts. Manitoba, Department of Agriculture, "Commodities transported in 1882," in *Report for the Year 1883*, 226.
34 *Manitoba Weekly Free Press*, 21 December 1878; Warkentin, *The Mennonite Settlements of Southern Manitoba*, 89.
35 M. Cooke, "The Capabilities of Manitoba and the Northwest for the Production of Cereals," prize essay in Manitoba Department of Agriculture, *Report for the Year 1881* (Winnipeg: Queen's Printer 1882), 74; *Sessional Papers of the House of Commons, Dominion of Canada*, 1886, 49, "Appendix 6, Report of the Select Standing Committee on Immigration and Colonization, of the House of Commons, 1–60," Mr [Jacob] Shantz's Evidence, "The Mennonite Immigration – Mode of Settlement – Success of Settlers – Repayment of Loan," 38.
36 Isaac R. Horst, *Up the Conestogo* (Mt Forest: I.R. Horst 1979), 232. For various reasons, planting flax on new breaking was not as critical in the small plots of the Great Lakes region, but nevertheless Chapter 3 shows how flax disappeared completely from most areas.
37 John Lowe, Deputy Minister of Agriculture, "Report on Canadian Flax Industries," in Canada, *Sessional Papers No. 8*, 58 Victoria, 1895, 153.
38 "The Flax Crop," *Daily Mail and Empire*, 19 August 1895, 8.
39 Early growers in the East claimed that "nothing can beat a piece of old lea" for growing flax, but otherwise there is little evidence that it was considered a sod-busting crop there. J.A. Donaldson, "Flax Culture," *Canada Farmer*, 15 April 1869; William Leslie, "Flax in and about Meadowvale," *Canada Farmer*, 1 July 1865, 195; William Saunders, "Flax," *Central Experimental Farm Bulletin* 25 (May 1896): 7.
40 Charles M. Daugherty, "Flaxseed Production, Commerce, and Manufacture in the United States," *Yearbook of the Department of Agriculture, 1902* (Washington, DC: Government Printing Office 1903), 421–38; C.P. Bull, "Flax," in L.H. Bailey, ed., *Cyclopedia of American Agriculture*, Vol. 2 (New York: Macmillan Company 1907), 294, 295.
41 S. Kennedy, "The Draught of Flax upon the Soil: A Comparison," *O.A.C. Review* 22, no. 8 (May 1910): 414.

42 William D. Ennis, *Linseed Oil and Other Seed Oils: An Industrial Manual* (New York: Nostrand 1909), 2.
43 Walter S. Barker, "Flax: The Fiber and Seed: A Study in Agricultural Contrasts," *Quarterly Journal of Economics* 31, no. 3 (May 1917): 521. In 1928 Federic A. Delano told a similar story of a crop moving west as the country developed. Federic A. Delano et al., eds, *Chemical, Metal, Wood, Tobacco and Printing Industries* (New York: Regional Plan 1928), 28.
44 Olmstead and Rhode, *Creating Abundance*. See also Carville Earle, "The Myth of the Southern Soil Miner: Macrohistory, Agricultural Innovation, and Environmental Change," in Donald Worster, ed., *The Ends of the Earth: Perspectives on Modern Environmental History* (New York: Cambridge University Press 1988), 175–210.
45 M.C. Urquhart shows that from 1909 to 1923 flax was the third most valuable grain, after wheat and oats, except for 1911 and 1912 when it was second only to wheat and 1921 when it dropped briefly below barley. Urquhart et al., *Gross National Product, Canada, 1870–1926*, Table 1.9, 32–3.
46 Lewis H. Thomas found that breaking sod was often followed by a crop of flax "to break up the sod and to clean out the wire worms." Thomas, "A History of Agriculture on the Prairies to 1914," in R.D. Francis and Howard Palmer, eds, *The Prairie West: Historical Readings* (Edmonton: Pica Pica Press 1985), 232; Urquhart et al., *Gross National Product, Canada, 1870–1926*, 66; Jeffrey B. Roet, "Agricultural Settlement on the Dry Farming Frontier, 1900–1920," PhD dissertation, Northwestern University; Russell, "The Far-from-Dry Debates," 493–521.
47 Bill Waiser, *The Field Naturalist: John Macoun, the Geological Survey, and Natural Science* (Toronto: University of Toronto Press 1989), 50; Doug Owram, *Promise of Eden: The Canadian Expansionist Movement and the Idea of the West, 1856–1900* (Toronto: University of Toronto Press 1980), 111.
48 "The Colonial and Indian Exhibition," *Minnedosa Tribune*, 18 June 1886, 1; "The Flax Growing Capabilities of Canada," *Brandon Sun Weekly*, 2 December 1886, 5.
49 This notion was popularized earlier in the United States by Charles Dana Wilbur. Donald Worster, *Dust Bowl: The Southern Plains in the 1930s* (New York: Oxford University Press 1979), 82; Waiser, *The Field Naturalist*, 49.
50 J. Castell Hopkins, *Canada: An Encyclopædia of the Country: Agricultural Resources and Development; Literature and Journalism; Chief Cities of Canada*, Vol. 5 (Linscott Publishing Company 1899), 494.
51 Peter Russell's recent synthesis challenges the simple picture of Canadian grassland farmers importing technologies such as dry farming from a supposedly homogeneous area like the Northwest states. Russell, "The Far-from-Dry Debates," 500.
52 See, for example, Michael Percy and Tamara Woroby, "The Determinants of American Migration by State to the Canadian Prairies: 1899 and 1909," University of British Columbia Economics Discussion Paper No. 79-02, January, 1979; Roet, "Agricultural Settlement on the Dry Farming Frontier, 1900–1920," 180–98.
53 Owram, *Promise of Eden*; David Spector, *Agriculture on the Prairies, 1870–1940* (Ottawa: Canadian Parks Service National Historic Parks 1983); Clinton Evans,

The War on Weeds in the Prairie West: An Environmental History (Calgary: University of Calgary Press 2002), 86.

54 Saskatchewan, Department of Agriculture, *Bulletin* 10 (2 July 1909): 4; Saskatchewan, Department of Agriculture, *Bulletin* 16 (27 May 1910): 4.
55 Canada, *Census of Canada, 1911*, Agriculture, Vol. 4 (Ottawa: J. de L. Taché 1914), xlvii; Saskatchewan, Department of Agriculture, *Bulletin* 26 (30 June 1911): 21, 5–7.
56 Don C. McGowan, *The Green and Growing Years Swift Current, 1907–1914* (Victoria: Cactus Press 1982), 21; Voisey, *Vulcan*, 46–7.
57 A.F. Mantle, "Hints for Flax Growers," Saskatchewan, Department of Agriculture, *Bulletin 24* (Regina 1911), 4, 9, 5.
58 William Duane Ennis, *Linseed Oil and Other Seed Oils: An Industrial Manual* (New York: D. Van Nostrand 1909), 2; Federic A. Delano et al., eds, *Chemical, Metal, Wood, Tobacco and Printing Industries*, Regional Plan (New York 1928), 28. To Whitney Eastman, crushing was related to frontier farming: "[A]s long as civilization kept pushing west and new land was made available, the flax crop continued to expand." Flax acreage began to decrease, therefore, "because of lack of available new land." Whitney Eastman, *The History of the Linseed Oil Industry in the United States* (Minneapolis: T.S. Denison 1968), 26, 73, 75.
59 John Bracken, *Dry Farming in Western Canada* (Winnipeg: Grain Growers Guide 1921), 94–5.
60 Victor S. Clark, *History of Manufactures in the United States*, Vol. 2 (New York: P. Smith 1929, republished 1949), 126.
61 W.B. Hurd and T.W. Grindley, *Agriculture, Climate and Population of the Prairie Provinces of Canada: A Statistical Atlas Showing Past Development and Present Conditions* (Ottawa: Dominion Bureau of Statistics 1931), 41.
62 "Alfred and Mabel Murch," in Miry Creek Area History Book Committee, *Shackleton, Abbey, Lancer, Portreeve*, Vol. 2 (Abbey: Miry Creek Area History Book Committee 2000), 1038–9; and "John and Rachel (Vistnes) Murch," in ibid., 1045–7.
63 GMA, "Joe Bellas' Homestead Shack, Carlstadt (later Alderson), Alberta," NA-4711-3; Canada, *Census of Canada 1911*, LAC, Manuscripts, Schedule 1, Medicine Hat, AB, Sub-district 35, 9.
64 GMA, "Cutting Flax on Canadian Wheatlands Fields," NA-587-18; GMA, Oliver S. Longman, "Pioneers and Pioneering in Western Canada" (1963), O.S. Longman Fonds, 76, 82; Harvey White, "The Big Four Farm," in *Little Town in the Valley: History of Flaxcombe and Surrounding School Districts* (Flaxcombe: Historical Society 1984), 32. That "summer, an additional 4,000 acres was cleared of rocks and broken." Verna Sawchuck, "Pleasant Memories of the Big Four – 1912–1914," in ibid., 36.
65 "C.S. Noble," *Lethbridge Herald*, 21 October 1912; GMA, C.S. Noble to Alfred Shepherd (of Guild & Shepherd W.S.), 18 October 1912, Charles S. Noble Fonds, Correspondence, 1912, M-7198-2. The estimates were typed in the spring before planting and were modified in pencil, presumably in the fall, to indicate that more flax had been sown in the newest property and that higher yields made the

crop worth $463,500. GMA, "Estimated Crop & Revenue," 17 May 1919, Charles S. Noble Fonds, Noble Foundation Financial, M-7198-65.

66 "A.C. Townley – The Apostle of Failure," *The Commercial West*, 23 March 1918, 22–3.
67 "Thomas Goldsmith," in E.M. Goldsmith and Grace D. Bradford McGibbon, *From Prairie Sod to Golden Acres: Chronicles of the Hoosier District* (Saskatoon: Modern Press 1956), 19.
68 Emmett Cronan, "Reminiscences of a Hoosier Grain Buyer," in ibid., 44–5.
69 Era Bell Thompson, *American Daughter* (St Paul: Minnesota Historical Society Press 1986), 100.
70 Paul Riegert, *From Arsenic to DDT: A History of Entomology in Western Canada* (Toronto: University of Toronto Press 1980), 95–7.
71 Evans, *The War on Weeds*, 95; James A. Young, "The Public Response to …," *Agricultural History* 62, no. 2 (1988): 124.
72 Ennis, *Linseed Oil and Other Seed Oils*, 2–9; Anatoly Marchenkov, Tatiana Rozhmina, Igor Uschapovsky, and Alister D. Muir, "Cultivation of Flax," in Alister D. Muir and Neil D. Westcott, eds, *Flax: The Genus Linum* (London: Routledge 2003), 76–7. "Canada, and especially the virgin soil of the North-West, is eminently suitable" for flax cultivation, one immigration agent promised, and the "Mennonites in Manitoba have for many years raised crops of flax." "Annual Report of the Emerson, Man., Immigration Agent (Mr. J.E. Têtu)," *Sessional Papers of the Dominion of Canada* 51 Victoria (4) 1888: 144–5.
73 Deborah Fitzgerald, *Every Farm a Factory: The Industrial Ideal in American Agriculture* (New Haven: Yale University Press 2003), 151.
74 Charles H. Clark, "Experiments with Flax on Breaking," Bureau of Plant Industry, USDA, *Bulletin No. 883* (20 September 1920), 2–3. Clark is presumably referring to flax in the Northern Great Plains, or to what he calls "the four principal flax-producing States" (North Dakota, Montana, Minnesota, and South Dakota). O.E. Baker, "The Agriculture of the Great Plains Region," *Annals of the Association of American Geographers* 13, no. 3 (September 1923): 109–67, 109.
75 The general belief of the Kansas Agricultural Experiment Station was that protection stimulated flax seed production in the United States, especially in the 1920s. "Growing Flax in Kansas," *Circular 133* (1927): 3–4; "Flax Production in Kansas," *Circular 173* (January 1934): 15; David Danbom, *Born in the Country: A History of Rural America* (Baltimore: Johns Hopkins University Press, 1995, 2006), 187–8; Eastman, *History of the Linseed Oil Industry*, 174–6.
76 Geoff Cunfer, *On the Great Plains: Agriculture and Environment* (College Station: Texas A&M University Press 2005), 30–1, 35–6.
77 David C. Jones argues that the general crop failures of the 1920s undermined the accepted practices of dry farming and led to a "shift in farming wisdom and a whittling of 'experts,' both professional and practical, down to size." "The Canadian Prairie Dryland Disaster and the Reshaping of 'Expert' Farm Wisdom," *Journal of Rural Studies* 1, no. 2 (1985): 141.
78 Noble to Shepherd, 18 October 1912, quoted in Voisey, *Vulcan*, 85; Mantle, "Hints for Flax Growers," 9–10.

79 John Bracken, "The Problem of Crop Production," Saskatchewan, Department of Agriculture, *Bulletin* 48 (1917): 6–13; Worster, *Dust Bowl*, 94; Cunfer, *On the Great Plains*, 89–92.

80 Production surged up 75 per cent in certain years in this decade, but given the lack of county-level data for those years, the decennial census is as close as we can get to the question of flax and sodbusting.

81 For more on the differences between representation and reality in historical nutrient cycles, see Donahue, *The Great Meadow*; Cunfer, *On the Great Plains*.

82 Bracken, "The Problem of Crop Production," 8.

83 Jeremy Adelman, *Frontier Development: Land, Labour, and Capital on the Wheatlands of Argentina and Canada, 1890–1914* (New York: Oxford University Press 1994), 61–70.

84 "Flax Production in Kansas," *Circular* 173 (January 1934), 14–15.

85 *Lakota Herald*, 30 March 1900.

86 F.H. Leacy, *Historical Statistics of Canada* (Ottawa: Statistics Canada 1983), M249–300; SA, A.F. Mantle to Thomas Thompson, 3 December 1910, "Flax, 1906–1945," File 1, Field Crops Branch, Ag 3. No. 38; Thomas Thompson to A.F. Mantle, in Mantle, "Hints for Flax Growers," 14.

87 Stone Diggers Historical Society, *Prairie Wool: A History of Climax and Surrounding School Districts* (Climax, SK: Stone Diggers Historical Society 1980), 98, 235.

88 *Lakota Herald*, 12 June 1903; Saskatchewan, Department of Agriculture, *Bulletin* 16 (27 May 1910): 4, 9–10; idem, *Bulletin* 29 (12 February 1912): 24; idem, *Bulletin* 35 (February 1913): 9, 70.

89 HLBP, H.L. Bolley, NDAES, to "Manager," American Linseed Company, 25 May 1918, Box 43, File 14.

90 John Bracken, *Crop Production in Western Canada* (Winnipeg: Grain Growers Guide 1920), 214; Saskatchewan, Department of Agriculture, *Bulletin* 17 (27 June 1910): 9; Longman, "Pioneers and Pioneering in Western Canada," 82–3.

91 Glenda Riley, *The Female Frontier: A Comparative View of Women on the Prairie and the Plains* (Lawrence: University Press of Kansas 1988), 135. Riley also describes a woman from Alberta who homesteaded in Montana and grew flax in the 1880s, Ibid., 20.

92 HLBP, E.H. Smith, American Linseed Company, to F.E. Gibson, County Commission, 10 May 1911, Box 39, File 6.

93 HLBP, E.H. Smith, American Linseed Company, to H.L. Bolley, 9 March 1911, Box 39, File 6.

94 HLBP, F.G. Orr, First State Bank to H.L. Bolley, 29 March 1911, Box 49, File 5.

95 HLBP, E.H. Smith, American Linseed Company, to H.L. Bolley, 4 May 1911, Box 39, File 6.

96 *Drumheller Mail*, 13 April 1916.

97 Walter S. Barker, "Flax: The Fiber and Seed: A Study in Agricultural Contrasts," *Quarterly Journal of Economics* 31, no. 3 (May 1917): 520–1; J.R. Leeson claimed it "is stated in a recent official document that 'in many instances a single crop (of seed) has paid for the land, in addition to the cost of breaking and planting.'" J.R.

Leeson, "Introduction," in Edmund A. Whitman, *Flax Culture: An Outline of the History and Present Condition of the Flax Industry in the United States* ... (Boston: Rand Avery Company 1888), 11.

98 Cora Hind, "Farm Problems," *Canadian Thresherman and Farmer* (February 1919): 36. Mantle actually claimed in a letter to Thomas Thompson that he would "not undertake to advise [farmers] to grow more flax." SA, A.F. Mantle to Thomas Thompson, 3 December 1910, "Flax, 1906–1945," File 1, Field Crops Branch, Ag 3. No. 38. In an undated summary of Mr McKay's presentation at a convention, we read that the Indian Head farm scientist felt that "if the government could stop the growing of flax in this country it would be a good thing – from the point of view of weed prevention." SA, "Flax, 1906–1945," File 1, Field Crops Branch, Ag 3. No. 38, 3.

99 HLBP, Archer-Daniels Linseed to H.L. Bolley, 18 June 1914, Box 39, File 11.

100 HLBP, H.L. Bolley to Archer-Daniels Linseed, 25 June 1914, Box 39, File 11; one month later Hind consulted Bolley on the use of green flax straw as a cattle feed. Much of the Prairie's crop had been caught in a frost on 9 August and was of limited use for seed. Cora Hind, *Winnipeg Free Press*, to H.L. Bolley, 19 August 1914, Box 41, File 3.

101 Saskatchewan, Department of Agriculture, *Bulletin* 26 (30 June 1911): 35, 69.

102 HLBP, "Letters from Farmers Regarding Co-operative and Extension Work Conducted by the Department of Botany, 1911–1914," Box 13, File 18, Institute for Regional Studies & University Archives, Fargo, ND; David Danbom, *Our Purpose Is to Serve: The First Century of the North Dakota Agricultural Experiment Station* (Fargo: North Dakota Institute for Regional Studies 1990); H.L. Bolley, "Growing Flax on Old Lands," North Dakota Agricultural Experiment Station, *Bulletin*, No. 40 (December 1910), 1.

103 Bolley believed during the First World War that unless the war stopped, the "spread" between flax and wheat prices would not narrow and that he should not promote flax when wheat prices were high. He figured he would be accused as trying to increase seed production in order to lower prices. HLBP, H.L. Bolley to Archer-Daniels Linseed, 8 April 1915, Box 39, File 11.

CHAPTER SEVEN

1 HLBP, Archer-Daniels Linseed to H.L. Bolley, 3 May 1911, Box 39, File 11.
2 The experiments had been performed at Iowa State College by Dr Otto Lugger, "Tales of First Fifty Years Rescuing the Flax Crop," in *Circular*, NDAC, May 1938.
3 Bill G. Reid, *Five for the Land and Its People* (Fargo: North Dakota Institute for Regional Studies, North Dakota State University 1989), 53.
4 Brian Black, *Petrolia: The Landscape of America's First Oil Boom* (Baltimore, MD: Johns Hopkins University Press 2000), 74.
5 HLBP, "Survival of the Fittest Garden," Photographs, "Flax – Experiments – Fargo," Negative 2705.

6 Alan L. Olmstead and Paul Webb Rhode, *Creating Abundance: Biological Innovation and American Agricultural Development* (New York: Cambridge University Press 2008), 136.
7 H.L. Bolley, "Growing Flax on Old Lands," North Dakota Agricultural Experiment Station, *Bulletin*, No. 40, December 1910, 1.
8 HLBP, H.L. Bolley to George B. Heckel, 22 April 1910, Box 42, File 18.
9 HLBP, H.L. Bolley to Archer-Daniels Linseed Oil, 10 March 1919, Box 39, File 11.
10 HLBP, Arthur E. Dillman, "'Plot 30' testimonial," 1936, Box 1, File 7.
11 Bill G. Reid, *Five for the Land and Its People*, 48.
12 Another treatment for halting the *Fusarium lini* was a formaldehyde solution sprayed on the seed before sowing, but it never completely eliminated the fungus. HLBP, H.L. Bolley to Cross (*Montreal Star*), 22 November 1934, Box 41, File 3.
13 LAC, Frank T. Shutt to Canadian Flax Mills Ltd, Toronto, 23 February 1908, Department of Agriculture, Dominion Chemist: Letterbooks, 1889–1933: T-1034, 591, p. 147.
14 HLBP, Farmer, Murray County, Minnesota, to H.L. Bolley, 25 September 1902, Box 41, File 16.
15 For example: SA, Kenneth Morrison, Spencer Kellogg and Sons, Winnipeg to A.F. Mantle, Saskatchewan, 24 August 1911, "Flax, 1906–1945," File 1, Field Crops Branch, Ag 3, No. 38; SA, Sherwin-Williams to Saskatchewan Minister of Agriculture, 25 July 1914, in ibid.
16 See, for example, SA, E.H. Smith, Western Linen Company, to A.F. Mantle, 29 April 1911; SA, A.F. Mantle to E.H. Smith, Western Linen Company, 16 May 1911; SA, W.B. Morrison, Dundee, Scotland, to A.F. Mantle, 25 March 1913; SA, George H. Campbell, Canadian Flax Mills, Toronto, to A.F. Mantle, 11 February 1913, "Flax: Straw Processing, 1906–1947," File 1, Field Crops Branch, Ag 3, No. 41.
17 Raymond A. Young, "Processing of Agro-based Resources into Pulp and Paper," in Roger M. Rowell and Raymond A. Young, *Paper and Composites from Agro-Based Resources* (Boca Raton: CRC Press 1997), 230–1; Charles F. Roland, *Flax* (Winnipeg: Winnipeg Industrial Bureau), 2–4.
18 For example: HLBP, F. Maclure Sclanders, Saskatoon Board of Trade, to H.L. Bolley, 20 April 1911, Box 41, File 3; A.E. McKenzie Company, Seedsmen to H.L. Bolley, 8 June 1917, Box 41, File 3; Charles H. Thornton & Company, Grain Brokers, Winnipeg, to H.L. Bolley, 11 August 1924, Box 41, File 3.
19 HLBP, J.W. Hirst to H.L. Bolley, 4 March 1903; J.E. Patton to H.L. Bolley, 25 November 1903, Box 39, File 6.
20 Tom Millhouse, "Piqua Once Was a Major Linseed Oil Producer," *Piqua Daily Call*, 24 January 1986.
21 HLBP, George A. Archer to H.L. Bolley, 6 March 1906, Box 39, File 9.
22 HLBP, C.M. Durbin, Sherwin-Williams, to H.L. Bolley, 30 November 1907, Box 44, File 16.
23 Ennis, *Linseed Oil and Other Seed Oils*, 216.
24 Paying for one's land with a single crop of flax was a motif often used by Archer-Daniels Linseed. HLBP, Archer-Daniels Linseed, *Circular*, 16 March 1911, Box

39, File 6; Hess & Small, Fredonia Linseed Oil Works to H.L. Bolley, 28 January 1910, Box 49, File 4.
25 HLBP, "Help Save the Flax Crop, Ask Us How," L. Simmons Hardware, Foxholm, ND, Photographs.
26 For example, HLBP, H.L. Bolley to Archer-Daniels Linseed Oil, 16 March 1910, Box 39, File 6; HLBP, H.L. Bolley to Archer-Daniels Linseed Oil, 3 January 1917, Box 39, File 11.
27 HLBP, H.L. Bolley to W.H. Kiichli, 16 May 1911, Box 44, File 16.
28 HLBP, Thornton to H.L. Bolley, 26 January 1910; H.L. Bolley to Thornton, 29 January 1910, Box 49, File 4.
29 HLBP, Durbin to H.L. Bolley, 3 March 1910, Box 44, File 16.
30 Eastman, *History of the Linseed Oil Industry*, 70; Alfred D. Chandler, *The Visible Hand: The Managerial Revolution in American Business* (Cambridge, MA: Belknap Press 1977), 506–7, 320, 368.
31 John W. Daniels claimed that part of his inspiration for this initiative was the railroad companies' practice of staffing agricultural experts, model farms, and free agricultural lectures – an effort to increase the farm production and population that generated their revenue. J.W. Daniels, "Report of the Paint Manufacturers' Committee," *Paint, Oil and Drug Review*, 6 October 1915, 31; Claire Strom, *Profiting from the Plains: The Great Northern Railway and Corporate Development of the American West* (Seattle: University of Washington Press 2003); Rex C. Myers, "Cultivating South Dakota's Farmers: The 1927 Alfalfa and Sweet Clover Special," *South Dakota History* 22, no. 2 (1992): 136–55.
32 HLBP, Archer-Daniels Linseed to H.L. Bolley, 14 March 1910; H.L. Bolley to Archer-Daniels Linseed, 22 March 1910; Archer-Daniels Linseed to H.L. Bolley, 23 March 1910, Box 39, File 9.
33 The linseed oil subscribers were American Linseed; Archer-Daniels Linseed; Minnesota Linseed; William O. Goodrich, Milwaukee; Sherwin-Williams, Cleveland; Hauenstein, Buffalo; Charles Nolan of National Lead, New York; Midland Linseed Oil Company; Northern Linseed Oil Company; Red Wing Linseed Oil Co; Hirst & Begley, Chicago; Spencer Kellogg, Buffalo; HLBP, Archer-Daniels Linseed to H.L Bolley, 15 August 1910, Box 29, File 9; HLBP, H.L. Bolley, "Report upon the investigation and crop improvement work provided for by the paint manufacturers' association fund, together with the outline of the expenditures against the fund from date of March 1st, 1911 to September 1st, inclusive," 1, Box 50, File 2; HLBP, C.T. Wetherill, "Flaxseed Developing Committee," Report, Atlantic City, 4 November 1910, 1.
34 HLBP, Archer-Daniels Linseed to H.L. Bolley, 2 August 1910, Box 39, File 9.
35 HLBP, H.L. Bolley to C.A. Brown, St. Anthony and Dakota Elevator, 15 September 1910, Box 49, File 4.
36 HLBP, L.P. Nemzek, "History of the Joint Flax Development Committee," in "In the Public Service: Report of Committee on Flax Development," 44th Annual Convention, Atlantic City, NJ (5–8 October 1931).
37 Bolley appreciated receiving Archer-Daniels Linseed circulars, claiming that otherwise he would rely completely on statistics gathered a year previously.

HLBP, H.L. Bolley to Archer-Daniels Linseed, 11 April 1911, Box 39, File 9; see, for example, Archer-Daniels Linseed to H.L. Bolley, 5 September 1910, Box 39, File 9.
38 HLBP, Archer-Daniels Linseed to H.L. Bolley, 15 August 1910, Box 39, File 9; H.L. Bolley to Archer-Daniels Linseed, 16 August 1910, Box 39, File 9.
39 HLBP, H.L. Bolley to J.W. Daniels, Archer-Daniels Linseed, 4 February 1911, Box 39, File 9.
40 HLBP, J.W. Daniels, "Report of the Paint Manufacturers' Committee," *Paint, Oil and Drug Review*, 6 October 1915, 31, Box 39, File 9.
41 "Flax Production in Kansas," *Circular* 173 (January 1934), 14–15.
42 Chandler, *Visible Hand*, 327.
43 HLBP, Archer-Daniels Linseed to H.L. Bolley, 16 September 1912, Box 39, File 10.
44 HLBP, Midland Linseed to Archer-Daniels Linseed, 19 June 1911; H.L. Bolley to Midland Linseed, 15 June 1912.
45 HLBP, e.g. Lowe Brothers, H. Lowe, Lowe Brothers, to H.L. Bolley, 9 July 1910.
46 Initially Bolley argued that "few farmers ever make a conclusive experiment" or give accurate information regarding acreage or yield, only whether or not their crop was a success; HLBP, H.L. Bolley to Archer-Daniels Linseed, 23 June 1911; HLBP, H.L. Bolley, "Preliminary Report of Flax Seed Development Work for Year Ending February 15th, 1915," 14 August 1914, in FDC Reports 1910–1915, Box 50, File 1, 7.
47 HLBP, W.M. Burns, Pittsburgh Plate Glass, to H.L. Bolley 15 April 1918, Box 44, File 5.
48 See, for example, HLBP, expense reports for 1911, 1913, Box 50, File 1.
49 HLBP, H.L. Bolley to C.T. Wetherill, 8 September 1913, 3.
50 HLBP, H.L. Bolley to Charles H. Clark, USDA Bureau of Plant Industry, 18 June 1913, Box 49, File 6.
51 HLBP, H.L. Bolley to Spencer Kellogg, 16 September 1911, Box 49, File 5.
52 HLBP, H.L. Bolley to Glidden Varnish, 24 May 1911, Box 49, File 5.
53 HLBP, L.P. Nemzek, "History of the Joint Flax Development Committee."
54 HLBP, H.L. Bolley to Archer-Daniels Linseed, 11 May 1911, Box 39, File 9.
55 For example, "we are flooded," Bolley claimed, with about twice the amount of correspondence they had before the FDC's support. HLBP, H.L. Bolley to Archer-Daniels Linseed, 31 January 1911, Box 39, File 9.
56 HLBP, H.L. Bolley to G.A. Archer, Archer-Daniels Linseed, 13 September 1912; H.L. Bolley to Archer-Daniels Linseed, 23 June 1911.
57 HLBP, H.L. Bolley to Archer-Daniels Linseed, 7 February 1911.
58 HLBP, Archer-Daniels Linseed to H.L. Bolley, 13 June 1911.
59 HLBP, Archer-Daniels Linseed to H.L. Bolley, 16 October 1912; H.L. Bolley to Archer-Daniels Linseed, 18 October 1912.
60 HLBP, H.L. Bolley to Archer-Daniels Linseed, 12 August 1912.
61 HLBP, H.L. Bolley, "Preliminary Report of Flax Seed Development Work," 15.

62 HLBP, Archer-Daniels Linseed to H.L. Bolley, 12 June 1915; Archer-Daniels Linseed to H.L. Bolley, 8 July 1915.
63 HLBP, Archer-Daniels Linseed to Thomas Cooper, NDAC, 29 September 1915.
64 HLBP, C.M. Durbin to H.L. Bolley, 19 August 1915.
65 HLBP, Archer-Daniels Linseed to H.L. Bolley, 3 March 1916; H.L. Bolley to Archer-Daniels Linseed, 6 March 1916, Box 39, File 11.
66 HLBP, H.L. Bolley to Archer-Daniels Linseed, 10 March 1919, Box 39, File 11.
67 See David B. Danbom, *The Resisted Revolution: Urban America and the Industrialization of Agriculture, 1900–1930* (Ames: Iowa State University Press 1979); and Roy Vernon Scott, *The Reluctant Farmer: The Rise of Agricultural Extension to 1914* (Urbana: University of Illinois Press 1971).
68 HLBP, H.L. Bolley to Archer-Daniels Linseed, 6 August 1919; Archer-Daniels Linseed to H.L. Bolley, 11 August 1919; H.L. Bolley to Archer-Daniels Linseed, 12 August 1919, Box 39, File 11.
69 HLBP, H.L. Bolley, "Report to Flax Development Committee, April 1921 (Crop 1921)," Box 50, File 1, 6.
70 HLBP, L.P. Nemzek, "History of the Joint Flax Development Committee."
71 HLBP, H.D. Long, "Flax Investigations – Seed Laboratory 1922," Box 50, File 1, 1.
72 HLBP, H.L. Bolley, "Preliminary Report of Flax Seed Development Work."
73 *USDA Statistical Bulletin*, No. 376, in Eastman, *History of the Linseed Oil Industry*, 58–9.
74 HLBP, H.L. Bolley, "Preliminary Outline Report, Division of Botany and Pure Seed for Flax Development Committee," 17 March 1927, Box 50, File 1, 2–3; Olmstead and Rhode, *Creating Abundance: Biological Innovation and American Agricultural Development*, 136.
75 Archer-Daniels had mills in Minneapolis, Chicago, Toledo, and by the early 1920s in Buffalo and New York, but their main mill was Minneapolis and presumably they shipped seed east or reported overseas imports in some other place. See, for instance, HLBP, Archer-Daniels Linseed, *Circular*, 27 September 1924; Archer-Daniels Linseed, *Circular*, 3 January 1930.
76 HLBP, American Linseed Company to H.L. Bolley, 18 July 1912; Louis Wommer, American Linseed Company, to H.L. Bolley, 11 September 1912; American Linseed Company to H.L. Bolley, 4 February 1913, Box 39, File 7.
77 The trend in false liberalism is described in Phillip Scranton, "Determinism and Indeterminacy in the History of Technology," in Merritt Roe Smith and Leo Marx, eds, *Does Technology Drive History? The Dilemma of Technological Determinism* (Cambridge, MA: MIT Press 1994), 143–68. Several of Bolley's correspondents, including Lyster H. Dewey of the USDA Bureau of Plant Industry, shared this sense that protection was undesirable but necessary. HLBP, Lyster H. Dewey to H.L. Bolley, 12 April 1909.
78 HLBP, H.L. Bolley to Archer-Daniels Midland Linseed, 18 July 1923; Archer-Daniels Midland Linseed to H.L. Bolley, 20 July 1923, Box 40, File 1.
79 HLBP, Archer-Daniels Linseed, *Circular*, 10 July 1923, Box 40 File 1.

80 HLBP, George B. Heckel, ed., "Fortunes in Flax," and "How a Farmer Was Made to See," *Paint, Oil, and Drug Review* ([n.d.] 1910), Box 42, File 18.
81 HLBP, Archer-Daniels Linseed, *Circular*, 23 September 1911.
82 Ibid., 20 September 1924.
83 HLBP, Charles H. Thornton to H.L. Bolley, 11 August 1924, Box 41, File 3.
84 Ralph H. Estey, *Essays on the Early History of Plant Pathology and Mycology in Canada* (Montreal & Kingston: McGill-Queen's University Press 1994), 109–11.
85 See, for example, HLBP, Howard Fraleigh to H.L. Bolley, 19 March 1934; Seager Wheeler to H.L. Bolley, 11 November 1913; Cora Hind to H.L. Bolley, 19 August 1914.
86 Joshua MacFadyen, "Long-Range Forecasts: Linseed Oil and the Hemispheric Movement of Market and Climate Data, 1890–1939," *Business History* 59, no. 7 (2017): 1–24.
87 Allensworth cautioned Bolley to "absolutely, travel first class. You will lose caste in Argentina if you should arrive second class." He also directed Bolley toward a Protestant church, where the "rector is a fine fellow and takes his Scotch and soda like a gentleman." HLBP, Allen P. Allensworth to H.L. Bolley, 15 March 1930, Box 63–13.
88 Eastman, *History of the Linseed Oil Industry*, 96.
89 Local industrialists were well aware of flax diseases and warned Allensworth to ensure only pure seeds were used in Bolley's experiments. Bolley sent seeds in advance to Buenos Aires (Pillitz), making certain the "box had more tags on it than it had seeds in it." HLBP, H.L. Bolley to Allen P. Allensworth, 24 March 1930, 1.
90 HLBP, H.L. Bolley, "Flax Seed Production in Argentina," Box 63, File 13.
91 Comision Nacional, República Argentina, *Tercer Censo Nacional, 1914, Explotaciones Agropecuarias* (Buenos Aires: L.J. Rosso y Cia 1919), Vol. 5, Chap. 12.
92 "By the mid-1890s ... colonization in the northern pampas ground to a halt," in Jeremy Adelman, *Frontier Development: Land, Labour, and Capital on the Wheatlands of Argentina and Canada, 1890–1914* (New York: Oxford University Press 1994), 69–70; and the expansion of cropland in Santa Fe and neighbouring provinces had ceased by 1914. Leslie Bethell, *The Cambridge History of Latin America*, Vol. 5, *c. 1870 to 1930* (Cambridge: Cambridge University Press 1986), 420.
93 Adelman, *Frontier Development*, 61–2, 201.
94 Daniel K. Lewis, *The History of Argentina* (Westport: Greenwood Press 2001), 63.
95 HLBP, H.L. Bolley, "Flax Seed Production in Argentina," Box 63, File 13. Scobie called the persistence of bag handling the great "peculiarity of the trade," James R. Scobie, *Revolution on the Pampas: A Social History of Argentine Wheat, 1860–1910* (Austin: published for the Institute of Latin American Studies by the University of Texas Press, 1964), 95.
96 In North America flax grew best on grey and dark grey chernozemic soils. Marchenkov, Rozhmina, Uschapovsky, and Muir, "Cultivation of Flax," 76–7; Adelman, *Frontier Development*, Soils Map.
97 HLBP, H.L. Bolley, "Observations on Flax Cropping in South America, Crop 1930–31: Outline Report," Box 63, File 13.

98 HLBP, H.L. Bolley, "Flax Seed Production in Argentina," Report, 18.
99 HLBP, H.L. Bolley to Cross (*Montreal Star*), 22 November 1934, Box 41, File 3.
100 Ibid., 4.
101 Ibid., 5.
102 Ibid.
103 Danbom, *Our Purpose Is to Serve*, 37.

CONCLUSION

1 AmeriFlax, "Flax: From Field to Fork," Video, published on 9 May 2011, https://www.youtube.com/watch?v=fAOhwSmAwiI (accessed August 2017); Camille D. Ryan and Alan McHughen, "Tomatoes, Potatoes and Flax: Exploring the Cost of Lost Innovations," in Stuart J. Smyth, Peter W.B. Phillips, and David Castle, eds, *Handbook on Agriculture, Biotechnology and Development* (Northampton: Edward Elgar 2014), 841–52.
2 AmeriFlax, "Just the Flax: For the Food Industry," *AmeriFlax: North Dakota Flax Producers* 1, no. 1 (June 2006): 4, http://www.ameriflax.com/UserFiles/Image/FoodNewsletter_June06.pdf (accessed August 2017).
3 Ankit Goyal, Vivek Sharma, Neelam Upadhyay, Sandeep Gill, and Manvesh Sihag, "Flax and Flaxseed Oil: An Ancient Medicine & Modern Functional Food," *Journal of Food Science and Technology* 51, no. 9 (2014): 1633–53.
4 Norman Salem Jr and Manfred Eggersdorfer, "Is the World Supply of Omega-3 Fatty Acids Adequate for Optimal Human Nutrition?" *Current Opinion in Clinical Nutrition & Metabolic Care* 18, no. 2 (2015): 147–54.
5 Goyal et al., "Flax and Flaxseed Oil."
6 Marzena Zając, Piotr Kulawik, Joanna Tkaczewska, Władysław Migdał, and Henryk Pustkowiak, "Increasing Meat Products Functionality by the Addition of Milled Flaxseed Linum Usitatissimum," *Journal of the Science of Food and Agriculture*, Epub 14 December 2016.
7 Lynette Boney Wren, *Cinderella of the New South: A History of the Cottonseed Industry, 1855–1955* (Knoxville: University of Tennessee Press 1995), 62.
8 Steve M. Wyatt, "Flax and Linen: An Uncertain Oregon Industry," *Oregon Historical Quarterly* 95, no. 2 (1994): 159–62.
9 Note that the future of these oilseed associations is uncertain, and just like the Flax Development Committee, they will likely be closely tied to the agendas of oilseed processors. "Flax Council of Canada to Shut Office," *Manitoba Cooperator*, 8 January 2018; Flax Council of Canada, website: https://flaxcouncil.ca/; AmeriFlax website: http://www.ameriflax.com/; Flax Institute of the United States, website: https://www.ndsu.edu/pubweb/~hammond/dept/flaxinst/; Jack Carter, "FLAXSEED as Functional Food for People ... and as Feed for Other Animals," website: https://www.ag.ndsu.edu/plantsciences/flax-institute/flaxseed.

INDEX

account books, 29, 43, 69, 84, 105; double-entry accounting, 41
adulterant oils and other oils, 191; cottonseed oil, 19, 191, 196, 197, 276; hemp oil, 191; menhaden oil, 191; other oils, 192, 196; petroleum oils, 191; rosin oil, 191; tung oil, 19, 191, 198
adulteration, 19, 155, 191–5, 198, 200, 204, 205
African-Americans, African-Canadians, 65, 96, 102–3, 106, 229
agricultural societies, 32, 37, 109, 113, 114, 117–20, 123
Ailsa Craig, ON, 93, 94
airplanes, 5, 134; Curtiss "Jenny" planes, 132, 133
Albany, NY, 142
Alberta Linseed Oil Company, 169, 171–3, 199, 205
alchemy, 5, 73, 126
alfalfa, 268
Allensworth, Allen, 266
Alma, ON, 76, 87, 88
Altona, MB, 210, 214
American Chemical Society, 199
American Civil War, 5, 19, 21, 32, 36, 39, 41, 63, 69, 112, 123, 131
American Cotton Oil Company, 184
American Linen Company, 80, 129, 241
American Linseed Company, 14, 173, 241, 244, 249, 250, 263; Flax Development Committee and, 254, 260; in the interwar period, 184–5, 197, 265; reorganized in 1898, 179. *See also* National Linseed
AmeriFlax, 277
Anderson, V.D. Company, 197
Andover, MA, 50, 80
Anishinaabe, 66, 92, 94, 97, 143
Anthropocene, 108, 206
Antrim County, Northern Ireland, 114
ants, 248
apothecaries, 145, 190
appliances, 48, 185
Archer, George, 176, 179, 245, 250, 263
Archer, William, 176, 250
Archer-Daniels and Archer-Daniels-Midland Linseed Oil Company (ADM), 14, 195–7, 202; acquires Midland Linseed Oil, 184; diversification, 197, 263; Flax Development Committee, 254–5, 259–61; interest in flax production, 241–2, 250; tariffs, 263–5; western expansion, 23, 173, 179, 263
Argentina, 25, 206, 264–8; flax imports from, 165–6, 174, 180, 183, 256, 275; flax production in, 25, 236–7, 262, 266–7, 270, 275; flax science and, 261, 265–7, 277
aridity, 25, 218, 219, 232
Arlington Farms, 109
arson, 59, 122
Arthur, ON, 83, 86, 87, 92, 93, 247

Assiniboine River, 109
Atlanta, GA, 129
automobiles, 164, 185, 187
Averill Chemical Paint Company, 150
Aylestock, Addie, 96
Aylestock, James William, 96, 103
azurite, 200

Baden, ON, 40, 59, 170, 181, 199, 209–10; fibre consumption in, 56; flax production near, 70, 74, 76, 79; linseed oil processing centre, 154, 159–61, 164, 166–8, 183–4
bagging, bags, sacks, 8, 10, 24, 40, 48, 52–4, 58, 123, 155
bail rope, 60
Bantjes, Rod, 113
Barber Brothers, 56, 112, 113
barium sulphate, 191
Barker, Walter S., 112, 115, 217, 241
barley, 213
Barron, Elwyn A., 127–9
baskets, 66, 92
beautification, rural, 147
Beck and Clair, 48
Beckert, Sven, 16, 115
bed cord (cordage), 13, 40, 53, 60
beets, 89, 98
Belfast, 119
Belgium, 3, 20, 51, 72–3, 80, 120, 127, 135, 180, 277
Bellas, Joe, 226, 227, 273
Belle River, ON, 125
Bethlehem, PA, 141, 142
Biart, Victor, 191, 200
Biggar, E.B., 211
Billings & Ingram Company, 48
biological innovation, 206, 207, 217, 218, 227, 228, 233
biology, 22, 194
Bismark, ND, 277
Black, Brian, 245
Blaw-Knox rotocel, 203
bleaching, 48, 112, 126, 127, 156

Body and Noakes Company, 172
Boisvert, Michel, 29
Bolley, Henry L., 129, 135, 195, 242, 244–70, 276–7
Bombay, India, 266
Boston, MA, 80, 108, 129; fibre customers in, 8, 48, 52, 56; paint in, 142, 145
Bosworth, ON, 82
botanical classification (plant taxonomy), 9, 73, 110, 230, 243, 247, 276
Bothwell, ON, 36
Bracken, John, 225, 232, 236, 240, 263
brakes, for fibre processing, 11, 47–9, 104, 116
Brandram-Henderson Lead and Paint, 170
Brantford, ON, 8, 97
Brazil, 111
breaking land, 6, 8, 138, 206, 207, 215–19, 222, 224–33, 236–43, 263, 273
breaking outfits, 227, 239–41
brewing, 172, 181
bricklayers, 48
Bridgeport, ON, 156, 157, 159, 160, 166, 209
British West Indies, 36, 96, 163
Brock District, ON, 67
Brockville, ON, 151
Brown, William, 182
Brown and Bauer, 211
Brown and Company, 54, 79
Buchanan, SK, 240
Buckeye Iron and Brass Works, 197
Buenos Aires, Argentina, 266, 275
Buffalo, NY, 7, 8, 14, 25, 28, 52, 165, 173, 177, 179, 183, 184, 202, 245
Burns, Daniel, 130
by-products, 22, 23, 45, 52, 69, 170, 181
Byrne, James, 85, 87, 89, 90

cabinetmakers, 52
Calcutta, India, 155, 166
Caledonia, ON, 93, 151
Cambridge (Galt), ON, 56
Cameron, Allan, 110, 111

Index

Cameron, E.D., 97
Cameron, Malcolm, 114
Campbell Farming Corporation, 230
Canada Company, 109
Canada Cordage (formerly Perine Brothers and Doon Twines), 61
Canada Linseed Oil Mills, 171
Canadian Bureau of Agriculture, 5, 118, 122
Canadian Department of Agriculture, 5, 109, 115, 132, 136, 241; Central Experimental Farm, 131, 148; Economic Flax Fibre Division, 107, 132, 133, 135, 136
Canadian Underwriters Association, 170
Canadian Wheatlands Company, 227
Candee, Richard M., 142, 143
cannabis, 136
canola, 6, 8, 31, 271, 272
capital, 24, 25, 27–8, 38, 41, 48, 63, 77, 122, 126, 172, 178, 180, 208, 224, 228, 237, 267, 275
capitalism, 7, 17, 26–8, 108, 217
Caradoc, ON, 136
carbonate of lime (chalk), 191
carding, 23, 102, 111, 126, 157
Carlingford, ON, 92, 93
Carlstadt, AB (later Alderson), 226
carpet industry, 19, 51, 154
carriage manufacturing, 8, 14, 25, 52, 140, 151–3, 170, 185
cartels, 185
cattle, 180–2, 272; beef cattle, 181–2; dairy products, 267, 272, 277
Cavendish, PE, 189
celery, 98
Chambers, M.S. Company, 48, 52
Charleston, SC, 32, 35, 36, 65
Charlottetown, PE, 153
Chatham-Kent County, ON, 125
chemistry, 8, 22, 31, 111–14, 124, 126, 140, 163, 185, 190–2, 194, 197–200, 202, 204, 205, 244, 248; agricultural, 199, 248; industrial, 113, 126, 148, 197–203, 205;

practical, lay, 31, 111, 112, 124; regulatory, 194, 195, 205, 244
Chicago, IL, 107, 175, 176, 250, 266
child labour, 32, 81, 82, 85–7, 96, 98–9, 133; in the mills, 101–3
Christian Island, ON, 136
Claussen, Chevalier (Peter), 111–13, 115, 116, 124, 126
cleanliness, 147, 186, 189, 203
Cleveland, OH, 40, 179, 197
climate, 6, 9, 14, 15, 223, 231, 268, 275
Climax, SK, 239
Cohen, Marjorie, 99
Cohen, S.J., 125, 127
Colchester, NS, 151, 152
Colonial and Indian Exhibition, 211
colour samples, 149, 150, 185
commodity blindness, 108
commodity history, 18, 30
commodity webs, 11, 17–25, 37, 64, 106, 161, 277
competition, linseed oil company, 88, 154, 161–3, 171–3, 178, 183–5, 192–3, 199, 204, 265, 268
Conestoga (Conestogo), ON, 40, 43, 45, 47, 48, 56, 59, 74, 77
Cooke, M., 215
cordage, 3, 13, 39, 48, 50–3, 56, 58–63, 101–2, 111, 115, 160
Corey, Bess, 240
cork. *See* linoleum
Cottingham, Walter, 170
cotton, 5, 11, 19, 31, 39, 68, 73, 81, 109–13, 115, 116, 127–9, 138, 179, 184, 246; cottonseed, 19, 144, 149, 191, 196, 197, 206, 276; famine (Lancashire famine), 5, 21, 32, 115
cottonization of flax, 111, 112, 114–16, 126
craftwork, 13, 28, 38, 48, 63, 142, 200
crash (textile), 56, 172
Credit River, 38, 39, 43, 58, 113, 115, 273
Cronan, Emmett, 229
Cronon, William, 17
Crowley, Terry, 83

Cunfer, Geoff, 232–3
Cuyahoga River, 8

Danbom, David, 269
Daniells, W.W., 181
Daniels, John W., 179, 263
Danysk, Cecilia, 88
Dashwood, ON, 45
Dayton, OH, 197
Deman, E.F., 72, 73, 80
Devoe, F.W., 150
de Vries, Jan, 81
Diamond, Jared, 18
disease, 15, 206, 230, 237, 239, 242, 243, 245–8, 251, 253, 255, 260, 265, 266, 268, 270; disease-resistant flax, 243, 244, 262; *Fusarium lini*, 242, 246–8, 250, 251, 268; wilt, 242, 246–8, 250, 251, 253, 259–62, 265, 268; wilt-resistant varieties, 248, 252, 258, 262, 269
Dodge, Charles, 77
Dominion Linseed Oil Company, 159, 160, 171, 205. *See also* Livingston Linseed Oil
Donaldson, John A., 57, 114–19, 121–4
Doon (Tow Town), ON, 39, 40, 43–5, 48, 51, 57–9, 61, 69, 74–5, 112, 122
Doon Twines, 49, 61, 62. *See also* Canada Cordage; Perine Brothers
Dorchester, Baron (Guy Carleton), 119
Dorion, A.A., 119
Drayton, ON, 45, 82, 90, 91, 96
Dresden, ON, 125, 130
dressers, flax fibre, 60, 209
Drewry, F.W., 172
driers, 14, 19, 141, 144–9, 156, 195
drought, 172, 219, 238, 261, 263
druggists, 146, 151, 156–9, 190–2, 194, 204
Drummond, Ian, 38
dry-farming, 223, 243, 249
Duluth, MN, 154, 166, 184, 249, 252, 253, 263
Duncan, Colin, 66

Dundas, ON, 53, 55, 56
Dupont, 254, 261
Durbin, C.M., 251, 254, 260
Dutton, ON, 90, 94
Dwight, Timothy, 99

Eastman, Whitney, 29, 202
Eby, Elias, 156, 157, 160, 209
Eby and Devitt, 157
Economic Flax Fibre Division. *See under* Canadian Department of Agriculture
Economy, NS, 151
Elgin County, 34, 118, 130
Elgin Flax Association, 36, 45, 72, 118, 119
Ellendorf, George, 60
Elliot and Company (Toronto), 149, 155, 156, 159, 181, 182, 190, 200
Elliott, Bonnie, 83
Elliott and Hunt, 43, 56
Elmira, ON, 104
Emerson, MB, 162, 209–11
English, Bud, 239
Ennis, William, 191–2, 196
Entre Rios, Argentina, 267
Erbach, Philip, 209, 210, 217
Erickson, Charlotte, 60
Erie Canal, 41, 154
Essex, ON (town), 104, 125–7, 130, 137
Estey, Ralph, 265
Evans, Sterling, 16, 38
Evenden, Matthew, 17
exhibitions, 118, 181
Eyck, Jan Van, 141

fabric, 10, 26, 40, 116, 127, 133
fallowing, 216, 234, 242
Fall River, MA, 80
Fargo, ND, 244, 246, 249
feed, livestock. *See* linseed meal and cake; oats: feed crop
fire, 37, 56, 58, 59, 94, 103, 125, 130, 164. *See also* arson; spontaneous combustion

Firey, Ernst, 60
First Nations flax workers, 14, 31, 32, 66, 85, 89–98, 103, 110, 125, 126, 131, 133–8, 143, 273, 274; Chippewa of the Thames First Nation, 66, 89, 90, 94, 136; Kettle Point First Nation, 98, 136, 137; Munsee-Delaware First Nation, 136; Munsee First Nations, 66, 91–5, 97, 126, 136; Oneida of the Thames First Nation, 89, 90; Saugeen First Nation, 89, 90; Six Nations, 89, 93, 97; Stony Point First Nation, 136, 137
First State Bank (Mott, ND), 241
fish, harmed by retting, 11, 35
Fitzgerald, Deborah, 66
Flaxcombe, SK, 227
flax-cotton, 19, 111–13, 116
Flax Development Committee (FDC), 245, 254–8, 261, 268, 269, 277
flax factorship, 66–7, 72, 77–80, 90, 110, 117, 160, 273–4
flax fibre industry, 6, 8, 13, 20, 23, 24, 33, 47, 61, 63, 81, 97, 159, 275, 276; processing, regional patterns of, 23, 24, 54, 61; processing, seasonal patterns of, 53, 54; production, regional patterns of, 22, 23, 41, 42, 67–72
flax fibre processing, stages of, 11, 35, 45; breaking, 10, 11, 78, 111, 123; hackling, 12, 51, 56, 57, 127; retting, 9, 11, 35, 48, 54, 73, 77, 80, 84, 86, 112, 115, 127, 130, 131; rippling, 9, 13, 20, 73; scutching, 10, 11, 13, 14, 45, 48, 50, 51, 68, 73, 74, 78, 101, 102, 104, 118, 120, 121, 164; spinning, 19, 24, 48, 53, 58, 60, 79; weaving, 60, 65, 67, 68, 99
flax harvest, 78, 79, 82, 83, 85–9, 91, 92, 94–9, 105, 133, 134, 136, 214, 274; gang harvest labour, 6, 80–2, 84–8, 90–2, 95–9, 104, 105, 126, 132, 133; puller, mechanical, 114, 124, 132, 136; pulling, manual labour, 9, 34, 35, 53, 73, 78, 81–90, 92–4, 96–9, 105, 126, 133, 211

flax-wool, 116
Floradale, ON, 38, 90, 91
Forbes Paper Company, 80
Ford, Henry, 164
Ford Canada, 132
formaldehyde soil treatment, 248, 251
Foxholm, ND, 253
Fraleigh, Howard, 98, 110, 135–7, 265
Fredonia, KS, 184
Fredonia Linseed Oil Company, 184, 251, 256
Fruitvale, CA, 41
fuel, flax straw as, 45, 113, 134, 228
Furtney, Christian, 77

Galt (Cambridge), ON, 56, 80
Gamble, John, 94
gang labour. *See under* flax harvest
Gasparri, Nestor, 16
Gates, Frederick T., 179
Geographic Information Systems (GIS), 29, 233; maps of flax production, 6, 29, 70, 76
Georgian Bay, 136
German-Canadians, 61, 82, 104, 162, 209, 211
Germany, flax in, 141, 180, 181
Gilford, Northern Ireland, 60
Glass, Samuel, 135
glazing, glass, 14, 141, 144–6, 150–4, 172, 190
Glen Allan, ON, 96
Goldsmith, Tom, 229
Gooderham & Worts, 39, 43, 54, 58, 80, 113, 116, 119, 122, 162, 182
Gooderich, William O., 184
Gould, Jay, 179
Graham, W.H., 117
Grand River, 38–40, 69
Great Exhibition, London, UK, 111, 124
Great Western Railway, 40
Green Gables, Cavendish, PE, 189
Greenwich, NY, 40, 60

Gretna, MB, 209–11, 214
Grey County, ON, 43, 47
Grisdale, J.H., 132, 135, 241
Guelph, ON, 48, 182
Gull Lake, SK, 239
Gunn Brothers (later Gunn and Murray), 94, 136

Halifax, NS, 152, 171
Halton County, ON, 43, 68, 69, 72, 116
Hamilton, ON, 8, 56, 152, 171
Harlem, MT, 239
Hart, A.H., & Company, 57
Harvey, Fernand, 29
Hay Township, ON, 45, 211
Heckel, George, 194, 197
hemp: categorized with flax, 73–4, 114; fibre production, 38, 60, 62, 107–9, 117–19, 219; hemp oil, 191; restricted, 136
Hendry, Charles, 75, 122, 161, 162
Hendry, Sarah, 59
Hendry, William, 43, 47, 59
henequen, 139, 249
Hensall, ON, 45
Hespeler, ON, 56
Hilborn, David, 78
Hilborn, Sara (Ferguson), 78
Hill, Peter, 95
Hind, Cora, 241–2, 265
Hind, Henry Youle, 5, 114
Hinnegan, Thomas Frank, 130
Hoffmann, J., 104
home production, homespun, 22, 26, 54, 61, 67–8, 74, 79, 99, 213
Hopkins, John Castell, 219
Hudson, Cleveland & Pittsburgh Railway, 40
Hudson River, 8, 179
Hutchinson, R.J., 132

India, 7, 54, 110, 119, 137, 180, 183, 206, 236, 264, 268
Indian agents, 85, 89–90, 93, 97, 136–7
industrial capitalism, 17, 26–7, 217

industrialization, 16, 32, 37, 38, 64, 66, 276
industrial knowledge networks, 17, 22, 28, 31, 41, 43, 44, 63, 77, 81, 113–15, 135, 154, 157, 164, 172, 173, 196, 204, 265–6, 277–8
information asymmetry, 277
Inglis. *See* Mair, Inglis & Company
Ingram, Henry, 41, 48
insurance, 37, 164, 169, 170, 178; Canadian Underwriters Association, 170
International Harvester Company, 250; of Canada, 225
Ipperwash land dispute, 137
Ireland: flax production, 6, 39, 131; industrial knowledge, 35, 60, 114–15, 236, 249, 277; linen industry, 3, 5, 20, 39, 60, 114, 127

Jackson, NY, 40
Jacques and Hay, 52
Jamaica, 35–6, 98
Janesville, WI, 80
japan. *See* driers
Johnson, Carl E., 169
Johnson's Pure Paints, 150, 190
Johnston Elevator Company, 241

Kalbfleisch planing and flax mills, 45
Kansas, 176, 183, 184, 232, 251, 256
Keachie, Anna Eliza, 43
Keith, William, 103, 106
Keller, Kenneth, 29, 99
Kellogg, Spencer and Company, 14, 173, 179, 184, 197–202, 205, 260, 264
Kettle Creek, 33–5, 38, 65, 69, 113, 130
Kiichli, W.H., 252
Kindersley, SK, 227, 231
King, William Lyon Mackenzie, 136
Kingston, ON, 153
Kirkwood, Alexander, 72–3, 114–18
Kitchener (Berlin), ON, 8, 43, 55–6, 59, 61–2, 70, 93, 140, 147, 164, 209
Kluzak, Emil, 239
Kristofferson, Robert, 28
Krug, Hartman, 61

Laboratory Letters (Spencer Kellogg), 197, 200
Lachine Canal, 7, 158, 170
Ladd, Edwin F. (Senator), 194–5, 205, 244
lake port linseed mills, 154, 176, 179, 183, 275
Lamb, Kathryn, 29
Lambton County, ON, 45, 98, 125
lampblack, 146
Lancaster, PA, 142
leather, 63, 141, 156
Leclerc, George, 119
Leeming Brothers, 119, 123
Lemp Studio (Tavistock, ON), 87, 95, 96, 100
Lewis, Robert, 171
Licht, Walter, 201
Liersch, E., 171, 184
lignans, 272
linoleum, 3, 9, 10, 14, 18, 19, 25, 33, 149, 154, 156, 186, 254, 264, 273
linseed meal and cake, 13, 155, 170, 171; consumption by livestock, 20, 22, 170, 180–2, 212; drawbacks on, 175, 182
linseed oil expellers, 202, 203
Livingston, James, 21, 59, 88, 104, 106, 154, 159–66, 170–1, 184, 196, 207–12, 241
Livingston, John, 21, 59, 154, 159–61, 164, 174
Livingston, Louisa (Liersch), 160
Livingston, Peter, 209
Livingston Linseed Oil (later Dominion Linseed Oil Company), 159, 160, 167–9, 172, 174, 208, 214
localism, 19, 54, 62, 129, 170, 181, 183, 196, 267, 273, 274
local knowledge, 48, 79, 214–15, 231, 237
Longman, O.S., 227, 240
Lund, Peter Wilhelm, 111
Lunn, Alice, 27
Lutz, John, 90
luxury goods, 8, 30, 33, 139, 196, 204, 213
Lyman, Clare and Company (Montreal), 14, 157, 158, 160, 170, 173, 183
Lyman, Elliot and Company (Toronto). *See* Elliot and Company

McCalla, Douglas, 144
McColl, James, 88, 89, 104
McCormick Company, 124, 225
McCracken, James, 110, 132, 134–7
McCracken, Morley William, 136, 137
McCredie, Austin L., 96, 130–2, 134, 135, 137
McCumber, Porter James (Senator), 255, 264
McDougall, William, 114, 116, 124
McGee, James, 45
McGowan, John, 88
McInnis, Marvin, 99
Mackenzie, William Lyon, 120
Macoun, John, 218, 219
McQueen, A.A., 97
Madawaska, NB, 68
Mahoning River, 8
Mair, Inglis & Company, 48
Maitland River, 38
Manitoba Department of Agriculture, 211, 214
Manitoba East Reserve, 162
Manitoba West Reserve, 21, 207, 209–15, 217, 227
Manning, J., 122
Mantle, A.F., 223, 224, 238, 239, 242, 249
marketing, 22, 54, 122, 140, 185, 186, 190, 195–7, 199, 204, 231; Clean-up, Paint-up, Fix-up campaigns (US), 8, 186, 189, 277; Clean up, Paint up, Modernize campaign (Vancouver), 186; "Crown Pure" linseed and paint, 190; Home Improvement Plan (Canada), 186; Old Dutch Process (National Lead), 196, 209; Save the Surface campaign, 187, 189, 277; "The Test Tells" (Spencer Kellogg), 199
Marlboro, VT, 41
Martin-Senour Company, 171
Maryborough Township, ON, 83, 96

Massie, Merle, 28–9
matrix of accumulation, 17, 64, 267
mattresses, 13, 115, 129
Maulson, G.J., 212
Meadowvale, ON, 80
Medicine Hat, AB, 169, 171–3
medicines, flax in, 13, 141, 142, 212, 271
Mennonites, 165, 184, 216–18, 274, 275; Manitoba (Russian), 8, 21, 162, 207–16; Ontario (Swiss or German), 41, 61, 77, 84, 148, 209
Merrick, John, 145
metalworkers, 48
meteorological forecasts, 265
Mexico. *See* Seasonal Agricultural Worker Program
Meyers, John, 84, 99
Miami River, 8, 176, 179, 276
Michigan, 28, 77, 88, 104, 135, 159, 161, 274
Middlesex County, ON, 45, 98, 136, 161
Midland Linseed Oil Company, 173, 184, 199, 254, 256, 260
Milverton, ON, 209
Minneapolis, MN, 8, 14, 21, 129, 162, 169, 173, 176, 179, 183, 184, 194, 255, 263, 265, 275
Montgomery, Lucy Maude, 189
Montgomery County, PA, 157
Montreal, QC, 7, 8, 14, 20, 21, 25, 28, 29, 49, 74, 119, 121, 151–3, 156–9, 161, 163, 165, 166, 170–3, 183, 184, 190, 211, 245
Moose Jaw, SK, 231, 240
Moravian Brethren, 141, 142
Morden, MB, 209, 210
Mornington Township, ON, 209
Morton County, ND, 259
Mott, ND, 241
Mount Aviv, MD, 65
Mudge, Benjamin Cushings, 107, 126–30, 137
Murch, John, 225, 227
Murray County, MN, 248

naphtha, 198
National Linseed, 178, 179, 192, 254. *See also* American Linseed Company
Naylor, Charles E., 126
Nelson, ND, 239
Nemzek, L.P., 261
NeuAnlage, MB, 214
Neustadt, ON, 43, 48, 55, 59
New Brunswick, 54, 68, 73, 109
New Dundee, ON, 43
Newell, Diane, 90, 98
Niagara Falls, NY, 52, 68
Nichol Township, ON, 100
Nith River, 40
Noble, Charles, 228–30, 232, 273
Nobleford, AB, 228
nodes (in commodity webs), 11, 18, 21, 22, 37, 106, 161, 165
North Brookfield, MA, 107, 126, 127
North Dakota Agricultural Experimental Station (NDAES), 194, 244–5, 254. *See also* Plot 30
Norval, ON, 43, 55, 58, 69, 80
Norway House, MB, 215
Nova Scotia, 64, 109, 114, 123, 151, 152
nutraceuticals, 271

oakum, 48
oats: competing crop, 6, 213, 238, 243; feed crop, 212; on new breaking, 216, 225, 232, 233, 241
Ohio, 6, 8, 28, 70, 169, 175, 176, 179, 183, 197, 199, 207, 217, 244, 250–2, 263, 273, 276
oilseed sector, 8, 16, 22, 25, 29, 31, 155, 185, 265, 271, 273
omega-3, 9, 271, 272
Ontario Agricultural College (OAC), 130, 132, 137, 168, 182, 217
Ops Township Agricultural Society, 117
Ottawa, ON, 8, 134, 152, 153, 199
Ottawa Valley, 114
outwork, 26, 60, 61, 63, 79

Owen Sound, ON, 159
Oxford Linen Mills, 107, 126–9

paint, 7–9, 14, 19, 25, 139–46, 156–9, 170–3, 208, 243–4, 273–4; consumption, 24, 140, 151–4, 163, 185–7, 189, 205, 231, 273–4; craft paint (not ready-mixed), 141–8, 154, 156–8, 171–3, 175, 180, 182–5, 200; ready-mixed, 7, 14, 140, 148–51, 163, 185, 196, 204, 273
painters and glaziers, 14, 19, 140–5, 150–4, 163, 186, 190–2, 195, 198, 200, 204
paint manufacturers, 14, 152, 194; lobby, 194; Paint Grinders' Association, 173, 193; White Lead Association, 173, 193
Palmerston, ON, 75, 76
Pampas, the (South America), 267, 273, 275
paper industry, 8, 45, 48, 52, 58, 80, 249
Paris, ON, 61
Pell, Edward, 141
Pembina Branch Railway, 162
Perine, Edward Graham, 61, 90
Perine, Hannah (Billings), 40
Perine, Helen Margaret (Hepburn), 61
Perine, Joseph Southworth, 39, 59
Perine, Moses Billings, 39–41, 45, 53, 59, 61, 79
Perine, William Danforth, 39, 41, 59, 77
Perine and Young (St Thomas, ON), 34–6, 113
Perine Brothers (later Doon Twines and Canada Cordage), 14, 33, 40, 51, 54, 66, 79, 113, 116, 121, 159, 162, 164, 255, 273
peripheries, peri-urban, 37, 43, 106
Perth County, ON, 68, 75, 76, 92, 161
Perth Flax and Cordage Company, 125
Philadelphia, PA, 20, 125, 142, 201
photographs as historical evidence, 84, 86–7, 95–102, 143, 201–3, 226
pigments, 141, 142, 144–7, 156, 157, 191–2, 200
pioneer villages, flax commemorated in, 26

Piqua, OH, 169, 197, 199
Pittsburgh, PA, 52
Pittsburgh Plate Glass Company, 151, 185, 187, 189, 200, 231, 250, 256
Plot 30 (North Dakota), 244–7, 253, 269
Pollan, Michael, 18
pollution, 11, 14, 25, 35
Portage la Prairie, 136
Port Colborne, ON, 157
Priester, P., 82
Prince Edward Island, 73, 108, 123, 153, 189
Printer, Robert, 60
Pure Food and Drugs Act (1906), 194–5, 204–5
purity, 20, 31, 204; of manufactured goods, 54, 190, 193–8, 202, 204; of raw materials, 20, 111, 155, 158, 200, 203; of seed for sowing, 210, 252, 268–9
putty, 13, 144–6, 149

Raibmon, Paige, 93
Rason, George E., 132
Raymond, A.H., 104, 125, 130, 137
Red Wing Linseed Oil Company, 173
Reynolds, Hezekiah, 145
Riga, Russia, 119–21, 266
Riga seed subsidy (Canada), 119–21
Riley, Glenda, 240
Riley, John, 28
Ripley, ON, 90
Robinson, William J., 129
Rockefeller, John D. Jr, 14, 179, 184
Rominger, Frederick, 104
Rosario, Argentina, 266
Rosenfeldt, MB, 210
Rosenort, MB, 213, 214
rotting. *See* flax fibre processing, stages of: retting
Ruddel, David Thierry, 29
rural photography, 85–7, 92, 95, 133, 169
Russia, 119, 121, 123, 180, 212, 230, 247, 277
rust (fungus), 185
rust (iron oxide), 248
Ryan, John, 112

Ryan, Nathan, 65, 66, 102, 103, 106
Rygiel, Judith, 68

saddlers. *See* leather
sails, 8, 18, 51, 273
St Boniface, MB, 159
St Catharines, ON, 115, 152
Saint John, NB, 151, 152, 171
St Marys, ON, 95, 131, 143, 144
St Thomas, ON, 32–8, 43, 45, 59, 65, 78, 100–3, 106, 113, 118, 121, 130. *See also* Kettle Creek
Salem, OR, 207
Sallows, Reuben R., 85, 86, 95–7
salt, 18, 20
Samson, Daniel, 113
San Mateo, CA, 41
Santa Fe, Argentina, 267
Sarnia, ON, 136, 151
Saskatchewan Department of Agriculture, 222, 223, 238, 240, 248
Saunders, William, 216
"Save the Flax Crop" campaign, 242, 244, 245, 252–4, 269
sawmills, 24, 40, 45, 57, 96, 156
scent, linseed oil, 191, 198
Schellenberg, David, 214
Schenck retting process, 112
Schmalz, Peter, 89, 143
Schnarr, Werner, 45
Schneider, Joseph, 147
Schneider's Creek, 40
Scranton, ND, 241
Scranton, Philip, 17, 64, 201
screw presses, 13, 142, 155, 202
Seaforth, ON, 104
Seasonal Agricultural Worker Program, 98
Selkirk, District of, 212
Selkirk, Earle of, 109
Seneca, Adam, 66, 94, 103
separation of consumption and production, 26, 27
Shackleton, SK, 225

Shantz, A.D., 161
Shantz, Jacob Y., 209, 215
Shepperd, John, 245
Sherwin-Williams Company, 150, 170–1, 179, 190, 251–4, 260
shipbuilding, 17, 130, 140, 145, 153
shives, 11, 45, 104, 126
Shoemaker, Jacob S., 157
shoemaking, 51
Shutt, Frank T., 199, 248
sign painting, 145, 152, 153
Silberfeld, MB, 214
Simays, E., 119
Simcoe County, ON, 134
Simpson, George (Governor), 109
sisal, 38, 61, 62, 249
slavery, 32, 35, 36, 65, 102, 273
Smith, Dove & Company, 50, 51, 80
Smith, E.H., 241, 249
Snyder, Christian B., 77, 90, 91, 148
sodium benzoate, 192
soils and soil exhaustion, 6, 208, 216–19, 223–4, 230, 236–7, 241–2, 245–8, 250, 258, 267–9, 274
solar energy, 9, 15, 35, 186
South America, 265–8, 270, 273, 275–6
South Dorchester, ON, 130
Southwestern Ontario, 31, 53, 63–9, 74, 76, 90, 96–9, 103, 125, 132, 135, 152, 159, 161–2, 166, 273–4, 276
Southworth, Joseph, 40
sowing, 8, 70, 77, 121, 210, 215, 216, 231, 233, 240, 242, 248, 262
soybeans, 8, 16, 154, 197, 263, 271, 272, 275
Spenst, Jacob, 214
spontaneous combustion, 169, 178
staples thesis, 29, 38, 108
Stephens, G.F., & Company, 172
Stepney Green, UK, 111
Sterling Debenture Corporation mail fraud, 107, 126, 127, 129
Strang, Robert, 161
Stratford, ON, 39, 54, 57, 66, 79, 91, 94, 117, 125

Strathroy, ON, 136
Streetsville, ON, 39, 54–6, 58, 59, 72, 116, 122
Strutt, Sir John, 227
subsistence, 22, 26, 28, 68, 212
Suffield, AB, 227
sulphate of baryta, 191
sulphate of lead and lime, 200
sunlight, protection from, 187, 204
Swift, Gustuvus, 27
Sylvester, Kenneth, 66

Taché, Joseph Charles, 120–2
tank cars, stations, 176, 178, 199
tariffs. *See* trade policy
Tavistock, ON, 82, 87, 88, 91, 95, 97, 99, 100, 104, 105
taxonomy. *See* botanical classification
telecoupling, 16, 17, 22, 23, 108, 273, 276, 277
Tilt, John, 43
Toronto, ON, 5, 8, 28, 36, 39, 52, 53, 56, 58, 68, 125, 132, 151–3, 155, 157, 159, 161, 163, 165, 166, 171, 181, 187, 190, 193
Toronto Exhibition, 158, 181
Tow Town. *See* Doon (Tow Town), ON
trade associations, 169, 172–3, 192, 204–5, 254, 269, 277. *See also* paint manufacturers
trade policy, 21, 56, 63, 110, 249, 263; drawbacks on meal, 175, 182; Emergency Act, 264; Fordney McCumber Act, 264; general tariffs, 8, 39, 56–8, 223, 224, 231, 244, 255, 261–5, 268, 269; Morrill Tariff, 56; National Policy, 163; Reciprocity (Canada–US treaties), 39, 40, 52, 56, 57, 63, 122, 265; Underwood Tariff, 182
Traill, Maulson, & Clark Company, 212
transition to capitalism, 26
triglyceride, 8, 23, 139
Troy, NY, 8, 24, 39, 41, 44, 48, 52, 57
Troy & Boston Railway, 40
Truro, NS, 152

trusts, 154, 169, 171–3, 178, 179, 184, 192, 244, 254, 263. *See also* cartels; trade associations
Truxes, Thomas, 20
Tulchinsky, Gerald, 43
Turkey, 141
turpentine, 145, 146, 148
Tweedsmuir histories, 87
Twin Cities. *See* Minneapolis, MN

Ulrich, Laurel Thatcher, 29
Ulster, Northern Ireland, 114
unionists, 32, 35, 36
United States Department of Agriculture (USDA), 77, 194, 216, 230, 246, 247, 251, 257, 265–7; fibre division, 77, 109
Upstate New York, 6, 64, 99, 273
urban industrial complex, 23–7, 30, 162, 170, 175, 195, 210, 255
Urquhart, M.C., 72
Uruguay, 25, 261, 266
USDA. *See* United States Department of Agriculture

value, acre of flax, 243, 254, 262–5, 269, 270
Vancouver, BC, 171, 186
Van Pelt, Hugh G., 181
varnish, 10, 14, 25, 139–43, 145, 146, 149–52, 154, 156, 158, 163, 171, 185–7, 189, 197, 237, 243, 254, 264, 269
Vessot, Charles, 134
Vickers, Daniel, 81
Victoria, Entre Rios, Argentina, 267

Waddelove, Joseph, 66, 92
Walker, Caroline (Howard-Gibbon), 35, 36
Wallaceburg, ON, 125, 130
Warren, Christian, 192, 194
Waterloo Township, 61
Welland Canal, 157
Wellesley (village), ON, 45, 160, 161

Wellesley Township, ON, 70
Wellington, Grey & Bruce Railway, 75
Wellington County, 43, 68, 74–6, 96, 100, 161
West Gwillimbury Township, ON, 134
Weston, ON, 45
west reserve. *See* Manitoba West Reserve
Wetherill, C.T., 257
wheat, competing crop, 6, 28–9, 80, 208, 213–19, 223, 226, 229, 237–8; on new breaking, 232–3
Wheeler, Seager, 265
Willowdale Experimental Farm, Willowdale, ON, 132–3
Wilmot Township, ON, 69, 70, 76, 79, 159, 161, 164, 209
Windsor, ON, 132
wings. *See* airplanes
Winkler, MB, 210
Winnipeg, MB, 153, 162, 171–2, 209, 212, 239, 241–2, 249, 263, 265, 266, 275, 277
Winnipeg Industrial Bureau, 249
Winnipeg Linseed Oil Company, 160, 171–2
Winnipeg Paint Company, 172

Winona, MN, 129
Wismer, E.A., 125
women, 14, 81, 82, 106, 194; activists in WCTU, 194; consumers, 79, 186, 212; farmers, 240; in knowledge and extension work, 242, 258; work in fields, 14, 81, 82, 85–7, 96, 98, 99, 106; work in mills, 40–1, 59–60, 61, 99, 102
Woodstock, ON, 94, 96, 159
wool, 23, 56, 74, 102, 116, 157
Woolwich Township, ON, 69, 70, 76, 151
Worster, Donald, 233
Wrenn, Lynnette Boney, 197
Wright, Ottis, 82
Wynn, Graeme, 19

Yale, MI, 77, 88, 104, 159, 161, 164
Yankton, SD, 176
Yantzi, Chris, 104, 105
yarn, 56, 60, 121, 154
Young, Alexander, 43, 59, 65, 78

Zurich, ON, 45
Zurich flax company, 45